A. Nowotny

Basic Exercises in Immunochemistry

A Laboratory Manual

Second Revised and Enlarged Edition

With 62 Figures

Springer-Verlag
Berlin Heidelberg New York 1979

ALOIS NOWOTNY, Professor of Immunology
University of Pennsylvania
4001 Spruce Street, Philadelphia, PA 19104/USA

ISBN 3-540-09453-9 2. Auflage   Springer-Verlag Berlin Heidelberg New York
ISBN 0-387-09453-9 2nd Edition   Springer-Verlag New York Heidelberg Berlin

ISBN 3-540-04666-6 1. Auflage Springer-Verlag Berlin Heidelberg New York
ISBN 0-387-04666-6 1st Edition Springer-Verlag New York Heidelberg Berlin

Library of Congress Cataloging in Publication Data. Nowotny, Alois. Basic exercises in
immunochemistry. Bibliography: p. Includes index. 1. Immunochemistry--Laboratory
manuals. I. Title. QR183.6.N68. 1979. 574.2'9'028. 79-14029.

Printing and bookbinding: Beltz Offsetdruck, Hemsbach/Bergstr.
2131/3130-543210

Preface of the Second Edition

First we wish to say a few words about the appearance of this
book, as compared to the first edition. The choice of printing
and paper was simply for financial reasons. This book is written
primarily for students who can barely afford to pay today's fees
for higher education and are even less able to buy books, some
of which are getting beyond the reach of the individual as well
as less-well-endowed libraries. We have tried very hard to keep
the price of this book low so that students can buy it.

As far as the content is concerned, we revised all the exercises which appeared in the first edition and omitted several
of these which could be replaced with better ones, or which were
found to be too complicated for student exercises. New references were added wherever they were needed, although in staying
with the philosophy of the first edition, only a very small percentage of the enormously increased literature has been cited.
As we said in the introduction to the first edition, this book
is not an exhaustive survey of the field. For such purposes
numerous other publications should be studied first. On the other hand, in this manual great emphasis has been placed on clar-
ity of description, completeness of methodological details and
reproducibility of the experiments.

Several old procedures were left in the second edition. We do
not think that methodologies (or other publications, for that
matter) should be ignored only because they were printed a few
decades ago. We have still included those which are useful both
for experimentation and for teaching purposes.

The 32 new exercises we have included were either developed
after the first edition left the press, or have become applicable since then for inclusion in a collection of basic exercises. Radioimmunoassays, enzyme-labelled antibody methods,
preparative thin-layer and affinity chromatographies, several
new variations of electrophoretic procedures, membrane studies,
immunocyte isolations and in vitro correlates of cell-mediated
immune reactions are only a few examples of new exercises which
we have included.

At the same time we realize fully that a great number of methodologies have been developed and published in various immunological and immunochemical laboratories which should have been
included in this book. The obvious limiting factors were time
and space. It would defeat the purpose of this basic manual to
try to achieve completeness, and it would be foolhardy for any-
one to attempt to cover the entire field alone.

This brings me to the point of thanking all who helped me in
this work. The deepest gratitude goes again to my wife, Anne
Nowotny, who untiringly tested and corrected so many of the descriptions and remained steadfast during this endeavor. My students, postdoctoral fellows and other research associates did

their share in elaborating several of the methodological details, without which the exercises simply would not function.
I express here once more my heartfelt thanks for their devoted
work. In addition to this, several descriptions were sent in
draft form to the scientists who pioneered the developments of
some procedures included in the new edition. Their comments,
corrections, and additions were extremely helpful and we greatly appreciate them. These scientists were:
    Dr. C. Boone, National Cancer Institute, Bethesda, MD; Dr.
G. Chen, University of Alabama, Birmingham; Dr. T.P. King,
Rockefeller University, New York, NY; Dr. F. Karush, University
of Pennsylvania, Philadelphia, PA; Dr. J. Lightbody, Wayne State
University, Medical School, Detroit, MI; Dr. J.J. Marchalonis,
Frederick Cancer Research Center, Frederick, MD; Dr. R.H. Purcell,
National Institute of Health, Bethesda, MD; Dr. L.A. Sternberger,
University of Rochester, Rochester, NY; and Dr. R.M. Zacharius,
U. S. Dept. of Agriculture, Philadelphia, PA.

Philadelphia, July 1979   A. NOWOTNY

Preface of the First Edition

This book intends to be neither a complete survey of the field nor an exhaustive source of references. For these purposes, the use of the extensive compilation *Experimental Immunochemistry* by E.A. Kabat and M.M. Mayer (1962), or the excellent methodological textbook, *Methods in Immunology* by D.H. Campbell, J.S. Garvey, E.E. Cremer and D.H. Sussdorf (1963), or the quite comprehensive series *Methods in Immunology and Immunochemistry* by C.A. Williams and M.W. Chase (1967) are more suitable. The main purpose of this manual is to provide students with a simple book which will introduce them to some frequently occurring problems in the three major sections of the immunochemistry of natural products. These are the isolation of the materials, the chemical analysis of the constituents and their structure, and, finally, the assays of the most important biologic and immunologic activities.

In this manual the exercises are simplified and several short-cuts are taken in order to fit them into the framework of a teaching course. The introduction to each exercise gives a brief and elementary explanation of the reaction on which it is based. "Materials and Equipment" lists all tissues or cells, chemicals, glassware, and special equipment which must be available to carry out the exercise, although the very common laboratory tools are usually omitted from the list. The "Procedure" attempts to give a description so simple that students with minimal experience in this and related fields should be able to learn and carry out the procedure. The section is divided into several steps, and these have been set up so that the exercise may be interrupted and continued on the following day, starting with the next step. In the few cases where this could not be done, it is up to the instructor to make the students aware of this circumstance. The "Evaluation" either gives simple ways of obtaining quantitative data from the experiment or refers to other procedure given in this manual by which such information can be obtained. The paragraph headed "Use and Limitations" gives a short survey of the applicability of the method. It is perhaps most important in methodological handbooks to point out clearly what may be expected from the procedure, the purposes for which it may be used, and the realistic values of the data obtained. The "References" list only a few which were considered to be either the most important or the simplest. Several valuable contributions were not mentioned because it would not serve the purposes of this manual to include them.

The figures in the text illustrate only a few pieces of equipment, tools, or manipulations. The drawings of some pieces of equipment are taken from catalogs of supply companies, with their permission. Needless to say, a number of equally reliable and usefull similar tools are available from other sources. Wherever applicable, efforts have been made to introduce homemade equipment.

Most of the procedures are described as they are routinely carried out in our laboratories. Some, which are not used frequently, were thoroughly investigated and tried out by us. Some of the exercises in this manual were used in graduate laboratory courses given at our school. In a few cases, handy modifications or recently developed improvements were learned from authentic experts in the corresponding field, and these have been incorporated as "personal communications".

Although the manual does not cover all possible fields of immunochemical research, we have tried to bring in a few examples from various fields, and to give a useful selection of exercises from different areas of immunochemistry which will enable the teacher to choose those which he considers to be the most relevant to his teaching program. In the last chapter suggestions are given regarding a logical sequence of exercises which should be followed by the students. All these start with the isolation of a certain immunologically active natural product, and include a few steps of purification followed by both chemical analysis of the material obtained and the measurement of its biologic or immunologic activity in various assays.

I must express my deep gratitude to my wife, Anna M. Nowotny, and to those of my co-workers who tested the exercises in the laboratory and who contributed to this manual with suggestions and sound criticism, and last but not least, to Mrs. Alice Stone who edited, proofread and typed the manuscript.

Philadelphia, July 1969   A. NOWOTNY

Contents

1 ISOLATIONS AND PREPARATIONS

Exercise No. 1	Fractionation of Serum Proteins by Ammonium Sulfate	1
Exercise No. 2	Isolation of Immunoglobulins by Gel Filtration on Sephadex G 200 Column	3
Exercise No. 3	Isolation of IgG by DEAE Cellulose	8
	Part A Column Chromatography	9
	Part B Batch-Type Operation	11
Exercise No. 4	Dissociation of Macroglobulins With 2-Mercaptoethanol	12
	Part A	13
	Part B	13
Exercise No. 5	Enzymatic Cleavage of IgG	14
Exercise No. 6	Chromatographic Separation of the IgG Fragments	16
Exercise No. 7	Separation and Isolation of H and L Chains of IgG	17
Exercise No. 8	Affinity Chromatography and Its Application to Antibody Isolation	19
Exercise No. 9	Cellulose Acetate Paper Electrophoresis of Human Serum	23
Exercise No. 10	Disc Electrophoresis in Polyacrylamide Gel	26
	Part A Separation of Serum Proteins	27
	Part B SDS Gel Electrophoresis of Erythrocyte Membrane Proteins	31
	Part C Molecular Weight Determination by SDS Gel Electrophoresis	32
Exercise No. 11	Separation of Human Hemoglobins by Electrofocusing	35
Exercise No. 12	Conjugation of Antibodies With Fluorescein and Rhodamine	41
	Part A Fluorescein	42
	Part B Rodamine	42
Exercise No. 13	Enzymatic Labeling of Proteins With ^{125}I	44
	Part A	44
	Part B	46
Exercise No. 14	Preparation of Peroxidase-Labeled Antibody	48
Exercise No. 15	Preparation of Soluble Dinitrophenyl Proteins	51
Exercise No. 16	Preparation of Dinitrophenyl- and Trinitrophenyl-Labeled Erythrocytes	54
	Part A	54
	Part B	56
Exercise No. 17	Isolation of Cell Membranes From Ascites Tumors	57

Exercise No. 18   Preparation and Use of Synthetic Liposomes   61
Exercise No. 19   Isolation of Murine Transplantation Antigen
                     Preparation   63
Exercise No. 20   Isolation of Bacterial H-Antigens   66
Exercise No. 21   Extraction of Bacterial O Antigens (Endotoxins
                     or Lipopolysaccharides)   69
                     Part A Trichloroacetic Acid Method   69
                     Part B Phenol-Water Procedure   71
                     Part C Chloroform-Methanol Extraction of
                              Endotoxic Glycolipid (EGL) From Rough
                              Mutants   72
Exercise No. 22   Preparation of ^{14}C-Labeled Endotoxin and
                     Endotoxic Glycolipid (EGL)   73
Exercise No. 23   Isolation of Bacterial Capsular Polysac-
                     charides, K Antigens   76
Exercise No. 24   Extraction of Teichoic Acid   78
Exercise No. 25   Extraction of *Pneumococcus* Type XIV Polysac-
                     charide   80
Exercise No. 26   Isolation of Phytohemagglutinin   82
Exercise No. 27   A and B Blood Group Antigen Preparations
                     From Erythrocytes   84
                     Part A Solubilization With Dilute NaOH   84
                     Part B The Use of Detergent   85
Exercise No. 28   Isolation of NN Blood Group Antigen From
                     Human Erythrocytes   86
Exercise No. 29   Isolation of Forssman Hapten From Erythro-
                     cytes   88
Exercise No. 30   Isolation of Ragweed Pollen Allergens   90
Exercise No. 31   Ultrasonic Disruption of Bacteria and Den-
                     sity-Gradient Centrifugation of the Products   93
Exercise No. 32   Vacuum Distillation and Freeze Drying   96

2   STRUCTURAL STUDIES

2.1   QUALITATIVE METHODS   99

2.1.1   Hydrolytic Procedures   99

Exercise No. 33   Determination of Optimal Hydrolytic Condi-
                     tions   99
                     Part A Carbohydrate Liberation   100
                     Part B Liberation of Amino Acids and Amino
                              Sugars   101
Exercise No. 34   Hydrolysis of Lipopolysaccharide With Acid   102
Exercise No. 35   Partial Hydrolysis of Lipopolysaccharide
                     With Ion Exchanger   104
                     Part A Discontinuous Method   105
                     Part B Continuous System   107

2.1.2   Separation and Isolation of the Split Products   110

Exercise No. 36   The Methods of Analytic and Preparative
                     Paper Chromatography   110
                     Part A Preliminary Runs   111
                     Part B Preparative Procedure   113

Exercise No. 37 Separation and Isolation of Peptides or
                   Oligosaccharides by High-Voltage Paper
                   Electrophoresis 117
Exercise No. 38 Demonstration of Ion Exchange Column Chro-
                   matography of Amino Acids 120
Exercise No. 39 Gas-Liquid Chromatography for Fatty Acid
                   Separation 124
Exercise No. 40 Carbohydrate Analysis by Gas-Liquid Chroma-
                   tography 127
Exercise No. 41 Gas-Liquid Chromatographic Analysis of
                   Amino Acids 131
Exercise No. 42 Demonstration of Thin-Layer Chromatography 135

2.1.3 Some Other Qualitative Methods for Structural
       Studies 140

Exercise No. 43 Qualitative Analyses of the Constituents of
                   an Unknown Natural Product 140
Exercise No. 44 Identification of N-Terminal and C-Terminal
                   Amino Acids 147
                   Part A N-Terminus 149
                   Part B C-Terminus 150
                   Part C Paper Chromatography of the Amino
                            Acids and Their Derivatives 151
Exercise No. 45 Staining of Gels for Carbohydrates 153
Exercise No. 46 Methylation of Carbohydrates 156

2.2 QUANTITATIVE ANALYTICAL DETERMINATIONS 159

Exercise No. 47 Dry Weight and Inorganic Ash Determinations 159
                   Part A Total Dry Weight 159
                   Part B Inorganic Ash 160
Exercise No. 48 Microdetermination of Nitrogen 162
Exercise No. 49 Microdetermination of Phosphorus 166
Exercise No. 50 Protein Determination by the Biuret Method 168
Exercise No. 51 Determination of Primary Amino Compounds
                   With Ninhydrin 169
Exercise No. 52 Carbohydrate Determination by Phenol-Sul-
                   furic Acid 171
Exercise No. 53 Ultramicrodetermination of Reducing Car-
                   bohydrates 173
Exercise No. 54 Enzymatic Determination of Carbohydrates .. 175
                   Part A Glucose 176
                   Part B Galactose 177
Exercise No. 55 Determination of Amino Sugars 178
Exercise No. 56 Determination of Heptoses and Pentoses 180
                   Part A Heptose 180
                   Part B Pentose 181
Exercise No. 57 Determination of Sialic Acid and 2-Keto-
                   3-Deoxyoctonic Acid Derivatives 182
                   Part A Sialic Acid 183
                   Part B 2-Keto-3-Deoxyoctonic Acid 184
Exercise No. 58 Chemical and Biologic Analysis of Compounds
                   Separated by Preparative Thin-Layer Chroma-
                   tography 185
                   Part A Chemical Analyses 186
                   Part B Measurements of Biologic Activities . 189

Exercise No. 59 Oxydation of Carbohydrates With Periodate.
                   Quantitative Spectrophotometric Determina-
                   tion 191
Exercise No. 60 Measurements of the Products of Periodate
                   Oxydation 194
                   Part A Formaldehyde Determination 194
                   Part B Formic Acid Determination 195
Exercise No. 61 Reduction of Carbohydrates With NaBH$_4$ 196
                   Part A Determination of the Degree of Poly-
                            merization 197
                   Part B Identification of the Reducing Termi-
                            nal Carbohydrate 198
                   Part C Investigation of the Glycosidic Link-
                            ages by Periodate and Borohydrate
                            Treatment (Smith Degradation) 200
Exercise No. 62 Quantitative Determination of Free Hydroxyl
                   Groups 203
Exercise No. 63 Lipid Determination With the Hydroxylamine
                   Method 205
Exercise No. 64 Qualitative and Quantitative Analysis of
                   Carboxylic Acids 207

3   IMMUNOLOGIC AND OTHER BIOLOGIC ASSAYS

3.1   ANTIBODY PRODUCTION 211

Exercise No. 65 Immunization and Adjuvant Effect 211
                   Part A Soluble Immunogen 212
                   Part B Particulate Immunogen 213
Exercise No. 66 Demonstration of Antibody Production at
                   Cellular Level (Immunoplaque Method)....... 215

3.2   ANTIGEN-ANTIBODY INTERACTIONS 217

3.2.1   Methods of Agglutination 217

Exercise No. 67 Bacterial Agglutination 217
Exercise No. 68 Hemagglutination and Its Inhibition 219
Exercise No. 69 Passive Hemagglutination and Its Inhibition 224
                   Part A Neter's Method 224
                   Part B Boyden's Procedure 226
                   Part C Carbodiimide Coupling Method 226
                   Part D Inhibition of Passive Hemagglutination 227
Exercise No. 70 Charcoal Agglutination 230

3.2.2   Precipitation Methods 232

Exercise No. 71 Double Gel Diffusion/Ouchterlony Method ... 232
Exercise No. 72 Immunoelectrophoresis 235
Exercise No. 73 Quantitative Immunoelectrophoresis/Rocket
                   Method 237
Exercise No. 74 Crossed Immunoelectrophoresis 240
Exercise No. 75 Counter Immunoelectrophoresis 243
Exercise No. 76 Semiquantitative Micro-Precipitin Assays .. 245
                   Part A Microtitration in Gel 245

Contents   XIII

                         Part B Determination of Optimal Antigen/
                                 Antibody Ratio by Gel Diffusion   248
Exercise No. 77   Quantitative Precipitation   249
Exercise No. 78   Quantitative Radial Immunodiffusion   251
Exercise No. 79   Radioimmunoassays for Cyclic AMP and Cyclic
                         GMP Determinations   254
                         Part A Cyclic AMP   255
                         Part B Cyclic GMP   258
Exercise No. 80   Solid-Phase Radioimmunoassay   261

3.2.3   Other Reactions of Antibodies and Immunocytes   263

Exercise No. 81   Complement Fixation   263
                         Part A Preparation of Reagents   265
                         Part B Hemolysin Titration   266
                         Part C Titration of the Complement   266
                         Part D Qualitative Complement Fixation
                                 (Determination of the Proper Antigen
                                 Dose for Quantitative Complement Fixa-
                                 tion Test)   268
                         Part E Anticomplementary Controls   269
                         Part F Quantitative Complement Fixation   269
Exercise No. 82   Total and Viable Cell Counts   272
Exercise No. 83   Direct and Indirect Staining of Bacteria
                         With Fluorescein-Labeled Antibodies   275
Exercise No. 84   Unlabeled Antibody-Enzyme Method to Localize
                         Antigens   277
Exercise No. 85   Isolation of Human Lymphocytes by Sedimenta-
                         tion   280
Exercise No. 86   Isolation of Adhering Leukocytes   282
                         Part A   283
                         Part B   284
Exercise No. 87   Rosette-Forming Cell Assay   285
Exercise No. 88   Passive Cutaneous Anaphylaxis   288
Exercise No. 89   Lymphoblast Assay   290
Exercise No. 90   T Cell-Mediated Cytotoxicity of Influenza
                         Virus-Infected Cells   293

3.2.4   Miscellaneous Biologic Reactions   297

Exercise No. 91   The Local Shwartzman Phenomenon   297
Exercise No. 92   Measurement of the Activity of the Reti-
                         culoendothelial System   299
Exercise No. 93   Enhancement of Nonspecific Resistance to
                         Infection   301
Exercise No. 94   Determination of Toxicity   303
Exercise No. 95   Measurement of Pyrogenicity   305

SUBJECT INDEX   309

1 Isolations and Preparations

Exercise No. 1

Fractionation of Serum Proteins by Ammonium Sulfate

Although several more sophisticated procedures for serum protein fractionation have been described since Kendall (1937) published the ammonium sulfate procedure for gamma globulin isolation, this is still the simplest and one of the most efficient methods if one does not wish to separate the different immunoglobulin fractions from each other. This method gives a preparation which contains almost all of the immunoglobulins together with other nonimmune serum globulins and with relatively low percentages of albumin. The material obtained is suitable for the preparation of fluorochrome- or ferritin-labeled immunoglobulins as well as for further purification. The procedure requires no elaborate equipment, and is carried out at room temperature.

Materials and Equipment

Rabbit serum
Saturated ammonium sulfate solution. (Mix 200 g $(NH_4)_2SO_4$ in approximately 200 ml water, dissolve at $50^\circ C$ and let cool to room temperature overnight. Use the supernatant.)
2 *N* sodium hydroxide
Barium chloride solution, 1%
Phosphate buffered saline (PBS). Dissolve in 1000 ml glass distilled water 1.09 g KH_2PO_4 + 2.14 g $Na_2HPO_4 \cdot 2H_2O$ and 9.0 g NaCl. This will give an 0.02 M, pH 7.0 phosphate buffer, made isotonic with NaCl. Another way to prepare if from stock solutions is by mixing 400 ml 0.02 M KH_2PO_4 (2.72 g per liter) and 600 ml 0.02 M $Na_2HPO_4 \cdot 2H_2O$ (3.56 g per liter) and dissolve in it 9.0 g NaCl
Pipette
Beaker, 50 ml
Graduate cylinder
Dialysis bag and jars
Centrifuge tubes, 50 ml capacity
Centrifuge
Magnetic stirrer and Teflon covered magnetic bar
pH meter

Procedure

1. Add 10 ml serum to a 50 ml beaker and place it on a magnetic stirrer. Under constant stirring of the serum, very slowly

add 6 ml saturated $(NH_4)_2SO_4$ from a pipette. Immerse a combination glass electrode into the suspension and, while continuing
the stirring, adjust the pH with 2 N NaOH to 7.4. Remove the
glass electrode, cover the beaker, and continue the stirring for
30 min. at room temperature. Transfer the contents to a centrifuge tube and spin it down at room temperature for 30 min. at
1000 g. Decant the supernate and save it for further analysis.
Dissolve the sediment in saline, make up the volume to 10 ml.
     2. This sediment still contains appreciable amounts of albu-
min which may be removed by repeated precipitation. Again use a
50 ml beaker and magnetic stirrer. Under constant stirring, as
before, slowly add 5 ml saturated $(NH_4)_2SO_4$ to the crude globulin solution. Adjust the pH and proceed as above. The precipitation may be repeated once more by dissolving the centrifuged
sediment in saline.
     3. Dissolve the final sediment in 10 ml NaCl solution and
dialyze it in the cold room against buffered saline until the
outer fluid is sulfate-negative. (Take a few ml dialysate and
mix with equal volume of 1% $BaCl_2$.) Change the outer fluid twice
daily, morning and evening. After the $(NH_4)_2SO_4$ has been removed
from the sample, centrifuge the contents of the bag at 1000 g
if a precipitate has been formed. Keep the clear supernate and
measure its volume in a graduate cylinder. Calculate the dilution which occurred during dialysis.
     4. The supernate obtained after the first precipitation should
also be dialyzed against pH 7.0 saline. Proceed as above. The
dilution of this preparation after dialysis will be higher, due
to the higher $(NH_4)_2SO_4$ content.
     5. Store the protein solutions in a freezer for further use.

Evaluation

Measure the protein content of the two fractions, and also the
protein content of the starting serum sample, by the biuret method described in Exercise No.50. Calculate the yield and percent recovery of the total proteins used for the fractionation.
     Investigate the purity of the two preparations by paper elec-
trophoresis or by immune electrophoresis as described in Exercises Nos.9 and 72.
     The serologic reactivity of the fractions obtained may also
be investigated. For this purpose, either passive hemagglutination (Exercise No.69) or the quantitative precipitin test (Exercise No.76 or 77) or the quantitative complement fixation test
(Exercise No.81) may be used.

Use and Limitations

While this method has been successfully applied to obtain human
or rabbit globulins, the same method will not give identical results if used for the fractionation of horse or other serum proteins.
     Components of the alpha and beta regions are believed to par-
ticipate in some immune pathologic reactions (such as skin-sen-
sitizing reagins). For their isolation, more refined procedures
are necessary, but the globulin preparation obtained here is a
good starting material.

Better purification of serum proteins can be achieved by the application of the ethanol precipitation procedure of Cohn et al. (1946). This method requires operation at or below $0^\circ C$ and very careful adjustment of pH and ionic strength. A modified procedure was described by Deutsch (1952).

It must be emphasized at this point that the immunoglobulins are much more heterogeneous than was believed originally. Results indicate that antibodies with high specificity to one single chemical determinant differ in the structure of their combining regions (Knight et al., 1966). There are no chemical procedures today which would isolate any of these immunoglobulins in a perfectly homogeneous form. Immunologic methods, using insoluble antigens as specific adsorbents were successfully used in a few experiments. Affinity chromatography (Exercise No.8) can be applied for this purpose. In the chemistry of biologic macromolecules it is generally true that any substance may seem to be homogeneous in the light of existing criteria, but may show heterogeneity with the application of more refined procedures. The known heterogeneity of immunoglobulins, discussed in detail by Fahey (1962), is only one of the typical examples. An excellent review on the heterogenetic of antibodies can be found in Kabat, 1976.

References

Cohn, E.J., Strong, L.E., Huges, jr. W.L., Mulford, D.J., Ashworth, J.N.,
    Melin, M., Taylor, H.L.: J. Am. Chem. Soc. *68*, 459 (1946)
Deutsch, H.F.: Meth. Med. Res. *5*, 284 (1952)
Fahey, J.L.: Adv. Immunol. *2*, 41 (1962)
Kabat, E.A.: Structural concepts in immunology and immunochemistry. 2nd ed.,
    p.223. New York: Holt, Rinehart and Winston Publ. 1976
Kendall, F.E.: J. Clin. Invest. *16*, 921 (1937)
Knight, K.L., Lopez, M.A., Haurowitz, F.: J. Biol. Chem. *241*, 2286 (1966)

Exercise No. 2

Isolation of Immunoglobulins by Gel Filtration on Sephadex G 200 Column

In contrast to adsorption, to distribution, or to ion exchange chromatographic procedures, Sephadex gel chromatography or gel filtration fractionates components of a mixture on the basis of their molecular size. These gels, which are able to function on this principle, are also called molecular sieves. Small molecules which may enter the pores of a highly cross-linked gel polymer, will be retained on the column, while large molecules will travel free in the liquid that fills the spaces between the swollen gel granules (void volume). For a detailed description of the gel filtration processes the students should consult specified literature (Flodin, 1962).

Flodin and Killander (1962) were the first to apply gel filtration to the separation of serum protein components. In this system, the largest globulin, the IgM (Mw=900,000), will obviously appear in the effluent as the first component. If the elution is continued with physiologic saline, the other smaller globulins will also leave the column, and finally the albumin,

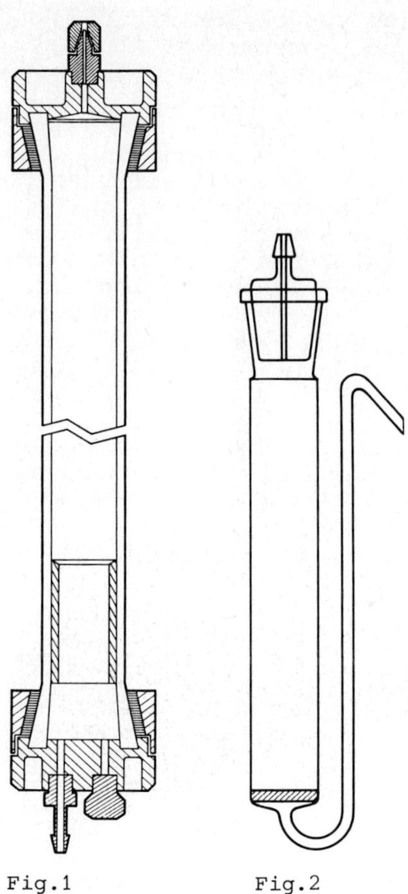

Fig. 1. "Sephadex" column (courtesy of Pharmacia Fine Chemicals, Inc., Piscataway, N.J. or Uppsala, Sweden)

Fig.1   Fig.2

Fig. 2. Regular chromatographic column

which has the smallest molecular weight among the major protein components of the serum, will appear. The version of Sephadex fractionation described here (Killander and Hogman, 1963) is adapted to the fractionation of 2 ml undiluted serum samples or to any other serum fractions containing a comparable amount of protein.

Materials and Equipment

Immune serum, or $(NH_4)_2SO_4$ fractionated products (Exercise No.1)
Sephadex G 200
Buffered saline: 0.05 M pH 7.3 phosphate buffer containing 2.2% NaCl and 0.02% sodium azide (to prevent bacterial growth)
40 ml 0.5 M Na$_2$HPO$_4$ + 10 ml 0.5 M KH$_2$PO$_4$ up to 500 ml + 11 g NaCl + 0.1 g sodium azide
Chromatographic column, 2 × 50 cm size (Fig.1 or Fig.2)
2000 ml graduated cylinder
18 × 150 mm test tubes
Automatic fraction collector (Fig.3)

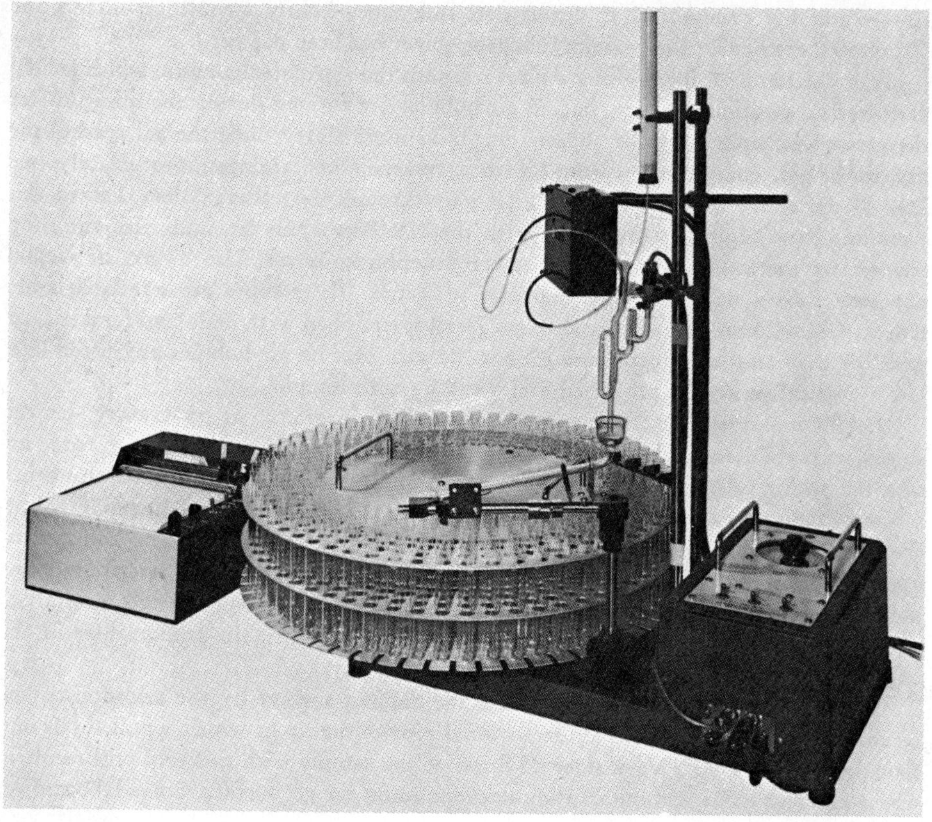

<u>Fig. 3.</u> Automatic fraction collector (courtesy of Canal Industrial Corp., Rockville, MD.)

    UV spectrophotometer, or
    UV flow analyzer with automatic recorder equipped with "event
    marker"

<u>Procedure</u>

    1. Take 30 g of dry Sephadex G 200 and disperse it in a graduated cylinder or similar long jar in 2000 ml buffered saline by slow agitation at cold room temperature overnight. The next day stop the stirring and wait approximately one hour. The larger granules will sediment and the smaller particles can be removed by decantation. Fill the cylinder with the same buffer, stir up the contents and allow it to sediment for another hour. Decantation and sedimentation must be repeated until the supernate becomes virtually free of slow sedimenting granules. This may take ten or more washings and the preparation of the gel may require a few days. But the proper preparation of the Sephadex gel is most important in gel filtration chromatography. If the granules are not completely swollen, or the filling of the column contains very fine particles, a very slow flow rate will

result. Application of pressure on the somewhat elastic gel bed
in the column will not increase the flow rate; on the contrary,
it may stop it almost completely. Sephadex preparations available today from vendors are rather uniform in particle size,
therefore the above described decantation and sedimentation may
not be necessary.

The gel particles may have tiny air bubbles trapped within.
These may form larger bubbles later, which would result in uneven flow in the column. To remove this trapped air, transfer
the properly watered and completely swollen slurry into a 500 ml
filter flask. Close it with a rubber stopper and apply approximately 10-16 mm Hg vacuum on it through its side tube. Gently
swirl the contents for 30 min., disconnect the vacuum tube; the
slurry has been degased.

Pour the swollen slurry along a glass rod into the assembled
column, which is held in a vertical position. Be sure that no
air bubbles are formed during the filling. After the gel settles,
siphon off the supernatant, add more slurry until the gel
reaches the desired height, connect the column with a reservoir
containing approximately one liter of the above buffered saline.
Let it pass through the column, then discard this washing. Now
connect the effluent with the UV flow analyzer, and with the
drop counter or the volumetric syphon of the fraction collector.
Use Tygon or Teflon capillary tubing with an inner diameter of
1 mm. These tubes are available from Kontes Glass Company (Vineland, New Jersey), equipped with male and female Luer ends for
easy and leakproof connections. Be sure that no air bubbles are
in the line. Use a medical syringe to fill the capillary tubing
with liquid.

2. Disconnect the reservoir and by very slight pressure, or
by gravity, let the supernatant fluid enter the gel. Here again,
great care must be taken not to force air into the packing. Using a pipette, apply 2 ml undiluted immune serum or any other
serum protein preparation on the top of the gel. If diluted serum is to be separated on the Sephadex column, the protein concentration of the sample must be adjusted to approximately 5-8%
protein content. It must also be stressed that serum fractions
obtained after DEAE filtration or salt precipitation must be
equilibrated by dialysis against the buffered saline used here
by dialyzing these protein preparations in the cold room for at
least two days before chromatography.

It is very easy to stir up the gel on the column surface by
the addition of the serum sample. It is advisable to use a special pipette the tip of which is pulled out in a flame and bent
at 90°. Very slow addition of the sample with this tool will
result in a nice uniform band without causing unevenness on the
gel surface. It is even better to place a fine Nylon screen or
a circle cut from glass fiber paper on the top of the column to
help even introduction of the sample into the column. After the
sample has been applied start collecting the fractions, taking
5 ml into each tube.

As soon as the sample enters the gel, apply 2 ml buffered saline on the top with similar care and let this flow into the
filling also. Now add approximately a 2 cm layer of swollen gel
slurry in buffered saline to the top of the column, and connect
it with the large reservoir. Approximately 300 ml buffered sa-

line must be run through the column to elute all proteins of the
applied sample. Do not use more than 200 mm water pressure.
    3. If a UV flow analyzer has been used, the automatic record-
er connected to it will show the adsorption of the chromatographic effluent. The maximum of the protein adsorption curves is at
280 nm, but sometimes lower wavelength, such as 255 nm has been
arbitrarily chosen for automatic monitoring of the effluent,
which has such a high protein content at certain phases that
full extinction would be the result at 280 nm, thus masking separation of the eluted fractions. If the laboratory is not e-
quipped with an automatic recorder, the collected samples must
be measured in a spectrophotometer at 280 nm wavelength.

Evaluation

Plot the optical densities against the tube number or take the
recorded chart and identify the protein-containing tubes by the
signals of the "event marker". An "event marker" is a signaling
device of better automatic recorders. It may be connected to the
fraction collector and when the collector turns and brings the
next tube to the effluent end of the column, the marker pen indicates this event on the margin of the recorder chart. With the
help of this mark on the recorded chart, it is easy to locate
those tubes which contain the eluted fractions. If the turntable
of the fraction collector is operated by a timer and the accurate chart speed is known, such an "event marker" is not necessary provided a perfectly even flow rate could be maintained.
    You will observe large zones separated by this procedure. Se-
parate the first peak and also pool the tubes under the second
peak into another fraction. Do the same with the components covered by the third peak. The first half of the first peak is
relatively pure IgM, while other parts of the elution curve cover mixtures of serum protein components. The third peak, which
is the largest, contains albumin in addition to other components.
The use of Sepharose 4b gel, developed by the A.B. Pharmacia,
separates three added fractions from the first peak obtained by
Sephadex G 200.
    Homogeneity and serologic reactivity may be investigated as
described in other exercises (such as 69,72,74,76,77,78,81).

Use and Limitations

Swelling of the Sephadex may be facilitated if carried out by
immersion of the homogeneous slurry, contained in a large beaker, into a boiling water bath for two hours. Removal of fine
particles is carried out as described above. Sterilization by
autoclaving of the gel, glassware, solvents, tubes, and all
tools is recommended to prevent bacterial growth in the gel.
Addition of 1% Na azide to the buffer will considerably slow
down microbial growth in the gel.
    As the results show, this procedure is far from being able
to provide us with homogeneous immunoglobulin preparations. The
collected fractions may be enriched in certain immunoglobulins,
but they are obviously still mixtures of a number of constituents. Combination of the DEAE ion exchange cellulose and the Se-

phadex G 200 gel filtration chromatographic procedures will give
us the IgG and IgM groups in a relatively purified form.
    There are a number of Sephadex derivatives. One of them is
Sephacryl S-200. It is prepared by cross-linking dextran with
N,N'-methylenebisacrylamide. These gel beads have a higher ri-
gidity and therefore a higher flow rate can be achieved in column chromatography by applying pressure to the system. The Sephacryl S-200 is quite resistant to a number of solvents, including 0.2 N sodium hydroxide, dimethylsulfoxide, chloroform,
tetrahydrofuran, or acetone. It can also be autoclaved repeatedly.
    Others were obtained by coupling covalently nonpolar func-
tional groups, such as octyl or phenyl substituents, to Sepharose
CL-4B. Such gels will retain nonpolar proteins by hydrophobic
interaction. Components of a protein mixture, for example, can
be separated based on their functional groups, on the differing
strength of their hydrophobic reactivity with functional groups
on the gel. This gel is also quite stable and can be used in
the presence of urea or guanidine HCl. It works in water as well
as in organic solvents.
    Inorganic molecular sieves which can separate molecules ac-
cording to their molecular dimensions have been known for a long
time. Zeolites are one of these and Eastman Company markets similar products. Very small glass particles with controlled inside
pore sizes are manufactured by Corning Glass Works, Corning,
NY 14830. Great advantage of this porous glass to Sephadex
preparations for molecular sieving lies in the fact that glass
can be easily cleaned, autoclaved and exposed to harsh chemicals. Furthermore, using column chromatography on controlled porous glass beads allows the operator to use pressure. A disadvantage of the use of porous glass is that the resolution one
can obtain is inferior to the fractionation seen using the proper Sephadex gel. Another series of products, suitable for gel
filtration are the polyacrylamide Bio-Gel P preparations manufactured by Bio-Rad Laboratories, Richmond, CA. 94804.

References

Flodin, P.: Dextran gels and their application in gel filtration. Uppsala:
   AB Pharmacia 1962
Flodin, P., Killander, J.: Biochim. Biophys. Acta *63*, 403 (1962)
Killander, J., Hogman, C.F.: Scand. J. Clin. Lab. Invest. Suppl.69, *15*, 130
   (1963)

Exercise No. 3

Isolation of IgG by DEAE Cellulose

Weakly basic cellulose ion exchanges, such as DEAE cellulose,
can be used for the separation of serum proteins according to
the procedure of Sober and co-workers (1956). This procedure is
especially useful for isolation of IgG antibodies, which leave
the column in a relatively pure form right after the void volume.

This method is slightly modified as described here in *Part A*, using a continuous gradient elution.

James and Stanworth (1965) elaborated a simple batch-type operation using DEAE cellulose for the separation of IgG from other constituents of sera. This is described in *Part B* of this exercise.

Part A. Column Chromatography

Materials and Equipment

Rabbit antiserum, dialyzed against 0.005 M pH 7 phosphate buffer
DEAE cellulose
0.1 N sodium chloride
0.1 N hydrochloric acid
0.005 M sodium phosphate buffer, pH 7
0.05 M sodium chloride containing 0.05 M sodium diphydrophosphate
Büchner funnel
Automatic fraction collector (Fig.3)
Filter flask, 500 ml
UV flow analyzer
Chromatographic colum assembly (Fig.1 or 2)
Automatic recorder
Density gradient device (Fig.4a)

Procedure

1. Prepare the DEAE cellulose as follows: Take 25 g DEAE cellulose, suspend it in 500 ml 0.1 N HCl, stir it for 30 min at room temperature, then filter through a Büchner funnel; apply vacuum to remove as much liquid from the cellulose as possible. Remove the cellulose cake, suspend it in 500 ml 0.1 N NaOH, stir slowly for 10 min, filter again as before. Suspend the cellulose in 500 ml 0.1 N NaOH, filter, but this time wash the cellulose in the filter with further 500 ml 0.1 N NaOH. The cellulose must be washed with distilled water until the filtrate is neutral. Be sure that all traces of NaOH have been removed. Final washing will be done using 500 ml 0.005 M pH 7 phosphate buffer. The pH of the effluent must be the same as the pH of the buffer. If this is not achieved after 500 ml, further washings are necessary. Suspend the cake in 200 ml of the above buffer and store it in this form. This suspension may be sterilized by autoclaving if necessary.

Transfer the cellulose into a 500 ml filter flask, close the flask with a rubber stopper and evacuate it well with a vacuum pump to remove the trapped air bubbles from the fibers. Keep it under vacuum for approximately 30 min.

2. Pour the slurry along a glass rod into a chromatographic column, size 25 × 500 mm. After a well-packed column is formed, connect the effluent end of the chromatographic column to a UV flow recording unit as described in Exercise No.2 and set the wavelength of the detector to 280 nm. Let the buffer enter the column bed, then turn off the stopcock at the bottom of the column. The serum sample has to be dialyzed against the initial

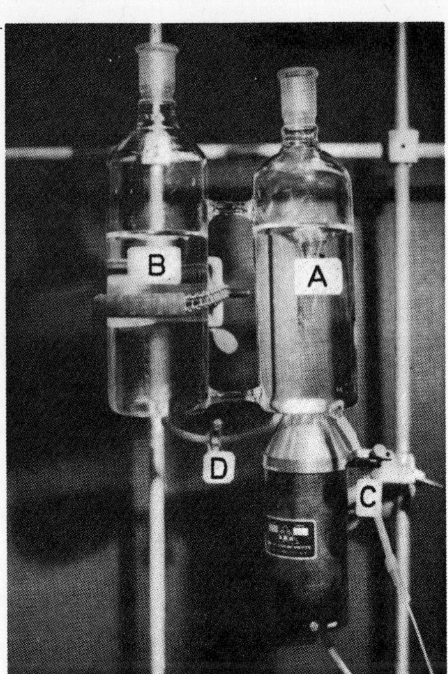

Fig. 4a. Linear gradient device constructed in the laboratory

buffer, i.e., 0.005 M pH 7 sodium phosphate, for at least two
days at approximately 5°C. Carefully pipette 5 ml dialyzed immune serum into the column. Let the serum also enter the cellulose bed, then varefully add 5 ml of the above 0.005 M sodium
phosphate buffer, pH 7.

   3. Take a gradient elution device (Fig.4a), close clamps C
and D, and fill into chamber A 400 ml 0.005 M pH 7 sodium phosphate buffer and into chamber B 400 ml 0.05 M NaCl dissolved in
0.05 M sodium dihydrogen phosphate. Start the magnetic stirrer
in chamber A. Open the clamps and start the elution of the proteins from the chromatographic column. Turn on the automatic
recorder connected with the UV flow analyzer. Collect samples,
obtaining 5 ml effluent in each tube.

Evaluation

The automatic recorder will draw a curve showing the protein
content of the chromatographic effluent. Such a pattern obtained
under conditions identical to those described here is shown in
Fig.4b.

   The content of those tubes which contain protein according
to the chart of the recorder may be tested for homogeneity in
the immunoelectrophoretic system. If rabbit antiserum has been
fractionated on the DEAE column, antirabbit goat serum may be
used in the immunoelectrophoresis. You will see that only the
first few tubes of the fraction are homogeneous protein preparations, while the others are mixtures of several different serum
proteins. Sterile handling of the entire procedure is possible
by autoclaving the glassware, the solutions, and the DEAE cellulose slurry.

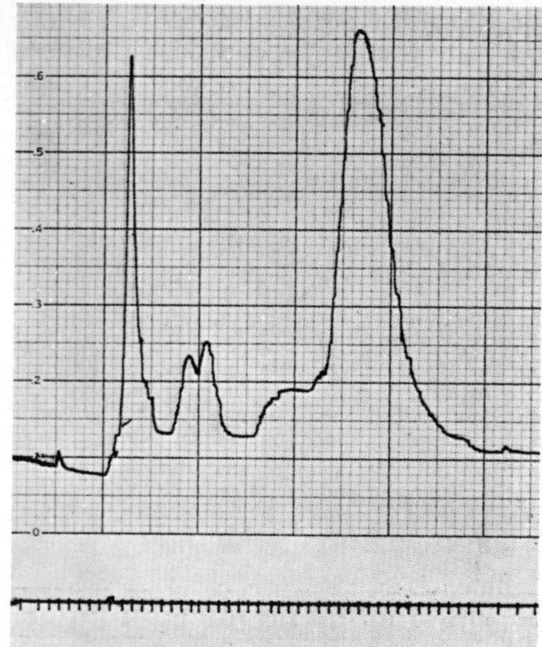

Fig. 4b. Typical serum protein elution pattern from DEAE chromatographic column. Note the signals of the "event marker" on the bottom

Part B. Batch-Type Operation

Materials and Equipment

The same materials will be needed as in *Part A*, with the exception of 0.005 M pH 7 phosphate buffer containing 0.3 M NaCl, and the same 15 ml centrifuge tubes.

Procedures

1. Prepare the DEAE cellulose as described in *Part A1*, and after final washing with 0.005 M pH 7 phosphate buffer, filter the cellulose on a Büchner funnel using vacuum.
2. The serum sample must be dialyzed against 0.005 M pH 7 phosphate buffer as described above in *Part A2*.
3. To 2 ml serum, add 5 g wet but well-packed DEAE cellulose as obtained above in Step 1. Stir the mixture at room temperature for 30 min.
4. Centrifuge the suspension at 1000 g for 15 min, decant the supernatant, wash the sediment by resuspending it in 2 ml of the above pH 7 phosphate buffer. Repeat the washing once more. Combine these with the first supernatant. This solution will be rich in IgG.
5. To elute other components from the DEAE cellulose, add 2 ml 0.3 M sodium chloride containing 0.005 M pH 7 phosphate buffer. Stir for approximately 30 min. Centrifuge, save the supernatant, and wash the cellulose twice with the same salt solution. This supernatant will be enriched in IgM.

6. If the cellulose residue is now washed by 0.8 M sodium chloride containing phosphate buffer, this will elute serum albumin and other globulins from the cellulose.

Evaluation

To evaluate the purity of the fractions obtained by batch-type operation regular or crossed immunoelectrophoresis procedures should be used, as described in Exercise No.72 and No.74 of this Manual.

Use and Limitations

The separation with the batch-type operation is very simple but the fractions obtained may be contaminated with other serum components. Repetition of the entire procedure with the isolated fractions will greatly improve the purity of the preparations.

It is important to emphasize the necessity of proper equilibration of the DEAE cellulose with the phosphate buffer. The dialysis of the serum samples against the same buffer for at least two days is equally important. These apply to both the column chromatographic as well as the batch-type operations.

It should be mentioned here, that a new DEAE cellulose in beaded form became available from Pharmacia Fine Chemicals (Piscataway, NJ 08854). This is sold under the trade name DEAE-SE-PHACEL. It gives a good flow rate in column chromatographic application, it does not release fine cellulose particles and it has a much higher capacity than the fibrous or the microgranular DEAE cellulose preparations, in both the column and in the batch-type operations.

References

James, K., Stanworth, D.R.: J. Chromatogr. *15*, 325 (1964)
Sober, H.A., Gutter, F.J., Wyckoff, M.M., Peterson, E.A.: J. Am. Chem. Soc. *78*, 756 (1956)

Exercise No. 4

Dissociation of Macroglobulins With 2-Mercaptoethanol

Deutsch and Morton (1957) reported that sulfhydryl compounds, such as 2-mercaptoethanol, are able to dissociate IgM macroglobulins, thus reducing their 19 S sedimentation constants to 6.5 S. The reactivity of the IgM with the corresponding antigens is also abolished. The breakdown products of the dissociated macroglobulins will be recombined if the 2-mercaptoethanol is removed by dialysis, but the conformation of the reaggregated macromolecules is said to be different from the native globulin. This recombination may be prevented by the addition of SH-blocking agents such as iodoacetate to the phosphate buffer in which the dialysis of the 2-mercaptoethanol and macroglobulin mixture is carried out.

   Two procedures are described here. The more elaborate procedure given first in this Exercise in *Part A,* follows the original
description of Deutsch and Morton. A very simple modification
of it is given in *Part B,* which can be applied for both isolated
macroglobulins and for dissociation of macroglobulins in whole
serum without previous fractionation. It can also be applied
for the dissociation of mammalian cell membrane components.

Materials and Equipment

   Rabbit immune serum obtained in Exercise No. 65
   IgM serum globulins obtained in Exercise No. 2
   1 *M* 2-mercaptoethanol (reagent grade) solution in water
   Iodoacetate
   0.2 *M* pH 7.4 phosphate buffer
   Dialysis bags and jars

Procedure

Part A

   1. Adjust the protein concentration of the IgM solution between 1.0% and 1.5%. Take 2 ml of this solution, or 2 ml undiluted immune serum, and mix with 2 ml of 0.2 *M* pH 7.4 phosphate
buffer. Add 0.5 ml 1 *M* 2-mercaptoethanol to the sample, mix
thoroughly, and let stand at room temperature for 48 h. Transfer
the preparations into dialysis bags and dialyze them in the cold
room for three days against 0.2 *M* pH 7.4 phosphate buffer, changing the outer fluid twice daily. The buffer must contain 0.02 *M*
final concentration of iodoacetate.
   2. After dialysis, measure the obtained volumes in ml and
calculate the changes in concentration if dilution occurred during this operation. As a control, prepare samples of untreated
IgM and immune serum by using phosphate buffer containing iodoacetate (the same composition as was used for dialysis), achiev-
ing the same final dilution with both treated and untreated serum
samples.

Part B

A simplified version of the 2-mercaptoethanol treatment has been
used more frequently. This treatment does not include the removal
of the 2-mercaptoethanol from the serum.
   1. Mix 0.2 ml of the above immunoglobulin solution or 0.2 ml
of an undiluted serum with 1.6 ml saline and 0.2 ml 1 *M* 2-mer-
captoethanol solution and let it stand at room temperature for
1 h.

Evaluation

Since 2-mercaptoethanol inactivates IgM and IgA, compare the
serologic reactivity of the preparations obtained in the following assays: charcoal agglutination (Exercise No. 70) may be used

if the antiserum used was obtained from animals immunized with
preparations containing bacterial O-antigens; passive hemagglutination can also be used (Exercise No. 69); immune electrophoresis (Exercises No. 72 and 74); or passive cutaneous anaphylaxis as described in Exercise No. 88. The easiest way to
demonstrate the effect of 2-mercaptoethanol on the structure of
the IgM fraction is by the passive hemagglutination method.

Use and Limitations

The dissociation with 2-mercaptoethanol has been widely used
recently for the inactivation of macroglobulins. The same treatment leaves the serologic reactivity of the smaller immunoglobulins virtually unchanged, although, from simple chemical consid-
erations, one must assume that structural alterations are going
on in IgG antibodies as well as in other proteins. Evidence shows
that serologic reactivity of other than macroglobulins will also
be affected by the above treatment (Adler, 1965). Addition of
0.1 or 0.01 molar 2-mercaptoethanol to commercially available
fetal calf serum preparations will greatly improve their usefulness in lymphoblast transformation assays or in other tissue
culture techniques where fetal calf serum is used as an essential
ingredient of the culture medium.

References

Adler, Frank L.: J. Immunol. *95,* 39 (1965)
Deutsch, H. F., and J. I. Morton: Science *125,* 600 (1957)

Exercise No. 5

Enzymatic Cleavage of IgG

In the study of active sites of natural products one of the main
goals is to obtain the smallest possible fragments of the macromolecule while maintaining biologic activity. Such an attempt
was made by Porter (1959) who, in his classical work, described
the cleavage of rabbit antisera by papain and the chromatographic
fractionation of the fragments obtained. In the present exercise
the work of Porter will be reproduced.

Materials and Equipment

    IgG globulin (preparation No. 1 obtained in Exercise No. 3)
    Crystalline mercuripapain (Commercially available from
     Worthington Biochemical Corp., Freehold, NJ 07728)
    Dextrane
    EDTA buffer. 0.15 *M* pH 7 phosphate buffer which contains
     10 m*M* cysteine and 2 m*M* EDTA, Na salt
    Dialysis bags and jars

Procedure

1. Dissolve 150 mg rabbit IgG in 10 ml EDTA buffer. If chromatographic effluent or filtrate of DEAE cellulose fractionation is used, its protein content should be between 10 and 30 mg/ml. In order to achieve this, the eluate must be concentrated. The easiest way is to fill a small dialysis bag with the eluate, then close both ends of the bag with strong knots. Lay the bag on a small tray or Petri dish and pack dry powder of high molecular weight dextrane around it. This will "suck out" the solvent from the bag, thus increasing the protein concentration inside. In approximately 3 to 4 h a 20 ml sample may be reduced to 5 ml. Predialyzed polyvinyl pyrrolidone or Carbowax may also replace dextrane. Another procedure to concentrate diluted solutions of biologically active and sensitive macromolecules, using a water-absorbing dried gel manufactured in granular form, is described in Exercise 8. If the protein concentration of the IgG solution is about 2%-3%, take 5 ml of this and mix it with 5 ml EDTA buffer.

2. Add to 10 ml of the buffer-dissolved immunoglobulin solution 1.5 mg crystalline mercury papain and let it stand with occasional shaking at 37°C for 16 h. In order to prevent bacterial growth during the incubation, a few drops of toluene must be added. Transfer the globulin plus enzyme mixture into a dialysis bag and dialyze against water in the cold room for two days. Store the globulin solution frozen or sterile-filtered.

Use and Limitations

Regular papain may also be used in this experiment, but crystalline mercuripapain is preferred because of its higher purity. Removal of the chelating agent EDTA from the reaction mixture converts the mercuripapain into an inactive dimer form, thus stopping the reaction.

A similar procedure can be used for the production of peptide fragments from other proteins, if the pH has been adjusted to the proper value. For the majority of proteins, the pH optimum of papain digestion is between 4 and 6. Exceptions are strongly basic or acidic substrates. It is recommended to measure the activity of the enzyme in the proper assay before its use to cleave IgG. Such a method is described by Smith and Kimmel (1960). The manual of the Worthington Biochemical corporation is also very useful in finding the proper assay procedures for the determination of enzymes.

References

Porter, R.R.: Nature (London), *182*, 670 (1958), and Biochem. J. *73*, 119 (1959)
Smith, E.L., Kimmel, J.R.: In: The enzymes, *4*, 133. Boyer, Lardy, Myrback (eds.), New York, N. Y.: Academic Press 1960

Exercise No. 6

Chromatographic Separation of the IgG Fragments

Carboxymethyl cellulose has been shown to be one of the most
useful ion exchangers in protein chromatography. The procedure
elaborated by Peterson and Sober (1956) as well as by Sober et
al. (1956) gives good resolution of different proteins. This
method has been applied by Porter (1959) for the separation of
breakdown products of IgG globulin.

Materials and Equipment

    Chromatographic column, 2 × 40 cm (Fig. 2)
    Filter flask
    Automatic fraction collector (Fig. 3)
    Vacuum pump
    UV spectrophotometer, or
    UV flow analyzer for chromatographic effluents combined with
      an automatic recorder and "event marker" (see Exercise No. 2)
    18 × 150 mm test tubes
    0.1 M sodium hydroxide
    1.0 M acetic acid
    Eluant buffers:
      (1) 2000 ml 0.01 M sodium acetate buffer pH 5.5
      (2) 0.9 M sodium acetate buffer pH 5.5.
      All buffers and diluents must be saturated with toluene
    Carboxymethyl (CM) cellulose
    Gradient producing device (see Fig. 4)
    Magnetic stirrer

Procedure

   1. Prepare 50 g carboxymethyl cellulose as follows. This ion
exchanger contains carboxyls as functional groups, therefore the
activation process is slightly different from that for DEAE. Stir
the CM cellulose into 500 ml 0.1 N NaOH and mix it slowly for
30 min at room temperature. Transfer it to a Büchner filter, remove
alkali by suction. Wash with distilled water until the filtrate
is neutral. Remove the cellulose cake from the filter and transfer it into 1000 ml 1 N acetic acid, stir gently for 30 min at
room temperature, then filter and wash as before until neutral.
Remove the cake again and suspend it in 500 ml 0.01 M acetate
buffer pH 5.5, stir and filter as above. Add to the cellulose,
without removing it from the filter, another 500 ml acetate buffer
and let half of this filter through by gravity, then apply suction.
Finally, remove the cake and suspend it in 500 ml fresh acetate
buffer and degas it as described in Exercise No. 3.
   2. Pour this slurry along a glass rod into the column, and by
repeated addition of further amounts, build up a column approximately 40 cm high. Force the liquid which is on top of the CM
cellulose column into the packing, using light pressure, but very
carefully avoid driving air into the packing. Place the column
in its final position, connect the Luer joints of the column and
the capillary Teflon tubing, be sure that the chromatographic
effluent goes through the UV flow analyzer, and also check to

see that the volumetric syphon or drop counter of the fraction
collector is working properly.
    3. If everything seems to funtion well, pipette 10 ml of the
dialyzed papain digest on the column, using slight pressure again
to force it into the CM cellulose. Carefully layer approximately
5 ml of the unused slurry on the top of this and start collecting
20 ml fractions.
    4. The first eluant should be 200 ml 0.01 M pH 5.5 acetate
buffer. From here on, a linear gradient will be formed from
0.01 to 0.9 M acetate buffer, increasing the molarity but not
changing the pH. A simple device for the production of a linear
gradient is shown in Fig. 4. In this exercise chamber A of this
devise is filled with 500 ml 0.01 M acetate buffer, and chamber
B with 500 ml 0.9 M acetate buffer. Let the magnetic stirrer run
vigorously and open connection D between the two chambers. Open
stopcock C of chamber A which leads the effluent to the top of
the column.

Evaluation

Continuous monitoring of the chromatographic effluent may be
carried out as described in Exercise No. 2.
    The serologic activity of the eluates obtained may be investi-
gated in different systems. The papain digested fractions do not
form precipitates but are active in inhibiting the corresponding
antigen-antibody reaction. For these purposes, the description
in Exercise No. 69 may be used, with the difference that the
antibody fragment should be incubated with the antigen-coated
red blood cell preparation for 30 min. This will block the anti-
gen-receptor sites, which have been passively transfered to the
red blood cell surface; therefore they will not react with
homologous antibodies. The quantitative precipitin test will also
be inhibited by those antibody fragments which have the reactive
sites. The antigens must be incubated with the obtained fractions
for 60 min at 37°C, and then the reactivity of the antigens with
the homologous antiserum may be investigated.

References

Peterson, E.A., Sober, H.A.: J. Am. Chem. Soc. *78*, 751 (1956)
Porter, R.R.: Biochem. J. *73*, 119 (1959)
Sober, H.A., Gutter, F.J., Wyckoff, M.M., Peterson, E.A.: J. Am. Chem. Soc.
    78, 756 (1956)

Exercise No. 7

Separation and Isolation of H and L Chains of IgG

Reductive cleavage of the -S-S- bonds in an IgG molecule will
liberate the H and L chains. If this cleavage is carried out
near neutral pH and with mild reducing agents such as dithiothreitol, no further degradation and denaturation of the split
products will occur. To block the sulfhydryl groups set free by

the reductive cleavage, alkylation with iodoacetamide has to be carried out. The H and L chains can be separated by gel electrophoresis in 8 M urea-formic acid buffer, or by column chromatography on Sephadex G 100 equilibrated with 8 M urea in pH 6 Tris-citrate buffer.

The method for dithiothreitol cleavage of the IgG molecule has been reported first by Gunewardena and Cooke (1966). The modification as described here has been elaborated by Karush and associates (1977) and communicated for inclusion in this manual by Ming-Ming Chua from the same laboratory.

Materials and Equipment

10 mg IgG

pH 8.0, 0.2 M Tris-HCl buffer. Mix 50 ml 0.2 N HCl and 56.4 ml 0.2 M (2.42 g in 100 ml) Tris

pH 6.0, 0.01 M Tris-citrate buffer. Add to 1000 ml 0.01 M Tris (1.21 g/l) 0.01 M citric acid solution (2.1 g/l) until pH 6.0 is reached

0.1 M 1,4-dithiothreitol. Dissolve 0.154 g in 10 ml distilled water. The dissolution has to be carried out by bubbling N_2 gas through the solution at room temperature for 60 min. Close the container. Prepare the solution fresh every time shortly before use

0.5 M iodoacetamide. Dissolve 0.93 g in 10 ml distilled water. This solution also has to be prepared fresh before its use

Sephadex G 100 column, 1.6 × 100 cm. Prepare this column as described in Exercise 2, but disperse the Sephadex in 8 M urea containing pH 6.0, 0.01 M Tris-citrate buffer. Take 1000 ml of the above buffer and dissolve in it urea crystals until the concentration reaches 8 M (480 g/1000 ml)

Procedure

1. Dissolve 10 mg IgG in 1.8 ml 0.2 M Tris-HCl, pH 8 buffer. Stir it to facilitate dissolution by bubbling N_2 gas through the buffer for about 10 min at room temperature.

2. Add 0.2 ml 0.1 M dithiothreitol (final concentration is 10 mM). Continue stirring with N_2 gas for 60 min.

3. Add 0.2 ml 0.5 M iodacetamide (final concentration of this is 50 mM). Continue stirring with N_2 gas at room temperature for 60 min.

4. Dissolve urea crystals in the reaction mixture to reach 8 M final concentration. In this experiment add 1.05 g.

5. Apply the entire reaction mixture to the Sephadex G 100 column. Start elution with pH 6.0, 0.01 M Tris-citrate buffer containing 8 M urea. Collect 5 ml aliquots and monitor the chromatographic effluent for proteins using a UV flow cell at 280 nm.

Evaluation

The first peak leaving the column right after the void volume contains the H chains, which are excluded from this gel. The L chains will be in the next peak. The ratio of H to L quantities is 2 to 1.

    If specific antisera to L and H chains are available, the
various gel diffusion and gel electrophoretic procedures described in Exercises 71, 72, and 74 can be used to determine the
homogeneity of the obtained fractions.

Use and Limitations

The H and L fractions may be contaminated by each other if the
chromatographic separation was not efficient. In such cases the
fractions should be rechromatographed using the same column. Of
course the column has to be regenerated by washing with 500 ml
Tris-citrate buffer overnight.
    The pure H and L chains can be used for structural analyses.
Amino acid sequence studies led to the discovery of constant,
variable and hypervariable regions in the L and H chains. (For
reviews on this topic see *Contemporary Topics in Molecular Immunology*
vols 2 (1973) and 4 (1975), Plenum Press.) They can also serve
as immunogens to produce specific antibodies to them, which are
extensively used in various immunochemical studies. The isolated
chains have been used to study structural changes associated
with their interaction (Lapanje and Dorrington, 1973). Hong and
Nisonoff (1966) prepared hybrid antibodies by combining isolated
H and L chains from two different antibodies and studied the
involvement of H and L chains in the affinity of immunoglobulins
to certain haptens.

References

Gunewardena, P., Cooke, K.B.: Biochem. J. 99, 8 p, (1966)
Hong, R., Nisonoff, A.: J. Immunol. 96, 602 (1966)
Lapanje, S., Dorrington, K.J.: Biochim. Biophys. Acta 322, 45 (1973)
Mitchell, K.F., Karush, F., Morgan, D.O.: Immunochemistry 14, 161 (1977)

Exercise No. 8

Affinity Chromatography and Its Application to Antibody Isolation

In affinity chromatography the chromatographic bed material contains a bound ligand. This ligand and the bed material form an
insoluble matrix. The ligand may be an antigen, or an enzyme
substrate or enzyme inhibitor, or other molecule which can react
specifically with certain molecules. The column filled with such
insoluble matrix containing a covalently bound antigen can be
used to separate the specific antibodies by passing serum through
this column. Enzymes can be also isolated from various extracts
if the covalently bound ligand is in an enzyme substrate (or
inhibitor) with which the enzyme can form a complex. Accordingly,
in affinity chromatography the separation is based on the reactivity of components.

To establish a covalent linkage between the ligand and the bed material, several procedures have been developed in the past, such as binding of proteins to p-aminobenzyl cellulose by the diazo reaction. A number of other methods have also been elaborated. One of the most efficient procedures is the cyanogen bromide (CNBr) method of Axen et al., 1967. Practically any compound which contains amino groups may be coupled to a CNBr activated Sepharose matrix. CNBr most probably reacts with the hydroxyl groups of Sepharose and they form cyclic and noncyclic imidocarbonates. The imidocarbonate groups react with amino groups on the ligand molecule.

Sepharose-bound enzymes will absorb antienzyme antibodies. Elution of the antibodies can be achieved by a variety of solvents. Low pH and higher ionic strengths usually dissociate antigen-antibody complexes. Slight elevation of the chromatographic temperature will facilitate the dissociation. In this exercise, 3 M ammonium thiocyanate will be used, according to the method developed by Anderson et al. (1975) to separate horseradish peroxidase enzyme from its antibodies.

Materials and Equipment

10 g Sepharose 4B-CNBr in lyophilized form. This amount is
  sufficient to prepare a 35 ml column. Available from Pharmacia Fine Chemicals, Inc., Piscataway, NJ 08854
0.1 M NaHCO$_3$ containing 0.5 M NaCl. Dissolve 8.4 g NaHCO$_3$ and
  29.2 g NaCl in 1000 ml. The pH of this solution is 8.2
1 M ethanolamine in water (61.1 g in 1000 ml)
0.001 M HCl
0.1 M acetate buffer, pH 4. This buffer also contains 0.5 M
  NaCl. Mix 18.0 ml 1 N NaOH and 100 ml 1 N acetic acid. Make
  up to 1000 ml. Dissolve 29.2 g NaCl in this solution
0.1 M pH 7.2 phosphate buffer. Mix 726 ml 0.1 M Na$_2$HPO$_4$ H$_2$O
  (7.8 g in 1000 ml) and 274 ml 0.1 M KH$_2$ PO$_4$ (13.6 in 1000 ml)
3 M ammonium thiocyanate (dissolve 228.36 g ammonium thiocyanate in 800 ml water and make up to 1000 ml)
0.1 M pH 8 borate buffer containing 0.5 M NaCl. To prepare
  this buffer mix 441 ml 0.1 N HCl with 559 ml 0.2 M sodium
  borate. The sodium borate is prepared by dissolving 12.4 g
  H$_3$BO$_4$ in 100 ml 1 N NaOH and made up to 1000 ml with distilled
  water
Chromatographic column, 15 × 200 mm (Fig. 2)
12-15 cm diam. medium glass filter
Adjustable speed motor stirrer with glass rod
3000 ml Erlenmeyer flask
1000 ml glass beakers
10 ml anti-horseradish peroxidase serum prepared in sheep or
  rabbits by the procedure described in Exercise No. 84
Lyphogel beads, available from Gelman Co., Ann Arbor, MI 48106
Phosphate buffered saline (PBS). See Exercise No. 1.
Horseradish peroxidase enzyme (Worthington, Freehold, NJ 07728)
0.1% bovine serum albumin dissolved in PBS

Procedure

1. Dissolve 200 mg horseradish peroxidase in 500 ml 0.1 M
pH 8.2 NaHCO$_3$ - NaCl buffer at room temperature.

2. Take a 3000 ml Erlenmeyer. Add 10 g Sepharose 4B-CNBr
powder in small proportions to 2000 ml 0.001 M HCl. Stir constantly but slowly with a blunt glass rod attached to a motor,
applying approximately 100-120 revolutions per minute. Do not
use a magnetic stirrer or other stirring devices which may break
up the beads. Continue the stirring for approximately 1 h at room
temperature or overnight in a cold room. After this filter the
swollen gel beads on a medium glass filter and wash with the
above HCl, using another 2000 ml and gentle vacuum to facilitate
the filtration. The gel beads are now ready to couple them with
protein. This should be carried out without delay.

3. Take a 1000 ml beaker and mix in it the enzyme solution
with the swollen, filtered Sepharose 4B-CNBr gel beads. Again
use a glass rod stirrer at low speed at room temperature for 2 h.
Filter the gel again, as in step 2 above.

4. Transfer the gel from the filter into a 1000 ml beaker.
To block the remaining active groups in the gel, add 500 ml 1 M
ethanolamine and stir slowly for 2-3 h at room temperature. Filter
as above.

5. The enzyme coupled CNBr-Sepharose gel can be washed on the
filter at this point to remove uncoupled ligand and excess
ethanolamine. For this purpose, 300 ml pH 4 0.1 M acetate buffer
containing 0.5 M NaCl should be used first. Add this to the gel
on the filter, apply no vacuum, stir the gel gently with a glass
rod and after about 10 min, by applying a gentle vacuum, filter
through the excess acetate buffer. Disconnect the vacuum again,
add 300 ml 1 M pH 8.0 borate buffer containing 8.5 M NaCl, stir
gently for another 10 min and vacuum filter as above. Repeat the
acetate and borate buffer washes alternately at least three times.
The last wash should be PBS. If the gel coupled to enzyme is not
to be used immediately, it should be stored in the cold room at
around +5°C. The gel should never be frozen.

6. To fill the column for chromatography, filter the gel again,
wash it once and disperse it in 100 ml PBS. Raise the effluent
tube to the top of the column. Fill it with the slurry by using
the same procedure and precautions as in every other Sephadex
type column chromatograpy (see Exercise No. 2).

7. Before use: a) wash the column with 100 ml 0.1% bovine
serum albumin dissolved in PBS, b) allow 500 ml PBS to go through
the column overnight in the cold room.

8. Antiserum to horseradish peroxidase can be prepared as described in Exercise No. 84. Dialyze 10 ml of this serum for three
days in the cold against PBS, then dilute it 1:3 with PBS. Pass
the total volume through the column, using a flow rate of not
more than 50 ml/h. Collect the effluent in one flask. Now wash
the column with a total of 200 ml PBS. The last milliliters should
not contain detectable amounts of protein, as determined by UV
spectrophotometry at 280 nm. If this is not the case, wash with
another 200 ml PBS.

9. The next step is the elution of absorbed antibodies. One
may wish to monitor the proteins leaving the column by using a
UV flow cell and collecting the effluent with a fraction collector,

as described in Exercise No. 2. Using the same flow rate as in
step 8 above, 200 ml 3 M NH$_4$SCN will flow through at room temperature.
    10. Dialyze the eluate for three days against PBS at cold
room temperature, changing the outer fluid twice daily. The use
of "hollow fiber" dialyzing systems, such as those available
from Amicon Corp., (Lexington, MA 02173) or chromatography on
Sephadex G-25 will speed up the dialysis procedure considerably.
    11. The immunoglobulins in the eluate can be concentrated by
precipitating them with ammonium sulfate at a 33% saturation
level (mixing 2 vol of dialyzed antibody solution with 1 vol of
saturated ammonium sulfate at room temperature) as described in
Exercise No. 1. Another procedure to concentrate the dialyzed
eluate uses Lyphogel. This consists of large beads of dry gel
which will pick up water and low molecular weight substances but
will exclude macromolecules, such as antibodies. The water uptake
capacity of the completely dry Lyphogel is more than five times
its dry weight. The uptake is fast and can be carried out in the
cold room without any damage to antibodies, and the recovery is
satisfactory. The swollen beads can be removed by vacuum filtration. Rinsing the filtered bead surface with a few milliliters
of saline (without vacuum), is recommended for better recovery
of concentrated antibodies. For the determination of the titer
in the individual tubes of the fraction collector or in the
pooled eluate, any of the procedures described in Exercises No.
69, No. 73 or No. 78 can be used. To establish the homogeneity
of the antibody preparation, immunoelectrophoresis should be
used. As antigen, a horseradish peroxidase preparation should
be used.

Use and Limitations

The enzymatic activity of the Sepharose-bound peroxidase can be
determined by using the procedure of Chance and Maehly (1955)
or by the more recent method described in the manual of Worthington Biochemical Products (1977).
    Such insolubilized or immobilized enzyme preparation can also
be prepared from other enzymes. If they are to be used only for
enzymatic reactions, the purity of the enzyme ligand is not always
critical. The activity of immobilized enzymes is usually reduced
as compared to the activity of free enzyme molecules. A great
advantage of such immobilized enzymes lies in the fact that they
can be removed from reaction mixtures by simple filtration or
centrifugation. One excellent application of immobilized lactoperoxidase coupled to CNBr activated Sepharose is for iodination
of proteins, as mentioned in Exercise No. 13. Several other
immobilized enzyme preparations are commercially available from
Worthington Biochemical Corp., Freehold, NJ 07728.
    Ammonium thiocyanate, a strong chaotropic agent, is frequently
used for the dissociation of antigen-antibody complexes, without
introducing irreversible changes in the molecular structures.
The effect of chaotropic ions on the dissociation of antigen-
antibody complexes was studied in detail by Dandliker and associates (1967). According to these studies, 2 M solution of NaSCN
at pH 6 was found to be one of the most efficient eluants. To
separate enzyme substrates or enzyme inhibitors, procedures similar to those described above can be used.

The antibody binding capacity of antigen ligand-containing
Sepharose is surprisingly high. Wofsky and Burr (1969) reported
successful isolation of 98% of all antibodies present in 370 ml
ascites fluid by passing it through a Sepharose-ligand system
having a total bed volume of only 16 ml. The amount of antibody
in the ascites fluid was 0.4 mg/ml. The applied flow rate was
50 ml/h. For the elution of the column, they applied 10^{-2} M
solution of the ligand and 0.1 M acetic acid. The eluted antibody
was purified by extensive dialysis.

Matrices other than Sepharose were also used, such as controlled pore glass beads. The glass surface may be treated so that
it will absorb a layer of ligand. The great advantage of the
porous beads as matrix is that such columns can work under relatively high pressure.

The applications of affinity chromatography in biochemistry,
immunochemistry, and immunology as well as in a number of related
fields is growing rapidly. Both DNA and RNA have been attached
to cellulose in order to purify other nucleic acids. Plant protein lectins, which have the capacity to bind certain carbohydrates, such as Concanavalin A, have been coupled to agarose.
Haptenic compounds such as dinitrophenyl group-containing peptides
were used to isolate highly specific antibodies. Whole cells and
subcellular organelles were also purified by affinity chromatography. Excellent reviews on the uses of affinity chromatography
were published. To mention only two, we refer here to the work
of Cuatrecasas and Anfinsen (1971), and Weetall (1973).

<u>References</u>

Anderson, N.G., Willis, D.D., Holladay, D.W., Caton, J.E., Holleman, J.W.,
    Eveleith, J.W., Attrill, J.E., Ball, F.L., Anderson, N.L.: Anal. Biochem.
    66, 159 (1975)
Anderson, N.G., Willis, D.D., Holladay, D.W., Caton, J.E., Holleman, J.W.,
    Eveleith, J.W., Attrill, J.E., Ball, F.L., Anderson, N.L.: Anal. Biochem.
    68, 371 (1975)
Axen, R., Porath, J., Ernback, S.: Nature London *214*, 1302 (1967)
Chance, B., Maehly, A.C.: In Methods of enzymology, Colowick, Kaplan (eds.)
    vol. II. p. 764, 1955
Cuatrecasas, P., Anfinsen, C.B.: Ann. Rev. Biochem. *40*, 259 (1971)
Danliker, W.B., Alonso, R., de Saussure, V.A., Kierszenbaum, F., Levison, S.A.,
    Schapiro, H.C.: Biochemistry *6*, 1460 (1967)
Weetal, H.H.: Separation and Purification Methods *2*, 199 (1973)
Wofsky, L., Burr, B.: J. Immunol. *103*, 380 (1969)
Worthington Biochemical Corporation, (Freehold, NJ 07728), Manual on: Enzymes,
    enzyme reagents and related biochemicals, p. 66, 1977

Exercise No. 9

Cellulose Acetate Paper Electrophoresis of Human Serum

Compounds with positive or negative net electric charges will
migrate towards the opposite poles in an electric field. This
phenomenon is called electrophoresis. If a mixture of different

components, such as serum proteins, or amino acids, is placed
in this electric field, the components will migrate toward the
oppositely charged electrodes. The direction and velocity of
their migration is determined mainly by their net electric
charge.

The electric field may be established in a glass cuvette filled with electrolyte or in a capillary system soaked with similar
electrolyte. The most popular capillary systems used for electrophoresis are filter paper and various gels. Paper electrophoresis
is a procedure in which the electric field is developed in a good
quality filter paper impregnated with a buffer. The paper acts
merely as a porous supporting material preventing remixing of
the components. Provided fairly low voltages (5-10 V per cm) are
used, very simple apparatus will suffice. The disadvantage of
the low voltage separation is that the required time for separation is long and the separations are not very sharp as a result
of diffusion.

The application of electrophoresis in a U-shaped cuvette filled
with electrolyte was developed by Tiselius (1937). This "Free
Boundary Electrophoresis" has been applied to the separation of
serum proteins. The much less demanding and much less expensive
paper electrophoretic separation of serum proteins was elaborated
by Turba and Enenkel (1950). The method described here is one of
the numerous modifications, using cellulose acetate paper strips
and microliter amounts of serum. For the staining of proteins the
acid fuchsin procedure of Nowotny (1952) will be applied.

Materials and Equipment

    Cellulose acetate strips 1″ × 6″ (available from Gelman Instru-
      ment Company, Ann Arbor, Mich. 48106). If regular filterpaper
      (Whatman No.1 or other similar quality) is used, cut 4 × 25 cm
      strips. These will fit into the equipment shown on Fig. 5.
      Apply 0.05 ml serum on regular filter paper strips
    Normal human serum
    Barbital buffer, pH 8.6, made by mixing 9.2 ml of 0.1 N HCl
      and 90.8 ml of 0.1 M sodium barbital with 100 ml water
    0.2% Acidic fuchsin in methanol : acetic acid : water = 5:1:4
    Methanol : acetic acid : water = 5:1:4 (without fuchsin)
    10% acetic acid in water
    Micropipettes
    Large beakers
    Petri dish
    Electrophoresis chamber (Fig. 5)
    Power supply
    Chromatographic drying oven

Procedure

1. Fill the electrophoresis chamber with the barbital buffer
pH 8.6, covering the electrodes with the electrolyte. Make connection between chambers having indentical poles with a soaked
piece of ordinary filter paper.

2. Mark the ends of the cellulose acetate strip with plus and
minus signs. Draw a line with a soft pencil across the cellulose

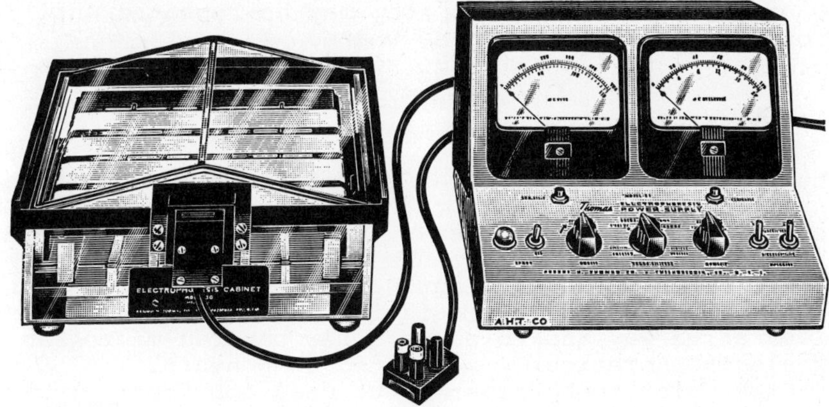

Fig. 5. One of the commercially available models for paper electrophoresis (courtesy of Arthur H. Thomas, Philadelphia, PA.)

acetate strip approximately 3 cm from the negative end. Write the number of sample and other data on the negative end.

3. Dip the cellulose acetate strip into a Petri dish containing the same buffer as the electrophoresis chamber. Put the soaked strip on a clean, dry, regular filter paper. By gently patting the cellulose acetate strip, let the dry filter paper take off the excess buffer. Place the wet strip in the electrophoresis chamber, immerse the negative end of it in the compartment containing the negative electrode, contact the positive end with the other pole.

4. Take 0.005 ml serum into a micropipette, wipe off excess serum from the outside. By gently touching the cellulose acetate strip along the pencil line several times, transfer the 5 μl serum to the paper strip, forming a starting line across the strip. Do not bring the serum close to the edges of the strip. Cover the chamber.

5. Connect the electrophoresis chamber with the power supply, turn on the main switch. Adjust the power to 200 V. Measure and record the current going through the strip. Let the electrophoresis proceed for 60 min. Turn off the main switch and wait a few seconds before touching the chamber; the capacitors of the power supply are still loaded.

6. Take the strip out with forceps, touching only the very end of it. Hang the strip with a wooden clamp in the preheated drying oven, let it dry for 5 min at 90°C.

Evaluation

The staining procedure is simple with acidic fuchsin. Pour 500 ml dye into a 1000 ml beaker and immerse the dried strip into the acidic fuchsin dye for 10 min. Remove it carefully with the forceps, without damaging the strip, and transfer it to the acidic methanol solution for 10 min to wash the dye from the paper. The separated protein bands will remain deep red. The washing can be improved by moving the beaker gently. Final washing is done in

10% acetic acid, and this step may be repeated until the background of the paper becomes white.
    Observe and identify the separated serum protein components.

Use and Limitations

The resolution of the paper electrophoretic method applied to
serum proteins is inferior to immune electrophoresis which
reveals many more components. The advantage of the paper electrophoretic method versus immune electrophoresis lies in the detection of weakly- and non-antigenic constituents. Another advantage
of the paper electrophoresis is that it is easy to use it in a
preparative way and to elute a few hundred microgram quantities
of electrophoretically separated components.
    Paper electrophoresis may be used for the analysis of differ-
ent proteins, enzymes, tissue extracts, animal poisons, amino
acids, polysaccharides, borate complexes of carbohydrates, inorganic ions, etc. Applying high voltage for a better and faster
separation is described in Exercise 37. Good reviews of the applications of paper electrophoresis were written by Block and
co-workers (1958) and by Heftmann (1961).
    Other electrophoretic procedures, which are much more fre-
quently used today than the paper electrophoretic techniques
are described in Exercises 10 and 11.

References

Block, R.J., Durrum, E.L., Zweig, G.: A manual of paper chromatography and
    paper electrophoresis. New York: Academic Press 1958
Heftmann, E. Chromatography. New York: Reinhold Publ. Corp. 1961
Nowotny, A.: Acta Phys. Acad. Sci. Hung. *3,* 469 (1952)
Tiselius, A.: Trans. Farad. Soc. *33,* 524 (1937)
Turba, F., Enenkel, H.J.: Naturwissenschaften *37,* 93 (1950)

Exercise No. 10

Disc Electrophoresis in Polyacrylamide Gel

Earlier electrophoretic procedures, such as paper electrophoresis
described in Exercise No. 9 of this Manual, were rapidly replaced by the much better resolving electrophoresis procedures
in gel. While regular paper or cellulose acetate paper electrophoresis detected a maximum of 6 to 7 components in a human serum
sample, electrophoresis in polyacrylamide gel separates 15 or
more components with ease. Particularly sharp electrophoretic
separation can be achieved by the so-called disc electrophoresis.
There are a number of physical-chemical explanations for the very
high resolution achieved by such procedures, but discussion of
these would be beyond the scope of this Manual. Those who wish
to understand the background of this phenomenon should read the
relevant chapter in the book by H.R. Maurer (1971) on disc electrophoresis. *Part A* of this Exercise describes the separation of
serum protein components by disc electrophoresis.

Further variations of polyacrylamide disc electrophoresis
exist. The procedures using sodium dodecyl sulfate (SDS) detergent are particularly valuable. With the use SDS one can separate
proteins normally insoluble in aqueous environments, for example,
proteins of cell membranes. Parts of these proteins are lipophilic,
residing in the bimolecular lipid layer of the cell membrane.
Detergents solubilize such proteins by hydrophobic binding of the
aliphatic regions of the detergent with the nonpolar region of
the polypeptide chain. Since the polar end of the detergent
remains to interact with water, it will solubilize the membrane
and liberate some components from it. *Part B* of this Exercise
describes such an experiment.

What made the SDS electrophoresis procedures particularly
useful was the fact that the presence of detergents in the electrophoresis system allows the estimation of molecular weights of
proteins. The reasoning is as follows: Most proteins have a great
number of nonpolar regions. The detergent molecules, such as SDS,
will bind in large numbers to nonpolar regions of protein molecules, will add a negative charge to them, and will mask the net
charge of the native proteins. The migration of such SDS saturated protein molecules will now depend almost entirely on the
dimensions of the molecules, and will be inversely proportional
to the molecular weight. Shapiro, Vinuela, and Maizel were the
first to describe molecular weight determination in SDS gel in
1967. *Part C* of this Exercise describes an experiment determining
the molecular weight of proteins. In all three parts of this
exercise the methods of Weber and Osborn (1969) were adapted

Part A. Separation of Serum Proteins

Materials and Methods

Protein solution for analysis. If serum is subjected to electrophoresis, no SDS dissociation is required. Dilute the
serum 1:5 using the buffer for gel. A 1 to 2 mg/ml protein-
containing solution should be prepared from other samples
Acrylamide solution: dissolve 22.2 g acrylamide and 0.6 g methylene-
bis-acrylamide in 100 ml water. Filter if necessary through
Whatmann No. 1 filter paper. Keep the filtrate in a cold room
in a dark bottle. (Warning: Acrylamide and its derivatives
are very poisonous in their monomeric form. The chemicals
irritate the skin, eyes, and respiratory system. It is a
neurotoxin which acts through inhalation and can cause headache,
nausea, vomiting, dizziness, blurring of vision, muscular
dysfunction and, if taken in large quantities, death. If it
gets on your skin, wash it off immediately. Once polymerized,
the acrylamide is no longer toxic.)
1.5% ammonium persulfate in water, freshly prepared for each
experiment
0.05% bromphenol blue dye in water
N,N,N', N' - tetramethyl ethylenediamine (TEMED)
Phosphate buffer for gel (pH 7.2): dissolve 7.8 g $NaH_2PO_4 \cdot$
H_2O and 38.6 g $Na_2HPO_4 \cdot 7H_2O$ in 900 ml water, adjust the
volume to 1000 ml

Phosphate buffer for electrophoresis: dilute above buffer
 with water 1:10
15% trichloroacetic acid in water. Prepare 10 ml
Staining solution: dissolve 1.25 g Coomassie Brilliant Blue in
 a mixture of 450 ml 50% methanol and 50 ml glacial acetic
 acid. Filter the dye solution if necessary through Whatman
 No. 1 filter paper. The destaining solution consists of
 100 ml acetic acid mixed with 100 ml methanol and 800 ml
 water, without the dye
Büchler gel electrophoresis apparatus (Fig. 6). Several other
 equally useful models exist but simple equipment can be constructed in the laboratory using the lower halves of two
 plastic bottles, as shown in Figure 7
12 electrophoresis tubes, 5 mm in diam and 80 mm long, open
 at both ends. These tubes must be washed before each use in
 chrome-sulfuric acid. The thoroughly water-washed tubes should

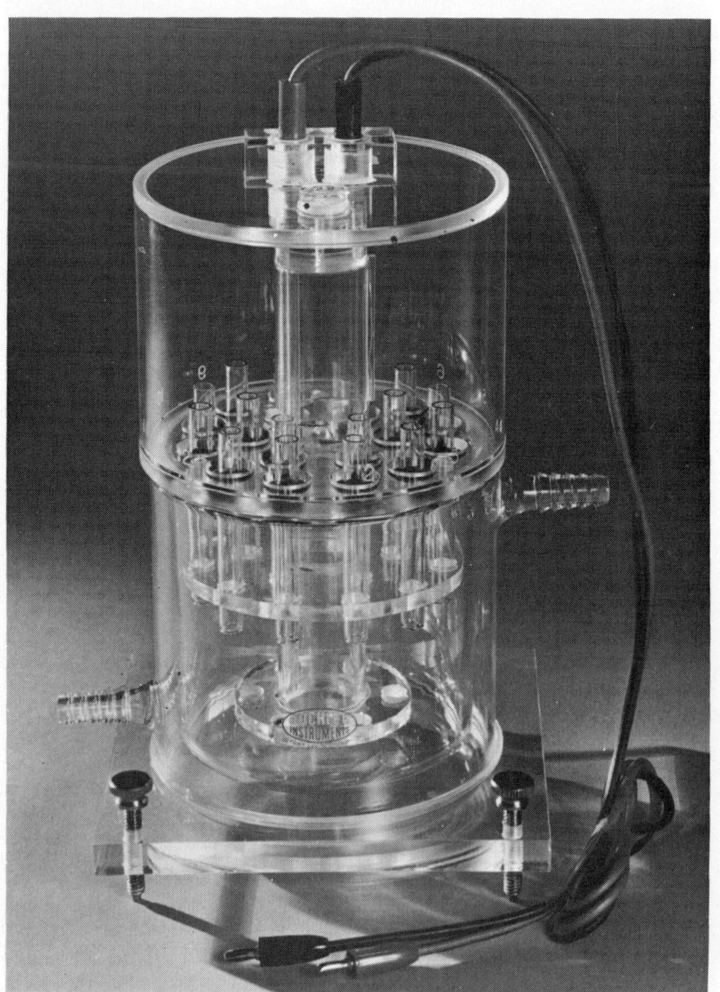

Fig. 6. Commercial
instrument for disc
electrophoresis
(Courtesy of Büchler
Instruments, Fort
Lee, NJ 07024)

<u>Fig. 7.</u> Home-made assembly for disc electro-
phoresis

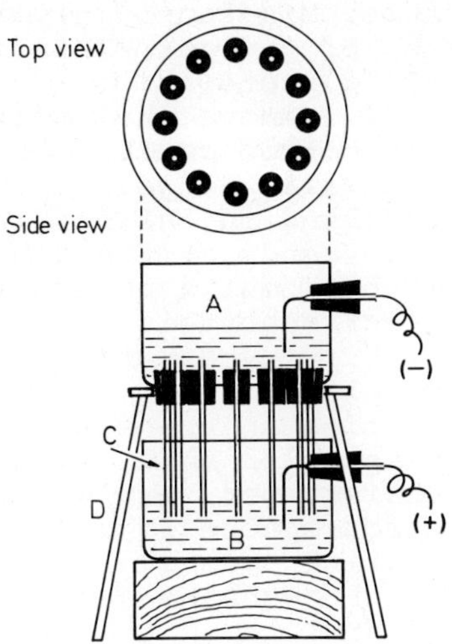

   be soaked in Kodak Photoflow 200 for a few minutes and
   washed again in twice-distilled water repeatedly. Dry them
   in a vertical position
Hypodermic needle and a 1 ml glass syringe
Disposable Pasteur pipettes
5, 25, and 50 µl Eppendorf or other type micropipettes
Illuminating box with opaque glass cover
Parafilm M (manufactured by American Can Comp., Greenwich,
   CT 06830)

<u>Procedure</u>

   1. Close the ends of the electrophoresis tubes with two layers
of Parafilm. Make it watertight. Place the tubes into a test tube
rack. The tubes have to stand perpendicularly.
   2. Prepare the following unpolymerized gel mixture: 50 ml gel
buffer (previously degassed under strong vacuum for about a
minute) mixed with 13.5 ml acrylamide solution. Degas the mixture
again briefly to avoid air bubble formation in the gel later.
Add 1.5 ml freshly prepared 1.5% ammonium persulfate solution
and 50 µl TEMED. Stir rapidly but gently for about 20 sec. Using
a fine-tipped 2 ml pipette, or better a Pasteur pipette calibrated
at approximately 2 ml, fill 2 ml of the nonpolymerized gel into
the tubes . As soon as all the necessary tubes are filled, a
drop of buffer must be placed very carefully on the top of each
gel. This must be layered by a Pasteur pipette or by a smooth-
working tuberculin syringe. The purpose of this is to provide a
perfectly flat gel surface. This interface will not be visible
before the gel polymerizes. Polymerization will take place in
about 25-30 min. After this, shake off the buffer overlay and
the polyacrylamide gels are ready for the experiment.
   3. For the electrophoresis of serum or for any immunoglobulin
preparation one should prepare the following mixture: 25 µl 40%
sucrose, 5 µl bromphenol blue, 25 µl buffer and 25 µl protein
solution (concentration 1 to 2 mg protein/ml) in a small test

tube. Mix thoroughly and then apply 50 µl of this mixture to the
top of the gel. This amount will approximately contain 16 to
32 µg protein.

4. Remove the Parafilm from the lower end of the tube. Fill
the bottom chamber of the electrophoresis tank with buffer so
that all the gels have approximately 1 cm of their lower end
immersed in the buffer. Sometimes there are bubbles under the
lower ends of the tubes. Remove these with a Pasteur pipette or
with a bent hypodermic needle (and fill the remaining space with
electrophoresis buffer).

5. Very carefully lay electrophoresis buffer on the top of the
samples without disturbing the layer. The high sucrose concentration in the sample will make this step easier. The bromphenol
blue color should stay in the sample layer as an indicator of
undisturbed sample application. If not all the grommets of the
electrophoresis unit are used, close these open grommets by in-
serting sealed glass tubes or glass rods. Electric current may
go through the sample-containing electrophoresis tubes only.

6. Turn on the power supply. Adjust the voltage so that the
current going through the system is approximately 5 mA/electro-
phoresis tube. Watch the bromphenol blue indicator moving down
the tube. When it is approximately 5-15 mm away from the end of
the tube, turn off the power supply. The migration of bromphenol
blue is a good indication of the approximate position of serum
albumin which is somewhat slower than the indicator dye. The
time required for such electrophoresis varies between 30 and
60 min, depending upon the number of tubes in the experiment.
More tubes require longer times.

7. Take twelve test tubes, 16 × 150 mm or similar size. Number
them as your electrophoresis tubes were. You are going to transfer
the gels into these tubes as follows: Take out the electrophoresis
tubes from the equipment, one by one. Remove the gels by forcing
water between the gel and the inside of the tube using a fine,
long hypodermic needle and any kind of syringe for this purpose.
Use a large Petri dish to catch the gel sliding out from the
tubes. The blue color of the indicator is still visible at that
point and this will identify the positive end of the gel. This
color will disappear during the forthcoming staining procedure,
therefore put the gels into the prenumbered test tubes so that
the positive ends of all gels will be in one direction.

8. The immersion of the gels in 15% trichloroacetic acid for
10 min before staining has been recommended by some authors for
protein stains. This fixes the proteins in the gel and eliminates
loss of proteins throughout the further steps of staining. Immersion of protein-containing gels into trichloroacetic acid will
make the protein bands visible since they will appear as white
bands in completely transparent gel. To stain the gels, rinse
them a few times with water, and fill the test tubes with Coomassie blue dye solution. Close them with rubber stoppers and let
them stand at room temperature for 60 min.

9. Complete destaining of the gels is important to make weak
protein bands visible. The destaining is carried out as follows:
Using a Pasteur pipette, remove the excess stain from the test
tubes containing the gels. This dye solution can be used again.
Fill the test tubes with the destaining solution described under
materials and methods. Leave it there at room temperature for

approximately 2 h. Remove and discard the destaining solution,
again using a Pasteur pipette, and refill the tube with fresh
destaining solution. Repeat this process every 2 h approximately
three times. Fill the tubes with fresh destaining solution and
let them stand overnight at room temperature. If the gels are
still slightly blue, repeat the destaining by soaking the gels
again overnight in a fresh solution. The gels can be stored in
10% acetic acid in water.

Evaluation

To have permanent records of the electrophoresis, photographic
procedures are the most suitable. Place the gels next to each
other in a flat-bottom glass or plastic dish and fill it with
distilled water. The starting lines of the gels should be lined
up exactly and the positive ends of all gels should point in one
direction. Mark the positive ends and the individual gels by
writing on the dish containing them. Take a picture by illuminating the gels from below, using the illuminating box. The enlarged
photographic print will be used to measure the migration in millimeters and comparing it to the migration of known standard
protein solution.

The separated protein components can also be isolated. Applying 100 or more µg proteins in one tube, two samples should be
run in parallel. One gel should be stored in a well-sealed empty
small test tube at 4°C. The other gel may be stained as described
above to localize the protein bands. The unstained gel can be
cut into 1 or 2 mm thick slices. These should be put into individual test tubes and suspended in 0.2 ml of 0.1% SDS solution
and incubated at 37°C for several hours. The solution can be
withdrawn and replaced with fresh SDS to complete the extraction.
The combined eluants can be lyophilized in a conical centrifuge
tube. Distilled water can be added to the dried residue to obtain
a an 0.1% SDS solution to which nine parts of ice-cold acetone should
be added. This precipitates the protein but the detergent stays
in solution. The precipitate can be collected by centrifugation
and the supernatant may be discarded. This procedure of elution
has been taken from the paper of Weber and Osborn (1969). The
precipitate can be dissolved in distilled water or saline for
further analysis or biochemical activity measurement provided
that a sufficiently sensitive assay is available for this purpose.

Part B. SDS Gel Electrophoresis of Erythrocyte Membrane Proteins

Materials and Equipment

The same equipment and the same chemicals are used here as
in Part A. The only exception is the buffer solution. SDS
buffers are used as follows: dissolve 7.8 grams $NaH_2PO_4 \cdot$
H_2O and 38.6 grams $Na_2HPO_4.7H_2O$ in 1000 ml distilled water
containing 0.2% SDS. This is the buffer used for the preparation of the gel. For electrophoresis the same buffer is
diluted 1:10 with 0.2% SDS dissolved in water.
For the dissociation of the membranes, the same phosphate buffer for gels as described in Part A is used, but with added
0.4% SDS and 0.4% 2-mercaptoethanol.

Procedure

1. To prepare a membrane sample for gel electrophoresis, lyse and wash human erythrocytes as described in Exercise No. 27. This membrane (stroma) suspension in distilled water should be adjusted to a 4 mg/ml dry weight concentration (See Exercise No. 47).

2. Mix 1 vol of the above stroma suspension with 1 vol of SDS and 2-mercaptoethanol-containing buffer for membrane dissociation as described above. Incubate the mixture at 37°C for 5 h.

3. Prepare the gel as in *Part A,* but use the SDS-containing gel buffer.

4. Take 25 µl of the dissociated membrane suspension, mix it with 25 µl 40% sucrose, 25 µl gel buffer and 5 µl bromphenol blue. Mix thoroughly, and apply 50 µl of this to the gel. Carry out the electrophoresis in the same manner as described in *Part A.*

5. Staining and destaining of the gel are the same as in *Part A.*

Evaluation

Again, photographic recording is recommended. Since SDS gel electrophoresis is very useful for molecular weight determination, the approximate molecular weight of the individual components in the stroma preparation can be estimated as described in *Part C.*

Preparative recovery of minute amounts of separated components is possible by using the method described in *Part A.* Additional procedures to detect separated components are described in the Use and Limitations section at the end of this Exercise.

Part C. Molecular Weight Determination by SDS Gel Electrophoresis

Materials and Equipment

The same materials, equipment, and chemicals are used as in *Part B.*

Obtain a few milligrams of the following six proteins: human serum albumin (MW = 68,000), catalase (MW = (60,000), ovalbumin (MW = 43,000), pepsin (MW = 35,000), trypsin (MW = 23,300), and lysozyme (MW = 14,300). It is quite essential that the purest available preparations be used for these molecular weight standards. Even twice crystallized commercial preparations frequently consist of several components.

Procedure

1. Standard samples from the above proteins should be prepared individually by dissolving 0.4 mg in 1 ml SDS and 2-mercaptoethanol-containing buffer as described in *Part B.* Incubate the samples at 37° for 5 h.

2. Mix 25 µl of these samples with 25 µl 40% sucrose, 25 µl gel buffer containing SDS and 5 µl bromphenol blue dye. Mix thoroughly and apply 50 µl of this to the gel.

3. Run one sample of each of the above standard protein solutions in every experiment analyzing protein mixtures with unknown
molecular weights.
4. The electrophoresis and staining procedures should be
identical with those described in *Parts A* and *B*.

Evaluation

Photographic recording must be carried out. On the enlarged print
of the photograph the migration of the components should be
measured in millimeters, taking the top of the gel as the starting point. An arbitrary internal standard must be chosen from the
six proteins and its relative migration will be taken as 1.00.
If highly purified and electrophoretically homogeneous lysozyme
preparation can be obtained, this may serve as the internal standard. If this is not available, either crystalline ovalbumin or
human serum albumin can be used. It is recommended that the unknown samples be run in at least two tubes. To one of them the
selected internal standard should be added by pipetting 25 mg of
this to the unknown 50 ml sample. By comparing the two unknowns,
one with the internal standard and one without it, it is easy to
identify the position of the internal standard. This distance,
as measured from the top of the gel, should be taken as relative
migration value of 1.00. The relative migration values of all
components, standards as well as unknowns, have to be determined.
Use semilog paper for the MW calibration curve of the standard
proteins. Plot the relative migration values obtained on the
linear scale as a function of their molecular weights on the log
scale given above. Use this calibration curve to determine the
molecular weight of unknown components. Figure 8 shows a plot
one can expect in running standard proteins in SDS gel electro--
phoresis, as described here.

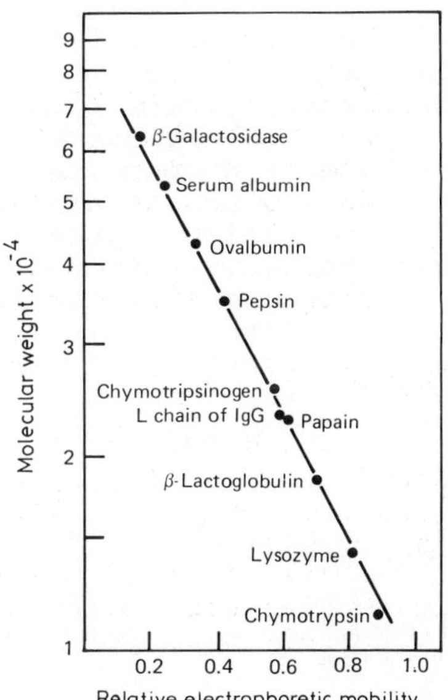

Fig. 8. Relative electrophoretic mobilities of
selected protein standards

By slicing the gel slabs into 2 mm thick discs (See Part A
of this Exercise) and counting them in a gamma counter, 125 I
isotope labeled membrane proteins (prepared as described in
Exercise No. 13) can be detected.

Use and Limitations

Many proteinaceous preparations can be analyzed by the procedure
described here. Ascites fluid, cerebrospinal fluid, gastric
juice, bacterial and plant extracts, various animal toxins, or
membrane preparations of normal and malignant cells, components
of viral particles can be studied very effectively.
On the preparative scale, only microgram quantities can be
obtained by the procedure described above. If the biologic activity of the separated components as expressed on a weight
basis is high, for example, that of some endotoxins or some enzymes, this quantity may be more than enough to identify the
active band among the separated ones.
The accuracy of molecular weight determinations by the SDS gel
method is ±10% for most proteins in a MW range of 15,000-150,000
daltons. Highly anisotropic molecules, or those with high lipid
or carbohydrate contents will show greater deviations from MW
values determined by more accurate physical-chemical procedures.
Antigenic components may be detected by embedding a non
stained polyacrylamide gel in melted (40°C) 1% agarose. After
it solidifies, cut an approximately 2 mm wide and 6 to 7 cm long
trough in the agarose, parallel to and not more than 4 or 5 mm
away from the polyacrylamide gel. Filling this trough with antiserum specific to a component present in the separated mixture,
the gel should be incubated for 60 min at 37°C and overnight at
cold room temperature in a moist chamber. Precipitin bands will
be developed where the antigen and the antibody met by diffusion.
Further instructions on similar double gel diffusion systems can
be found in Exercise No. 74.
The crossed immunoelectrophoretic method (Exercise No. 74)
can be used for SDS gel electrophoresis of cell membrane antigens.
Excellent separation of the erythrocyte membrane components has
been shown in the publication of Bjerrum and Bog-Hansen (1976).
The staining procedure described here will detect only proteinaceous components. Glycoproteins and other carbohydrate-
containing molecules can be detected by the periodic acid Schiff
(PAS) staining method. The procedure for this is given in Exercise No. 45, separately.
We should also mention that the gel concentration described
in these experiments is 10%. Frequently lower concentrations are
needed, particularly while working with dissociated membrane
components.
If one makes comparisons between two samples separated into
components by disc electrophoresis, it is very important to apply
from both samples the same amount of protein, expressed in weight
units. Use the Kjeldahl method for the quantitative protein content determination (Exercise No. 48).

References

Bjerrum, O.J., Bog-Hansen, T.C.: In: Biochemical analysis of membranes.
   Maddy, A.H. (ed.) London: Chapman and Hall 1976
Maurer, H.R.: Disc electrophoresis and related techniques of polyacrylamide
   gel electrophoresis. Berlin, New York: de Greuyter 1971
Shapiro, A.L., Vinuela, E., Maizel, J.V.: Biochem, Biophys, Res. Comm. *28,*
   815 (1967)
Weber, K., Osborn, M.: J. Biol. Chem. *244,* 4406 (1969)

Exercise No. 11

Separation of Human Hemoglobins by Electrofocusing

The isoelectric point (pI) of a protein is the pH at which the
number of positive and negative charges of the protein molecule
are equal, which means that the net electric charge of the molecule is zero. The procedure of electrofocusing is the separation
of proteins based on differences in their pI.

In the electrofocusing equipment a continuous pH gradient is
produced in a supporting gel. The pH usually varies from 3.5 to
10, but narrower pH intervals can also be taken for greater
resolution. To obtain such a pH gradient within one gel slab,
ampholytes can be used. Ampholyte substances can accept as well
as give protons, which means that they can act both as acids and
as bases. Amino acids and small peptides are ampholytes. Large
protein molecules can also behave similarly. A commercially
available preparation called Ampholine was developed by Vesterberg
and Svensson in 1966. They prepared synthetic aliphatic amino-
carboxylic acid mixtures with a molecular weight between 300 and
600. This Ampholine can be separated from proteins by simple
dialysis. It does not interact with most proteins and it has a
good buffer capacity. Its adsorbance at 280 nm is very small,
thus making UV scanning for proteins in the gel possible. For
the selection of the Ampholine products consult LKB catalog
No. 1809.

If a dextran gel, such as Utrodex or Sephadex G75 Superfine
(or, for analytical purposes, a polyacrylamide gel) prepared with
Ampholine is subjected to electrophoresis, a close to linear pH
gradient will develop, with a low pH at the anode. The sample to
be fractionated can be simply mixed into the entire gel, even
before the pH gradient develops. The components of the sample
will migrate in the electric field to the zone where their net
electric charge is zero. At that particular pH, they will not
move further, and they will be even sharply focused into a thin
line. Hence the name "electrofocusing". Samples can also be applied to a selected zone of the gel, either before or after the
development of the pH gradient. In "Use and Limitations" of this
Exercise, alternate methods of sample application are discussed.

In this Exercise we describe the use of the LKB Multiphor
No. 2117 electrofocusing instrument for the separation of normal
adult human hemoglobin in Ultrodex gel.

Materials and Equipment

Human hemoglobin solution (see below)
Ultrodex dextran gel (available from LKB Instruments,
  Rockville, MD 20852)
Ampholine (LKB catalog No. 1809-126) for linear pH gradient
  between 5.0 and 8.0
Whatmann No. 1 or similar quality filter paper
1 N H_3PO_4
1 N NaOH
Coomassie Blue Dye
10% Trichloroacetic acid
Scale with 1-500 g capacity
LKB Model 2117 Multiphor electrofocusing equipment (avail-
  able from above company)
Magnetic stirrer with Teflon-coated stirring bars

Procedure

1. Take 0.5 ml citrated whole human blood, centrifuge it to
sediment the red blood cells, and discard the supernatant. Wash
the sedimented cells three times with physiologic saline and
discard the washings. Add 2.5 ml distilled water to the cells
to hemolyze them. Centrifuge at 5000 g for 30 min to sediment
the red blood cell membranes. Lift approximately one ml from
the supernatant.

2. Preparation of the gel bed is somewhat circumstantial.
Take 5 ml of Ampholine No. 1809-126 and dilute it up to 100 ml
using distilled water. Weigh 4 g of Ultrodex and add it in small
portions to the constantly stirred 100 ml Ampholine solution, as
prepared above. Homogenize the suspension for a few minutes with
a magnetic stirrer.

3. Prepare the tray for the gel. Place two long square rods
along the longer ends of the tray, 10.5 cm apart from each other.
Cut ten Whatman No. 1 strips, 1.5 × 10.5 cm. Soak eight of these
in 10 ml Ampholine solution, prepared as above, without Ultrodex.
Place four layers of the strips at each of the two narrower ends
of the tray provided by the Multiphor electrofocusing instrument.
These strips will prevent the gel from running off the tray.
Determine the weight of this assembly. Gently stir the prepared
Ampholine-Ultrodex slurry and pour it in to the tray. Using a
scale, determine the weight of the tray containing the slurry.
From this value deduct the weight of the empty tray which was
determined earlier, to determine the weight of the slurry poured
into the tray. This weight must be reduced by evaporation, according to the instructions of the manufacturer given on the label of
the Ultrodex package.

4. The evaporation can be carried out if a small fan is mounted
about 3 ft above the tray and a light stream of air blown onto the
surface of the gel (See Figure 9). The air stream must be gentle
enough not to cause waves on the surface of the gel. Check the
rate of evaporation by weighing the tray about every 15 min.
Discontinue the evaporation when the water loss reaches the limit
given by the Ultrodex package labels. Overdrying will result in
the formation of cracks in the gel and this should be avoided.

Fig. 9. Concentrating the gel for isoelectric focusing. (LKB-Produkter AB,
Bromma, Sweden)

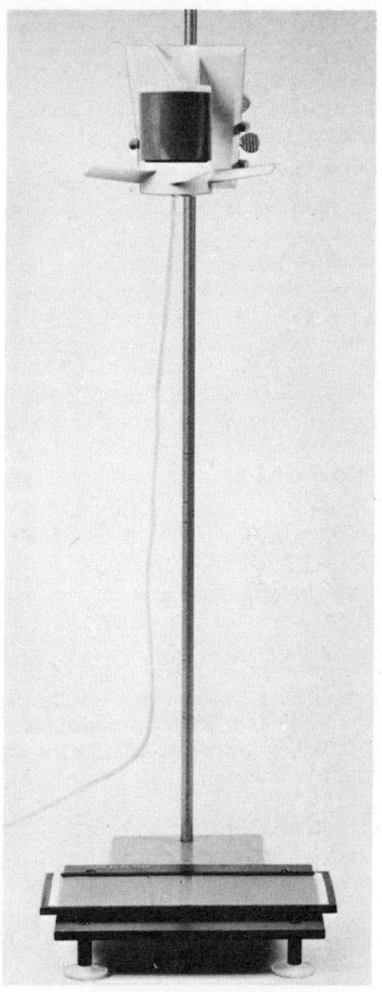

5. Transfer the tray to the cooling plate of the Multiphor
electrofocusing instrument. It is advisable to carry out the
entire experiment in the refrigerator or cold room. The cooling
plate should be connected either to running tap water if it is
sufficiently cold, or to a circulating cooling liquid pumped
from a refrigerated bath. The optimal temperature for such an
operation is between 5° and $10^{\circ}C$. To make good contact between
the gel-containing tray and the cooling plate surface, pipette
a few milliliters of water containing approximately 1% diethylene
glycol or glycerol on the cooling plate surface. This liquid
film will provide an efficient heat transfer between the tray
and the cooling plate.

6. There are two more 1.5×10.5 cm paper strips left. Take
one and dip it into 1 N phosphoric acid solution and place it
on top of the four strips at the anode side of the tray. Take
the last paper strip and soak it in 1 N NaOH. Place this strip
on top of the other four strips at the cathode side of the tray.

7. Cut a 20 × 5 mm small paper wick from Whatman No. 1 or
heavier filter paper. Take a forceps and dip this paper into the
hemoglobin solution. Shake off excess hemoglobin and carefully
place this small strip on the gel surface, parallel with its
shorter side. The position of the sample-containing paper strip
is relatively unimportant, which means that it may be close to
the center as well as to the anode or cathode of the electrophoresis. The sample on the filter paper will migrate to the
zone where the pH corresponds to the pI of the hemoglobin components. To demonstrate this phenomenon, prepare three such
samples on paper strips and place one in the center of the plate,
another approximately 5 cm toward the positive end, and the third
approximately 5 cm from the center toward the negative end of
the gel.

8. Turn on the power supply and apply a potential difference
to the two ends of the tray so that approximately 8 W is used.
(W = volts × amperes.) Under these conditions and at 10°C good
separation can be achieved overnight.

Evaluation

The hemoglobin components are clearly visible without staining.
Photographic procedures can be used to have permanent records
of the separation.

If colorless protein components are separated by electrofocusing, two procedures can be used for their detection after
separation. A so-called print technique has been developed which
is achieved by placing a dry filter paper on top of the gel.
A 10 × 24 cm filter paper will cover the gel surface sufficiently
well. Leave the filter paper strip in place for about 5 min so
that it absorbs small amounts of the separated protein components.
It is important to avoid trapping air under the filter paper,
which would prevent making a good print. Dry this filter paper
in an oven at about 110°C for 10 to 15 min. Soak the dried paper
in 10% TCA for 15 min. These procedures will fix the proteins
picked up by the filter paper to its fibers. Staining of the
filter paper strip with 0.2% Coomassie brilliant blue can be
carried out. Dissolve this dye in a mixture of methanol : water :
acetic acid = 50 : 50 : 10. Staining for 1 hour is sufficient.
After this, wash the paper in methanol : water : acetic acid =
50 : 50 : 10 to remove excess dye from the paper, until the background color completely disappears.

Another way to evaluate the electrophoretic separation is
achieved by dividing the gel into narrow strips, parallel to the
shorter ends of the gel slab. This can be done easily by pressing
a fractionating grid through the gel bed. This grid is provided
by the manufacturer. The pH of the gel within each divider can
be measured with a surface glass pH electrode, available from
LKB. With the aid of a spatula, scoop the gel out of the grid-
divided compartments. Transfer the gel into small filters provided with a glass wool plug or filter paper support, and wash
the gel suspension with small amounts of a suitable solvent. Each
grid compartment must be handled separately. The filtrates may
be subjected to chemical analysis to detect proteins, carbohydrates, nucleotides or other components by chemical procedures.
Similarly, if biologic or biochemical activities of the separated

components are known and can be measured, such analysis of the
eluates should be carried out in order to locate active components among the separated fractions. This analysis, combined with
chemical determinations, will give very valuable information
about the possible chemical nature of the active material separated by electrofocusing.

Use and Limitations

The procedure described here is eminently suitable for preparative purposes. It is, however, frequently required that smaller
gels be prepared, saving both on rather expensive chemicals and
on material to be separated. In such a case, homemade trays can
be used very well to replace the tray provided with the Multiphor
apparatus. It should be also emphasized that the gel serves only
as a matrix, which prevents rapid diffusion and remixing of the
separate components; other inert gels or slurries can also be
used as matrix.
    Before the rather recent appearance of horizontal electro-
focusing equipment on the market, vertical columns were used for
analytic as well as preparative electrofocusing. Such instruments
were made available by a few manufacturers, such as LKB, or ISCO,
(Lincoln, NB 68505) or MRA Corp. (Boston, MA 02121).
    The sample application described in this experiment is satis-
factory for relatively small amounts. If larger quantities are
available, one could mix the sample with the Ampholine-containing
dextran gel in a beaker, before pouring the gel. Under such conditions, as already mentioned above, the sample will be evenly
distributed before electrofocusing, and will be focused by the
procedure into rather sharp lines at the end of the fractionation. This is a very convenient procedure but cannot be used for
every compound. If longer contact of some more sensitive components with undesirable pH ranges should be avoided, this procedure of applying the samples cannot be used. Under these conditions, it is recommended to use the LKB sample applicator, which
cuts a channel in the gel. This channel can be refilled with a
sample-containing gel slurry. Removal of the applicator will
make contact between the sample and the rest of the gel. See
Figure 10 for the procedure.
    Ampholine does not interfere with most proteinaceous sub-
stances and for most biologic or chemical analyses their removal
is not necessary. It may be that the presence of Ampholine will
interfere with the assays used to detect separated components.
In such a case, dialysis must be used to separate it from the
eluates.
    Ampholine-containing polyacrylamide gels can also be used for
electrofocusing. The separation in polyacrylamide gel can also
be very sharp and small plates can also be prepared for electrofocusing. The separation of micro-quantities can be carried out
in Ampholine-containing polyacrylamide gels on microscope slides
for analytic purposes. For preparative purposes dextran gels
are more suitable as supporting media. A great advantage of the
polyacrylamide gel electrofocusing system is that it can be combined with immunologic procedures to detect antigenic components
separated by electrofocusing, provided that specific antiserum
to these components is available. Procedures described in Exer-

<u>Fig. 10.</u> Sample application for isoelectric focusing. (Courtesy of LKB-
Produkter AB, Bromma, Sweden)

cise No. 74 for crossed immunoelectrophoresis can be applied here
to enhance further the resolution.
    Analytic electrofocusing technique was first described for
the analysis of immunoglobulins by Awdeh et al. in 1968. The
horizontal electrofocusing system was developed by Vesterberg
in 1972. Gel electrofocusing techniques were applied for the
separation of hemoglobins, urinary proteins, serum components,
cell membrane preparations, and a number of other biologic materials. Abraham and Bakerman isolated Rh antigens from human
erythocytes and purified it by electrofocusing (1976). The Acta
Ampholinae Literature Reference List, which is available from
LKB representatives, is an excellent compilation of literature
references describing the use of electrofocusing.

<u>References</u>

Abraham, C.V., Bakerman, S.: Science Tools *24*, 22 (1976), LKB publication.
Awdeh, Z.L., Williamson, A.R., Askonas, B.A.: Nature London *219*, 66 (1968)
Vesterberg, O.: Biochem. Biophys. Acta *257*, 11 (1972)
Vesterberg, O., Svensson, H.: Acta. Chem. Scand. *20*, 820 (1966)

Exercise No. 12

Conjugation of Antibodies With Fluorescein and Rhodamine

Labeled antibodies provide one of the most indispensable tools
for the localization of antigens in tissue sections, in smears,
or in microbial preparations. The rather high specificity of the
antigen-antibody reaction permits discovering antigens in situ
without isolating them from the surrounding components. For these
purposes, the antibody produced in vivo against those sought for
antigenic macromolecules has to be labeled with a tracer.

Several tracers are known and widely used in histology, cyto-
logy, diagnostic microbiology, parasitology, embryology, and in
ultrastructure studies, in addition to immunology. The most common are fluorescent dyes, which reach an excited state after
absorbing a certain quantum of energy. This energy can be im-
parted to the dye molecules by UV irradiation. The amount of
absorbed energy introduces changes in the kinetic and potential
energy of a few electrons. If this excited molecule is stable
enough, the return to lower energy level occurs through emission
of radiation. The wavelength of this radiation is different from
that of the absorbed energy. If such a dye is illuminated by UV
light, it will radiate energy on the wavelength which is shifted
into the visible range. If such a dye is covalently bound to
antibodies, it maintains this property. Reaction between antigens
and antibodies results in the "fixation" of such dye molecules
to tissue sections, smears, etc. which contain antigens. These
preparations investigated under a microscope equipped with UV
illuminator and the proper filter system will show a bright
fluorescence. This basic phenomenon was used for the first time
by Coons and associates (1941).

The procedure described here is based on the work of McKinney
and co-workers (1964) as modified by Cherry and associates (1966).
It consists in two steps: fractionation of the globulins from
the immune serum and conjugation of the fluorescein dye with the
proteins obtained.

Materials and Equipment

    Serum globulins (from Exercise No. 1)
    Fluorescein isothiocyanate (FITC) and Rhodamine isothiocyanate
     (RB-200)
    0.2 *M* sodium phosphate, dibasic
    0.1 *M* sodium phosphate, tribasic
    0.01 *M* phosphate buffer, pH 7.5
    Phosphate-buffered saline. Dissolve 8.5 g NaCl in 1000 ml pH
     7.5. 0.01 *M* phosphate buffer
    Dialysis bags and jars
    Magnetic bar and stirrer

Procedures

Two procedures are described here using (A) the yellow green
fluorescent FITC, and (B) the orange color producing RB-200 for
the conjugation with proteins.

Part A. Flourescein

The first step is the determination of the amount of proteins in
the dialyzed preparation. For this purpose, the quantitative
assays described in Exercise No. 48 or in Exercise No. 50 may be
applied.

In the following procedure the fluorescein/protein (F/P)
ratio can be varied by simply changing the reaction time.

1. Add 2.5 ml O.2 M Na_2HPO_4 to 10 ml serum fraction containing
approximately 1% protein. Mix with a magnetic stirrer.

2. Dissolve 2.5 mg pure FITC in 2.5 ml O.2 M Na_2HPO_4 and add
2.5 ml water. Add this solution as soon as it has been prepared
to the protein solution, slowly, under constant stirring. This
addition should be prolonged to approximately 15 min. Adjust the
pH to 9 with O.1 M Na_3PO_4. Add physiologic saline to make the
total volume 20 ml.

3. Do not stir from this point on, and let the reaction
proceed overnight. If a lower F/P is desirable because of high
nonspecific staining observed with the conjugate, the reaction
time may be shortened to a few hours.

4. The purification of the conjugated proteins from unreacted
fluorochrome may be carried out by dialysis against phosphate-
buffered saline in the cold room for three days. Change the outer
fluid at least twice daily.

Another procedure which is frequently used applies chromatographic separation of the conjugate from free dyes on a Sephadex
column. This procedure is described here, according to Killander,
Ponten and Roden (1961). A 2 × 50 cm chromatographic column is
prepared as described in Exercise No. 2, but using Sephadex G 25
instead of G 200 washed with phosphate buffered physiologic
saline. Apply 20 ml filtered conjugate on the column and let drain
into the gel. Carefully add 20 ml buffered saline and as soon as
this amount also enters the column, start the elution of the conjugate with additional buffered saline. The conjugate will leave
the column right after the void volume has been collected, while
the unreacted fluorochromes will be eluted only after the void
volume of buffered saline has been driven through several times.
This procedure leaves the column regenerated and ready for repeated use.

Part B. Rhodamine

If the above 20 mg/ml protein concentration is used, add to each
10 ml protein solution, under constant stirring, 6 ml of the
Rhodamine isothiocyanate solution containing 1 mg/ml in O.2 M
Na_2HPO_4. Continue the stirring of the mixture for 2 h in the
cold room. The purification of this RB-200 protein conjugate may
also take place through dialysis or by separation on a Sephadex
column.

Evaluation

Place a drop of the conjugates on a filter paper strip. Dry it
and observe the fluorescence under an ultraviolet lamp. A good
conjugate must show intensive color after 10- or 100-fold

dilutions in this assay. The specificity of the staining has to
be investigated in proper control experiments.

Use and Limitations

A very useful monograph has been written by R.C. Nairn (1964)
about the basic principles, methodology, and application of
fluorochrome-labeled antibodies. Students are referred to this
book, the last paragraph of which briefly reviews the most outstanding achievements in the use of fluorescent antibody techniques up to 1964.

By using one of the modified applications, preparations containing two or more antigens may be "stained" in two or more
different colors if their antibodies are conjugated with different fluorochromes. For example, the sites of IgG and IgM production may be differentiated if the anti-IgG antibodies were labeled
with the green FITC and the anti-IgM antibodies were conjugated
with the red RB-200.

The application of the proper fluorochrome-to-protein ratio
is essential in these experiments. The optimal ratio should
result in maximal fluorescence with the minimal number of labeled
molecules per antibody molecule. Overloading of the protein molecule not only does not enhance the fluorescence, but leads to
denaturation of the protein and to changes in its serologic
reactivity.

Labeling under the above conditions allows obtaining a wide
range of different F/P ratios. It has been found by Cherry and
associates (1966, Lewis et al., 1964) that each antibody-antigen
system has its own F/P ratio where the nonspecific staining is
low while the specific reaction produces a bright fluorescence.
Tables for the preparation of different F/P ratios have been
published by McKinney et al. (1964).

The most serious limitation of the fluorescent antibody techniques is due to the frequently observed nonspecific fluorescence
of the preparation. This nonspecific staining of preparations is
due mostly to the following factors: (1) The labeled antibody
preparation contains free, unreacted fluorochromes, which react
nonspecifically with the microscopic preparation. (2) Proteins
may be overloaded with fluorescent dyes. (3) Presence of fluorescein-labeled nonimmunoglobulin proteins in the investigated
preparation may be due either to incomplete removal of these
after the antibody-antigen complex formation, or to nonspecific
binding of these proteins to constituents of the microscopic
preparation. Ideal procedure for the fluorescent antibody method
would be the use of specific and homogeneous immunoglobulins for
the conjugation with fluorochromes at an optimal F/P ratio. While
this condition cannot be achieved with ease, there are relatively
simple procedures described which eliminate most of the above-
mentioned sources of error. If foreign antigens are being located
in tissue sections of a certain animal species, it is important
to absorb the fluorochrome-labeled protein solution with acetone
powder made from the organs of animals. Such liver powder and
similar preparations are commercially available. The absorption
of the labeled protein solutions with organ powder will reduce
the nonspecific binding of fluorescent proteins by the tissue
sections. This has been recommended by Coons and Kaplan (1950).

Certain human, animal, or plant tissue sections in a microscopic preparation exhibit autofluorescence. In order to make
labeled antibodies visible, it is necessary to select a fluorochrome which emits a wavelength different from the tissue section. In most cases the selection of Rhodamine B 200, which
shows a red fluorescence, overcomes this difficulty, because the
autofluorescence of the tissues usually has a different color.

Tissue sections or leukocyte preparations, which sometimes
show intensive autofluorescence, may be treated with Flazo Orange.
This is applied after the conjugation has been carried out with
fluorescein-labeled antibodies. The Flazo Orange counterstain
will mask the autofluorescence while allowing the specific green
fluorescence of the labeled antibodies. This procedure has been
described by Hokenson and Hansen (1966).

References

Cherry, W.B.: Personal communication (1966)
Coons, A.H., Creech, N.J., Jones, R.N.: Proc. Soc. Exp. Biol. Med. *47,* 200
    (1941)
Coons, A.H., Kaplan, M.H.: J. Exp. Med. *91,* 1 (1950)
Hokenson, E.O., Hansen, P.A.: Stain technology *41,* 9 (1966)
Killander, J., Ponten, J., Roden, L.: Nature London *192,* 182 (1961)
Lewis, V.J., Jones, W.L., Brooks, J.B., Cherry, W.B.: J. Appl. Microbiol.
    12, 343 (1964)
McKinney, R.M., Spillane, J.T., Pearce, G.W.: J. Immunol. *93,* 232 (1964)
Nairn, R.C.: Fluorescent protein tracing. Baltimore, Md.: Williams & Wilkins
    Co. 1964

Exercise No. 13

Enzymatic Labeling of Proteins With ^{125}I

A major disadvantage of the earlier protein iodination procedures
lies in the use of some harsh chemicals which can introduce unwanted and irreversible changes in the structure of the labeled
preparations. To circumvent this, iodination by peroxidase en-
zyme-containing or peroxide generating systems was applied by
numerous authors since the first short report on the feasibility
of this by Keston (1944). It has been used for immunoglobulin
labeling by Marchalonis (1969) using lactoperoxidase, hydrogen
peroxide, and iodide. The procedure described here in *Part A* is
based on this procedure. *Part B* describes the procedure used by
Dr. Ah-kau Ng (NIH, National Cancer Institute, Bethesda, MD)
to iodinate exposed proteins on human tumor cell surfaces, based
on the publication of Phillips and Morrison (1971).

Part A

Materials and Equipment

IgG solution (from anti *S*.typhi O901 rabbit or other O-anti-
    sera as prepared in Exercise No. 65 and fractionated as

described in Exercise No. 3). A 5 mg/ml solution in saline
will be used for this Exercise
Lactoperoxidase enzyme (available from Calbiochem, La Jolla,
CA 92037, or from numerous other companies). 2 mg/ml solution in saline should be prepared
Hydrogen peroxide: an 8.8 mM solution will be needed. This
solution will contain 29.9 mg H_2O_2 in 100 ml distilled
water, and must be freshly prepared and kept in a well closed
container
$Na^{125}I$ solution, diluted to 4 mCi/ml in PBS. Carrier and re-
ductant free $Na^{125}I$ is available from New England Nuclear
Corp., Boston, MA 02118
Gamma radiation counter
Disposable polystrene tubes, 5 ml capacity
0.02 *M* pH 7.0 phosphate buffered saline (PBS) (See Exercise
No. 1)
Radiologic hood with charcoal filter
5 m M cystein. HCl solution

Procedure

WARNING: Since you are working with isotopes, the safety regulations must be followed very closely. In case of any doubts or
problems, contact your radiation safety officer immediately.

1. Mix 200 µl IgG solution (contains 1000 µg IgG) + 20 µl
lacto-peroxidase solution in a polystyrene tube with 20 µl $Na^{125}I$
(80 µCi). Use a Vortex.

2. Add 5 µl H_2O_2 solution and Vortex again vigorously for
5 min. Transfer the polystrene tube to a rotary mixer and allow
it to agitate for an additional 25 min at room temperature.

3. Stop the iodination reaction by adding 2 ml 5 m*M* cysteine.
HCl solution to the mixture. Mix thoroughly for 5 min.

4. To remove small molecular weight components, dialyze the
contents of the tube against 100 ml PBS for three days in a cold
room. Change the outer fluid twice daily.

Evaluation

Cone and Marchalonis (1974) mixed the labeled rabbit IgG with
normal rabbit serum and subjected it to cellulose acetate strip
electrophoresis and to immunochemical analysis. Accordingly, one
can mix 10 µl labeled rabbit antienzyme IgG with 1 ml normal rabbit serum, and subject an aliquot to electrophoretic separation,
as described in Exercise No. 9. The strip may be stained for
proteins as described in the same Exercise. To detect the isotope-
labeled protein components among the separated serum fractions,
now transfer the strip in a dark room to a Kodak BB54 X-ray film,
the same size as the cellulose acetate strip. Line them up and
hold them together between glass plates. Wrap the plates in
aluminum foil and keep them in the dark for one or two weeks.
Develop the film with Kodak B19 developer and fix it. Dry and
observe the film. Compare the position of radioactivity with the
position of stained bands on the cellulose acetate strip. If the
labeled proteins are not denatured, they should migrate from the
point of application toward the negative pole, together with the
unlabed IgG.

One can carry out immunoelectrophoresis, using *S.typhi* O901 LPS (1 mg/ml) as antigen and the labeled anti-LPS IgG preparation. As comparison, nonlabeled, purified rabbit anti-LPS IgG could be run on the same plate. For details see Exercise No. 72. If the stained gel is dried on the microslide, autoradiography can be carried out as described above. The labeled IgG should react with the LPS antigen in the same way as the unlabeled sample.

The specific activity of the labeled IgG preparation can be determined by adding various known amounts of IgG to counting vials and determining the cpm values.

Use and Limitations

In all enzymatic iodination procedures, the activity of the enzyme must be uninhibited. Trace amounts of peroxidase inhibitors may prevent iodination.

The H_2O_2 concentration in all systems must be kept low but sufficient. High H_2O_2 contents seem to have an adverse effect on the enzymatic activity, and may lead to iodination of the enzyme itself.

The procedure given here has been elaborated for IgG iodination. Other proteins may require slightly different conditions. For example, it may be necessary to change the pH of the reaction mixture, or the protein to $Na^{125}I$ ratios, or the duration and temperature of iodination. All these experimental details must be elaborated in order to find the optimal conditions for every protein. As a starting point, the procedure described here can be used without modification. If the results are not satisfactory, further procedural changes can be tried.

David and Reisfeld (1974) elaborated an iodination procedure using Sepharose-bound lactoperoxidase. This procedure offers simplicity and reproducibility. Exercise No. 8 in this Manual describes a procedure for the preparation of Sepharose bound enzymes.

Part B

The purpose of this Exercise is to label those components of cell membranes which are exposed on the surface. The great advantage of the lactoperoxidase procedure is that the large molecular weight enzyme (M.W. = 78,000) cannot penetrate the cells and will not label components other than those exposed on the cell surface.

Materials and Equipment

    0.05% Trypan blue in PBS
    0.03% H_2O_2
    The same reagents as in Part A will be used here.

Procedure

1. Wash the cells to be iodinated three times with cold PBS. Determine cell counts and percent viability using 0.05% trypan blue PBS as described in Exercise No. 82. The cell suspension

must have a viability greater than 95%. Suspend the washed cells in PBS to give a solution containing 1×10^8 cells/ml.

2. To 0.05 ml cells, add 0.1 mCi Na^{125}I, 20 µg lactoperoxidase and 10 µl H_2O_2. Incubate the mixture at 30°C temperature for 5 min with constant stirring.

3. Terminate the iodination by the addition of 10 ml cold phosphate buffered saline. Pellet the cells at 4°C at 1000 g for 15 min. Wash the cells twice in PBS.

Evaluation

Determine cell counts and percent viability in 0.05% trypan blue - PBS.

Determine total ^{125}I incorporation into the cells with a gamma radiation counter. Calculate the % recovery of total radioactivity used for iodination of the cells.

Use and Limitations

To ensure labeling of only exterior surfaces of cells, high cell viability (>90%) must be maintained throughout the iodination procedure. Since lactoperoxidase catalyzes iodide incorporation into the exposed tyrosine and histidine groups on proteins, the proteins to be labeled should contain some of these amino acid residues.

Recent evidence shows that the alteration of protein structure by the above treatment does not affect the serologic activities of immunoglobulins (Vitetta et al., 1971), histocompatibility antigens (Henning et al., 1976), and tumor associated antigens (Snyder et al., 1977). Taylor (1974) used ^{125}I-labeled rabbit antibodies to human immunoglobulins to detect immunoglobulin production in paraffin embedded, formalin-fixed lymphoreticular neoplasms.

References

Cone, R.E., Marchalonis, J.J.: Biochem. J. *140*, 345 (1974)
David, G.S., Reisfeld, R.A.: Biochem. *13*, 1014 (1974)
Henning, R., Milner, R.J., Reske, K., Cunningham, B.A., Edelman, G.M.:
   Proc. Natl. Acad. Sci. USA *73*, 118 (1976)
Keston, A.S.: J. Biol. Chem. *153*, 335 (1944)
Marchalonis, J.J.: Biochem. J. *113*, 299 (1969)
Phillips, D.R., Morrison, M.: Biochem. *10*, 1766 (1971)
Snyder, H.W., Jr., Stockert, E., Fleissner, E.: J. Virol. *23*, 302 (1977)
Taylor, C.: The Lancet, Oct. 5, p 802, (1974)
Vitetta, E.S., Baur, S., Uhr, J.W.: J. Exp. Med. *134*, 242 (1971)

Exercise No. 14

Preparation of Peroxidase-Labeled Antibody

The first procedure binding antibody to an enzyme was published
by Nakane and Pierce (1966). These authors established covalent
linkage between the two components by p,p'-difluoro-m,m'-dinitro-
diphenyl sulfone. A peroxidase enzyme was used to detect the
bound antibody both at light and electron microscopic levels.
Nakane and Kawaoi later (1974) developed a new procedure to bind
peroxidase to antibody by oxidizing the carbohydrate moiety in
the enzyme with periodate, thus forming aldehyde groups which
react readily with primary amino groups in the antibodies or
Fab' fragments. The periodate treatment of the enzyme does not
interfere with its enzymatic activity and a high percentage of
the oxidized enzyme can be bound to antibodies without inter-
fering with the reactivity of the antibody. This procedure is
given here, based on the description of Nakane and Kawaoi (1974).

Materials and Equipment

   Horseradish peroxidase (available from Sigma Chemical Co.,
    St. Louis, MO 63178).
   Anti-mouse IgG. (Commercially available from Cappel Labs.
    Inc., Downingtown PA 19335)
   1% Dinitrofluorobenzene dissolved in ethanol
   0.3 M $NaHCO_3$. Dissolve 2.52 g in 100 ml water
   0.01 M Na_2CO_3/$NaHCO_3$ buffer, pH 9.6. Mix 320 ml 0.01 M Na_2CO_3
    (1.06 g in 1000 ml) and 680 ml 0.01 M $NaHCO_3$ (0.84 g in
    1000 ml)
   0.08 M sodium metaperiodate ($NaIO_4$) in water. Dissolve 1.71 g
    $NaIO_4$ or 2.14 g $NaIO_4 \cdot 3H_2O$ in 100 ml water
   0.16 M ethylene glycol. Dissolve 9.93 g CH_2OHCH_2OH in 100 ml
    water
   Sodium borohydride ($NaBH_4$). Dissolve 50 mg in 10 ml distilled
    water
   Phosphate-buffered saline (PBS). (See Exercise No. 1)
   10 and 25 ml Erlenmeyer flasks
   Tiny magnetic spin bar (8 mm)
   Magnetic stirrer

Procedure

   1. Dissolve 5 mg peroxidase in 1 ml 0.3 M $NaHCO_3$. Add 0.1 ml
1% dinitrofluorobenzene and stir with a magnetic stirrer at room
temperature for 1 h.
   2. Add to the mixture 1 ml 0.08 M $NaIO_4$ in distilled water.
Continue mixing the solution for 30 min at room temperature.
   3. To stop the oxidation, add 1 ml 0.16 M ethylene glycol.
Continue mixing for 1 h at room temperature.
   4. Dialyze the mixture against 0.01 M carbonate-bicarbonate
buffer, pH 9.6, in the cold room for 48 h, changing the outer
fluid twice daily.

   5. Transfer the contents of the dialysis bag into a 25 ml
Erlenmeyer flask. Add 5 mg IgG (anti-mouse IgG prepared in goat)
dissolved in 2 ml carbonate-bicarbonate buffer. Mix the solutions
with a magnetic stirrer for approximately 3 h at room temperature.
   6. Add 2 ml $NaBH_4$ solution. Mix the solutions and transfer the
flask to the cold room overnight.
   7. Dialyze the mixture at $4°C$ against PBS. Dialysis should
last for two days, changing the outer fluid twice daily. If a
precipitate forms, remove it by centrifugation.
   8. If further purification is required, the dialyzed sample
should pass through a Sephadex G100 column (2×50 cm size) as
described in Exercise No. 2. The first protein peak leaving the
column after the void volume contains the enzyme-labeled IgG
preparation. The chromatographic effluent should be monitored
at 280 nm.

Evaluation

The preparation you have now is an anti-IgG immunoglobulin covalently linked to peroxidase. The dilution of this preparation
may be adequate in this form for detecting antibodies on cell
surfaces or in histological sections, but for some purposes it
may be desirable either to dilute it or to concentrate it. In the
latter case, the complex should be precipitated at cold room
temperature by mixing it with equal volumes of saturated ammonium
sulfate. Centrifuge the precipitate sharply at 5000 g for 30 min.
Discard the supernatant and redissolve the sediment in PBS so
that the desired protein concentration will be achieved. For the
determination of the protein concentration, the procedure in
Exercise No. 48 or No. 50 can be used.
   The enzymatic activity of the preparation has to be determined
next. If the immunoglobulin has an inert enzyme attached to it,
it will react with the antigen but this reaction will not be
detectable due to the lost activity of the enzyme. Moreover, if
the preparation contains a mixture of active and inactive enzyme
complexes, the inactive ones will compete for receptors and will
considerably reduce the number of detectable sites. The enzyme
activity must be determined and the procedure of Chance and Maehly
(1955) can be used for this measurement.
   The amount of enzyme used for the complexing being known
(5 mg), the total activity of the starting material can be measured and calculated. After the activity of the immunoglobulin-
enzyme complex has been determined, one must calculate the percent
of enzyme activity present in the complex. This should not be less
than 75% of the starting activity.
   It is also necessary to measure the antibody reactivity in the
complex. If the reactivity of the antibody is lost, no attachment
will take place. We know the amount of antibody we complexed with
the enzyme and now we must compare the reactivity of the complex
with the reactivity of the same amount of free immunoglobulin.
For this purpose, probably the simplest method is the semiquantitative procedure described in Exercise No. 76. To carry this out,
adjust the mg/ml concentration of a free immunoglobulin solution
so that it will contain the same amount of immunoglobulin we added
to the periodate-oxidized peroxidase. Make 12 double dilutions
from both preparations, using PBS and the microtitrator kit, as

described in Exercise No. 76. Make two microdiffusion plates with
the same pattern as shown in Fig. 56, but this time punch 4 mm
diameter holes in it on both sides of the central trough. Number
the wells as indicated in above Figure. Take the slides and fill
the holes with the dilutions of the free anti-immunoglobulin,
starting from the highest dilution end, filling hole No. 12 with
it. The holes in the other microslide will receive dilutions of
the complex, again starting in hole No. 12 with the highest dilution. Fill the central trough on both slides with a 1 mg/ml mouse
IgG solution. Let the diffusion pattern develop in a moist chambe
at cold room temperature. The following day read the highest
dilutions of both preparations which still give visible precipita
with the IgG. There should not be more than two or three well
differences between the two preparations.

Use and Limitations

The addition of dinitrofluorobenzene to the peroxidase before
oxidation with NaIO$_4$ was done to block the alpha and epsilon
amino groups of the enzyme. If free, these groups will react with
the aldehyde groups generated by periodate oxidation, thus resulting in linking of several enzyme molecules together. Blocking
of these groups of the enzyme with dinitrofluorobenzene will not
interfere with the enzymatic activity provided that HF, which is
a reaction side product, is removed from the preparation by
dialysis.

The addition of sodium borohydride is needed to stabilize the
complex. Nakane and coworkers elaborated the optimal NaBH$_4$ concentration which will stabilize the complex but will not interfer
with the enzymatic or antibody activities. According to Nakane
and Kawaoi, the above procedure will yield a preparation in which
99% of the IgG is labeled with peroxidase and approximately 70%
of the peroxidase will be coupled to IgG. Regarding the optimal
antibody to enzyme ratio in the complex, a 1 : 1 ratio would be
desirable. If this is higher than 1 : 4, the preparation cannot
be used for immunohistochemical labeling. The procedure given
here, which follows the method of Nakane and Kawaoi, approaches
the ideal antibody to enzyme ratios.

A very recent modification of the here described method has
been elaborated by Wilson and Nakane (Proceedings, VIth International Conference on Immunofluorescense and Related Staining
Techniques, 1978, Elsevier-North Holland publ., in press). In
step 3 (see Procedure above) they dialyze the IO$_4$ treated enzyme
against 1 mM pH 4.4 acetate buffer at 4°C overnight, and keep
the pH of the preparation low until just before the conjugation
reaction with IgG or its Fab' fragment. This procedure eliminates
the use of amino-blocking agent 1-fluoro-2,4 dinitrobenzene.
Before conjugation, the pH of the IO$_4$-treated enzyme is raised
to 9-9.5 by the addition of 20 µl 0.2 M Na$_2$CO$_3$ buffer, pH 9.5
and immediately mixed with 8 mg IgG (or 5 mg Fab') dissolved in
1 ml 0.01 M Na$_2$CO$_3$ buffer, pH 9.5. The reaction mixture is stirre
for 2 h at room temperature, finally reduced by the addition of
0.1 ml freshly prepared NaBH$_4$ (4 mg/ml) solution. The mixture is
transfered now to a cold room for 2 h, before chromatographic
purification on Sephacryl G-200 column.

Using enzyme-labeled antibodies, mouse immunoglobulin can be detected on lymphocyte surface, on histological sections of tissues, or on any particles to which IgG has been attached, either through immune specific or nonspecific reactions. Avrameas and Uriel used such preparations in immunodiffusion (1966). Ultra-thin sections of cells reacted with the enzyme-labeled preparation and osmium tetroxide can be viewed under the electron micro-scope, as described in Exercise No. 84. The same procedure can be used to attach enzymes not only to antibodies but to any other proteins, such as antigens, viral particles, protein fragments and synthetic polypeptides with free primary amino groups. The book *Immunoenzymatic Techniques,* edited by Feldman et al. (1976), describes the principles and techniques as well as the applica-tion of these enzymes to cell biology. Among many other applica-tions, it describes the use of lectin-coupled enzymes in detect-ing cell-surface lectin receptors.

<u>References</u>

Avrameas, S., Uriel, J.: C. R. Acad. Sci. (Paris) *262,* 2543 (1966)
Chance, B., Maehly, A.C.: In: Methods in enzymology, Colowick and Kaplan
    (eds.) Vol II, p. 764, 1955
Feldmann, G., Druet, P., Bignon, J., Avrameas, S. (eds.): Immunoenzymatic
    techniques. Amsterdam-Oxford: North-Holland, 1976
Nakane, P.K., Kawaoi, A.: J. Histochem. Cytochem. *11,* 1084 (1974)
Nakane, P.K., Pierce, G.: J. Histochem. Cytochem. *14,* 929 (1966)

Exercise No. 15

Preparation of Soluble Dinitrophenyl Proteins

The method described in Exercise No. 44 yields an insoluble dinitrophenyl (DNP) protein which is suitable for chemical analy-sis, but cannot be used for immunobiologic assays. Eisen (1964) prepared DNP protein derivatives which are readily soluble in water or saline using a mild procedure. Using dinitrophenyl sulfonate instead of dinitrofluorobenzene, more selective dinitro-phenylation could be achieved. This reagent is water soluble; its removal requires simple dialysis (or gel filtration). The number of DNP groups introduced on a protein can be measured photometrically.

The importance of this method in immunochemistry is that the DNP groups give a new immunologic specificity to the carrier protein. Antibodies formed against such a DNP protein will have antibodies against this determinant group.

The general principles of the soluble DNP protein preparation will be given here based on the work of Eisen (1964).

<u>Materials and Equipment</u>

Human serum albumin (HSA)
Dinitrophenyl sulfonate, Na salt, recrystallized (DNPS)

Potassium carbonate
Physiologic saline, pH 7.4
10% Sodium hydroxide
10% Hydrochloric acid
Amberlite IRA 400 in Cl⁻ form
Büchner funnel (Fig. 11) with filter paper
Chromatographic column, 20 × 400 mm (Fig. 2)
Suction flask (Fig. 11)
Spectrophotometer
pH Meter
Magnetic stirrer and bars

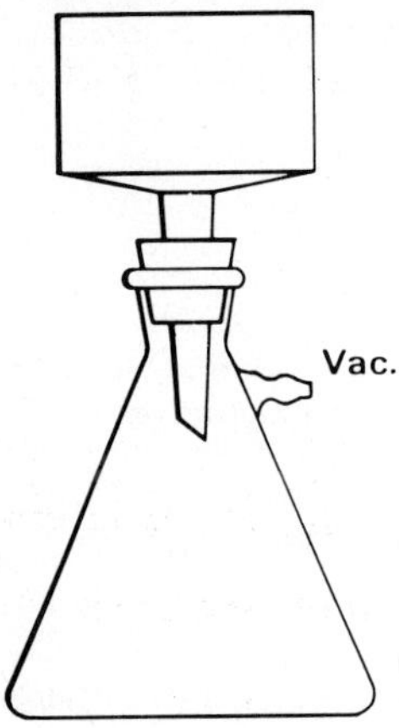

Fig. 11. Filter flask with Büchner funnel

Procedure

1. Dissolve 100 mg human serum albumin and 100 mg K_2CO_3 in 50 ml distilled water. Add 100 mg DNPS. Stir the solution in a dark cabinet overnight, at room temperature.

2. Prepare a 20 × 400 mm ion exchange column, using Amberlite IRA 400 in Cl⁻ form. For the preparation of freshly activated ion exchangers, stir the resin for 15 min, first in 10% NaOH, filter and wash until the filtrate is neutral, then add 10% HCl. After 15 min, filter and wash it on a Büchner funnel until it becomes neutral. The same process may be repeated with NaOH and HCl. Fill the column with this slurry. After the resin settles, pass the soluble DNP protein mixture through the column, using 200 ml pH 7.4 physiologic saline to elute all the proteins. The unreacted DNPS and the free dinitrophenol formed will be adsorbed by the column. Collect the effluent.

Evaluation

Measure the protein content/ml of the preparation as described in Exercises No. 48 or 50. From the protein content of the DNP protein, calculate the molarity of the solution (M_{HSA}), knowing that HSA has a molecular weight of 69,000. (For example, a 0.69% albumin is a 10^{-4} molar solution.) Determine the DNP group content

of the protein spectrophotometrically. Add 0.1 ml of the DNP-
HSA conjugate to 9.9 ml 0.1 N NaOH. An intense yellow color will
be visible. Measure the absorbance (OD) of this solution at 360 nm.
Assume that the yellow color of the DNP protein is due entirely
to ε-DNP lysine and note that the molecular extinction of this
lysine derivative in dilute NaOH is 17,530 (= absorbance of a
1 molar DNP lysine solution).

To calculate the molarity of DNP lysine in your preparation
($M_{conj.}$), multiply the absorbance measured at 360 nm by 100 (dilution) and divide it by the molecular extinction of DNP lysine

$$M_{conj.} = \frac{OD \times 100}{17,530}$$

In order to find the number of DNP lysine groups per molecule
protein (N)

$$N = \frac{M_{conj.}}{M_{HSA}}$$

Use and Limitations

The preparation of DNP- and TNP(trinitrophenyl)-tagged erythrocytes is described in the next Exercise.

A few examples are given here for the use of DNP protein
conjugates.

Rabbits may be immunized by this DNP-HSA and the serum obtained
may be used to demonstrate hapten specificity and carrier specificity.

Combination of DNP with another protein will yield a conjugate
which will react strongly with antiserum produced to the unrelated
DNP-HSA conjugate. Measurement of the precipitated antibody with
HSA alone will show the amount of antibodies produced against
the HSA carrier.

Using equilibrium dialysis as described by Karush (1950) or
Eisen (1964), the affinity of DNP lysine to the corresponding
antibody may be measured quantitatively.

The specific precipitate formed between DNP protein and homologous antisera may be dissociated by DNP lysine which displaces
the DNP protein conjugate.

References

Eisen, H.: Meth. med. Res. *10*, 94 and 106 (1964)
Karush, F.: J. Amer. Chem. Soc. *72*, 2705, 2714 (1950)

Exercise No. 16

Preparation of Dinitrophenyl- and Trinitrophenyl-Labeled Erythrocytes

The introduction of new immuno-determinant groups, such as
dinitrophenyl (DNP) and trinitrophenyl (TNP), into whole cells
became possible through the use of the DNP and TNP sulfonic
acids. The chemical reaction through which a covalent linkage is
established between the protein and the DNP or TNP hapten is a
relatively mild nucleophilic substitution. Electron donating
functional groups, in the carrier molecule, such as SH^- or NH_2^-
or OH^- groups will react with electron accepting sulfonic acid
radicals on the nitrated benzene ring, thus forming a stable
covalent bond between the hapten and the carrier. DNP and TNP
sulfonic acids are soluble in water or saline and the reaction
will proceed smoothly at room temperature or at slightly elevated
temperatures. The treatment of a carrier molecule or cell with
these reagents will not introduce considerable unwanted changes
in their structure.
   Labeling of proteins with DNP was elaborated by Eisen et al.
(1953) based on the work of Sanger (1946) which is also the basis
of Exercise No. 15 in this Manual. TNP coupled to hemocyanin has
been used as a more potent immogen than DNP by Rittenberg and
Amkraut (1966). The procedures described here in *Part A* are
based on the method of Eisen et al. (1953) and applied to DNP
labeling of sheep red blood cells. *Part B* is based on the work
of Rittenberg and Pratt (1966) who introduced TNB haptens to
erythrocytes.

Part A

Materials and Eqiupment

    10 ml sheep blood
    2,4-dinitrobenzene sulfonic acid (DNBS), sodium salt. Avail-
      able from ICN Nutritional Biochemical Corp., Cleveland,
      OH 44128
    DNP lysine (available from Sigma Chemical Co., St. Louis,
      MO 63178
    DARCO charcoal (available from Fisher Scientific Co., King
      of Prussia, PA 19406
    1 *N* NaOH
    1 *N* HCl
    0.02 *M* pH 7.2 phosphate-buffered saline (PBS) See Exercise No.1
    Boiling chips
    Reflux condenser with 24/40 ℑ joint
    1000 ml round flask with 24/40 ℑ joint and the corresponding
      glass stopper
    Water or oil bath
    Büchner funnel, 8-10 cm diameter and vacuum filter flask,
      500 ml capacity with filter paper (Fig. 11)
    Test tube shaker (Model 2997-M 10 A.H. Thomas, Philadelphia,
      PA 19105)

Procedure

1. Recrystallization of the DNBS sodium salt may be necessary, unless a highly purified commercial product is available.
   To proceed with the recrystalization, pour 300 ml ethanol into the 1000 ml flask and add 5 g DNP-sulfonate sodium salt together with a few boiling chips. Dissolve the DNPS by boiling the ethanol under refluxing on a boiling water bath (or on an oil bath adjusted to 95°-100°C). Take the flask out of the bath, add approximately 2-3 g DARCO charcoal to the solution. Close the flask with a glass stopper and shake it vigorously. Heat it to boiling again, using the reflux condenser. The mixture should be now filtered as quickly as possible through two layers of Whatman No. 1 or similar filter paper using a Büchner funnel and vacuum. The first few milliliters of filtrate may contain charcoal granules, refilter these until the filtrate is clear. Do not change the filter in the Büchner funnel. The crystals will be formed as the solution cools. Transfer the filter flask to a cold room and keep it there overnight. Collect the crystals on filter paper, again using a Büchner funnel and vacuum. Transfer the crystals into an open dish and place them in a vacuum desiccator over $CaCl_2$ and dry them under vacuum overnight. The collected filtrate may give you further crystals on the next day, these may be collected again by filtration and added to the previously harvested preparation.

2. Take 10 ml sheep blood, centrifuge it at 1000 g and wash the sedimented red blood cells three times with PBS. Suspend the packed cell sediment with 1 vol PBS.

3. Dissolve 5 mg recrystallized DNPS in 5 ml PBS and add 2 ml of this solution to every ml RBC suspension, which has been prepared in the previous step. To protect the reactants from light, wrap the reaction flask in aluminum foil. Incubate the cells at 37°C for 2 h under constant but gentle stirring.

4. Wash the cells with at least 20 vol of PBS, sediment at 1000 g. Repeat the washing three more times. The cells are now ready for immunization or for detection of antibodies.

Evaluation

The DNP to total membrane protein ratio can be determined if the conjugated membrane is isolated. Use the procedure described in Exercise No. 27 to isolate the membranes from an accurately measured 0.2 ml sedimented DNP-labeled RBC suspension. The washed membranes should be suspended in 2 ml 1 N NaOH at 50°-60°C for 60 min. The optical density of the solution should be read at 360 nm against an identically treated but not conjugated RBC membrane sample, also dissolved in 1 N NaOH. Prepare a calibration curve from DNP lysine standard dissolved in 1 N NaOH and use this curve to determine the approximate DNP : mg protein ratio. The protein content of the membrane preparation can be determined by the Kjeldahl nitrogen determination procedure. The DNP groups on the proteins will not contribute significantly to the percent of nitrogen determined by this procedure.
   Another way to establish the degree of substitution with DNP (or TNP) groups on the membrane is to determine the substituent concentration as above, and determine the dry weight of the

membrane preparation (without NaOH). Express the percent DNP or
TNP content of the dry weight. Exercise No. 48 gives you the
procedure for the Kjeldahl determination and Exercise No. 47
gives you the method for dry weight measurement.

Part B

Materials and Equipment

> 2,4,6-trinitrophenyl sulfonic acid (TNPS) available from ICN
> Nutritional Biochemicals Corp., Cleveland, OH 44128
> ε TNP lysine, available from Sigma Chem. Co., St. Louis, Mo
> 63178
> 1 *N* HCl
> 0.1 *M* Na barbital buffer with NaCl, pH 7.3. Mix 570 ml 0.1 *M* Na
> barbital (20.6 g/1000 ml) and 430 ml 0.1 *N* HCl. Dissolve in
> this 9 g NaCl
> 0.2 *M* cacodylate buffer, pH 6.9. Mix 250 ml 0.2 *M* Na cacodylate
> (42.8 g Na cacodylate.3H$_2$O per 1000 ml water) with 47 ml
> 0.2 *N* HCl. Add 703 ml water
> Glycyl-glycine (available from ICN Nutritional Biochem. Corp.,
> Cleveland, OH 44128. Dissolve 100 mg in 10 ml distilled water
> Magnetic stirrer

Procedure

1. If necessary, recrystallize 5 g TNPS in 10 ml 1 *N* HCl.
Dissolve the TNPS completely in hot HCl. If the TNPS does not
dissolve completely, filter it while hot, using filter paper.
Crystals will be formed. Collect them on a small Büchner funnel
and dry in a desiccator over KOH pellets.

2. Dissolve 60 mg crystallized TNPS at room temperature in
20 ml cacodylate buffer in a 100 ml Erlenmeyer flask. Use a
magnetic stirrer. Add slowly, dropwise, 6 ml washed, 1 : 1 diluted
erthrocyte suspension (see *Part A*). Continue stirring for 15 more
min at room temperature.

3. Add 50 ml Na barbital-NaCl buffer to the flask, stir 1 or
2 min. Centrifuge the suspension at 1000 g for 15 min. Discard
the supernatant. Wash the sediment three times with the following
mixture: 50 ml Na barbital buffer + 2 ml glycyl-glycine. Gently
resuspend the cell sediment, stir it well for a few minutes and
centrifuge as above. Discard the colored supernate. Repeat the
washing at least three more times. The last washing fluid should
be colorless, indicating that the excess TNPS has been removed.
The cells are ready for use. They can be stored at +5°C for
several days without noticeable change.

Evaluation

Use the same procedures as in *Part A* for the milliequivalent
TNP/mg membrane dry weight determinations. The TNP group has an
absorption maximum at 348. The calibration curve for TNP milliequivalent determination can be obtained by using εTNP lysine.
The molar extinction coefficient of εTNP lysine is 15,400.

Use and Limitations

DNP- or TNP-labeled proteins can be linked to CNBr-activated
Sepharose as described in Exercise No. 8, and used for the isolation of DNP or TNP specific antibodies by affinity chromatography.
The affinity of immunoglobulins to DNP or TNP hapten-carrying
molecules can be determined by equilibrium dialysis. Changes in
the affinity of immunoglobulins to these determinants during
various schedules of immunization can be followed. The role of
the carrier molecules in the immune response to DNP or TNP can
be studied. DNP- and more recently TNP-labeled soluble proteins
and erythrocytes have been used extensively in various studies
investigating diverse aspects of immune reaction mechanisms.
Paul et al. (1970), Tigelaar et al. (1971), Henney and Ishizaka
(1970), and Naor et al. (1974, 1975) are only a few of the many
examples.

References

Eisen, H.N., Belman, S., Carsten, M.E.: J. Am. Chem. Soc. *75*, 4583 (1953)
Henney , C.S., Ishizaka, K.: J. Immunol. *104*, 1540 (1970)
Naor, D., Morecki, S., Mitchell, G.F.: Eur. J. Immunol. *4*, 311 (1974) and
    Eur. J. Immunol. *5*, 220 (1975)
Paul, W.E., Katz, D.H., Goidl, E.A., Benacerraf, B.: J. Exp. Med. *132*, 283
    (1970)
Rittenberg, M.B., Amkraut, A.A.: J. Immunol. *97*, 421 (1966)
Rittenberg, M.B., Pratt, K.L.: Proc. Soc. Exp. Biol. Med. *132*, 575 (1966)
Sanger, F.: Biochem. J. *40*, 261 (1946)
Tigelaar, R.E., Vaz, N.M., Ovary, Z.: J. Immunol. *106*, 661 (1971)

Exercise No. 17

Isolation of Cell Membranes From Ascites Tumors

Various methods are described in the literature for the isolation
of membranes of nucleated cells. The procedure described in this
exercise is based on the method of Boone et al. (1973) with minor
modifications as applied to the isolation of TA3-Ha murine ascites
tumor membrane in the author's laboratory (Nowotny et al., 1976).

Materials and Equipment

   TA3-Ha transplantable tumor cells. Since the tumor grows in
     allogenic recipients, practically any mouse strain can be
     used. The cells harvested from the ascites fluid should be
     washed in Hank's balanced salt solution twice, using centri-
     fugation at 2000 g for 20 min. 2 ml packed sediment will be
     used for this Exercise
   Stock solution I for Tris-Mg buffer pH 7.4. Dissolve 121.0 g
     Tris (hydroxymethyl) aminomethane and 9.52 g $MgCl_2$ in a total
     of 800 ml glass-distilled water. Add to this under constant
     stirring 10 *N* HCl until the pH reaches 7.4. Fill up with
     distilled water to 1000 ml

Stock solution II for Tris-Mg buffer pH 8.6. Prepare the same
   800 ml Tris-Mg solution as above and add slowly 10 *N* HCl
   until the pH reaches 8.6. Make it up to 1000 ml with distilled water
Buffer A. Take 1 vol of Stock I and mix it into 99 vol of
   physiologic saline
Buffer B. Take 1 vol of Stock II and mix it with 99 vol of
   saline
Buffer C. Add 1 vol of Stock II to 99 vol of 0.25 M sucrose
   solution. (Dissolve 85.579 g sucrose in 1000 ml water)
60%, 40%, and 35% (W/W) sucrose solutions, prepared in Buffer
   B, as follows: place 60 g sucrose in a tared beaker on a
   top-loading balance and make this up to 100 g by adding
   glass-distilled water. Prepare the 40% and 35% W/W solution
   in a similar fashion
Hank's balanced salt solution. (Available from CIBCO, Grand
   Island, NY 14072, or from many other companies)
Dounce type homogenizer with a B type pestle (Bellco Glass
   Co., Vineland, NJ 08360)
Refractometer for sucrose concentration determination. (Bausch
   & Lomb or other products can be used)
Spinco centrifuge with SW 25.2 rotor
Refrigerated regular laboratory centrifuge
Ultrasonic equipment (Model W 350, Branson Sonic Power Co.,
   Danbury, CT 06810)

Procedure

1. To each ml packed cell sediment add 5 ml Buffer A at +4°C
and keep it there for 5 min while stirring it with a glass rod.
Transfer the cell suspension immediately after this to the Dounce
instrument and homogenize it by exactly ten strokes, applying
them slowly, avoiding bubble formation. This is sufficient to
break the cells. Transfer the contents of the homogenizer to a
beaker and immediately adjust the sucrose concentration to 45%
by mixing 1 vol cell homogenate with 3 vol 60% sucrose prepared
in Buffer B. Stir the homogenate thoroughly. The liberated membrane ghosts will disintegrate in the Tris-Mg buffer within a
few minutes, unless sucrose has been added to it.
2. Pipette 15 ml of this suspension in each of the 50 ml
SW-25.2 cellulose nitrate centrifuge tubes. Carefully layer on
the top of this 15 ml 40% sucrose solution, and on top of this
pipette 15 ml 35% sucrose solution. The last layer is 5 ml Buffer
C. Centrifuge the tubes at 20,000 rpm for 10 min at +4°C in the
Spinco centrifuge.
3. The sediment at the bottom of the tube will contain nuclei,
the intact whole cells to be found at the interface of the 40%-
45% sucrose concentration. The isolated cell membrane fragments
will collect at the 35%-40% interface. That band is collected
with a pipette and identical bands from several centrifuge tubes
can be pooled. Dilute this pool with Buffer B to a 10% sucrose
concentration. This sucrose concentration has to be controlled
by a refractometer. It will take approximately a three- to four-
fold dilution of the collected band to reach the 10% sucrose
final concentration. The collected samples should now be centrifuged in polycarbonate centrifuge tubes at +4°C at 5000 g for

30 min. The pellet is washed twice more with Buffer C, using
10 ml for each centrifugation. Discard the supernatant. The pellets of the isolated membranes can be resuspended in the same
Buffer C.

Evaluation

Electron microscopy can be used to study the purity of the membrane fractions. The preparation can be sedimented again and
fixed in 3% glutaraldehyde-1% osmium tetroxide and embedded in
Epon resin. Sections can be made from this preparation and stained
with various procedures to render them visible under the electron
microscope.

Other assays include biochemical determinations of enzyme
activity. Several enzymes are membrane-bound in nucleated cell
preparations and their activity can be determined. ATPase can
be determined by the method of Wallach and Ullrey (1964). Other
frequently used enzymatic "markers" of membrane preparations are
5'-nucleotidase, as measured by the procedure of Heppel and
Hilmoe (1969), or alkaline glycerophosphatase, as determined by
the procedure of Morton (1969).

If the membrane contains immunogens, either transplantation
antigens or virus induced membrane antigens, various immunochemical procedures can be used to detect and quantitatively
determine their amounts.

Use and Limitations

The same procedure can be used to isolate membranes from spleen
cells, peripheral blood samples, liver cells, and various tumors.
It is sometimes more difficult to tease apart solid tumors than
spleens or livers to obtain single cell suspensions. On the other
hand, some tumors shed membrane fragments into their surroundings
during growth. Such membraneous preparation has been isolated
from ascites fluid of TA3-Ha tumor-bearing mice by a modification
of the here described procedure (Nowotny et al., 1976).

The numerous subcellular elements make it significantly more
difficult to isolate nucleated cell membranes in a pure and
biologically still active form than to obtain cell membranes of
human erythrocytes. The procedure starts with efficient disruption of the cells. Repeated freeze-drying has been used for the
rupture of cells, this procedure breaks all membranes in the cell
rather extensively. N_2 decompression has been used by several
laboratories for the rupture of the cells. This procedure gives
good yield in active form and provides preparations ready for
relatively easy further purification (Wallach and Kamat, 1964;
Manson and Palm, 1968).

The use of the Dounce homogenizer, which breaks the cell
membrane without damaging more than a few percent of the nuclear
membrane, is much more gentle. Warren et al. (1966) used 0.001 M
$ZnCl_2$ to "harden" the cell membrane and facilitate its separation
from cell nucleus and protoplasm. Boon et al. (1969) used $MgCl_2$
as divalent cation for similar purposes. Finally, some laboratories
prefer the use of hypotonic saline, which swells the cell and
will initiate the formation of pseudopods. Gentle homogenization
of such swollen cells will yield rather pure membrane microvesicles,

although in a lower yield than the more drastic procedures. The
microvesicles and membrane fragments obtained can be further
purified by centrifugation. Neville (1960), whose method is
widely used, homogenized liver cells in 0.001 M $NaHCO_3$ and separated the fragments on discontinuous sucrose density gradient.
Wallach and Kamat (1964) preferred polysucrose (Ficoll) because
of its lower osmotic activity. The advantages and disadvantages
of the most frequently used methods were critically evaluated
and reviewed by Wallach and Winzler (1974).

Although a preparation as described here will be considerably
pure, as can be judged from the electron microscopic picture, it
still may contain ribosomes, mitochondrial elements, fragments
from the endoplasmic reticulum, and possibly other sub-cellular
components. Further purification of this preparation can be obtained by using a continuous sucrose density gradient of sonicated
ghost supernatans. A few seconds (5-10) sonication of the preparation described above will give rather uniform microvesicles.
Several of these will contain trapped fragments of other subcellular organelles. If the continuous sucrose density gradient is
made from 50% sucrose on the bottom to 20% sucrose on the top,
and the membrane preparation is layered in Buffer C on top of
the 20% sucrose layer, 12-18 h centrifugation at $+4^oC$ at 20000
rpm will give further purification. The purest membrane preparation, according to Boone and associates (1973), was found in the
zone which corresponded to 30%-33% sucrose content.

References

Boone, C.W., Ford, E., Bund, E., Stuart, C., Lorenz, D.: J. Cell Biol. *41*,
    378, 1969
Boone, C.W., Gerber, P., Brandschaft, P.B.: J. Nat. Cancer Inst. *50*, 841,
    1973
Heppel, L.A., Hilmoe, R.J.: In: Methods in enzymology. Colowick, S.P., Kaplan,
    N.O. (eds.): Vol. II, 'p. 546. New York: Academic Press
Manson, L., Palm, J.: In: Advances in transplantation. J. Dausset et al. (eds.)
    Munksgaard, 1968
Morton, R.K.: In: Methods in Enzymology. Colowick, S.P. Kaplan, N.O. (eds.),
    Vol. II, 1955 p 533. New York: Academic Press
Neville, D.M.: J. Biophys. Biochem. Cytol. *8,* 413 (1960)
Novotny, A., Butler, R.C., Grohsman, J., Keebler, C.: Ann N.Y. Acad. Sci.
    276, 106, 1976
Wallach, D.F.H., Kamat, V.B.: Proc. Natl. Acad. Sci. USA *52*, 721 (1964)
Wallach, D.F.H., Ullrey, D.: Biochem. Biophys. Acta *88,* 620 (1964)
Wallach, D.F.H., Winzler, R.J.: Evolving strategies and tactics in membrane
    research, Berlin-Heidelberg-New York: Springer 1964
Warren, L., Glick, M.C., Nass, M.K.: J. Cell Physiol. *68,* 269, 1966

Exercise No. 18

Preparation and Use of Synthetic Liposomes

Liposomes are small spherical bodies prepared from lipids in the
laboratory. Their wall consists of a concentric lipid bilayer
(or several concentric bilayers) in which the lipid molecules
are highly oriented. Liposomes prepared in water will have the
lyophilic end of the lipid molecules oriented towards the water
phase, in the same way as lipids are oriented in cell membranes.
According to Bangham (1968), the liquid crystal or liposome is
a preferred phase structure of many biologic lipids in the presence of water or salt solutions.
   There are several ways in which liposomes can be prepared.
One may dissolve the lipid in organic solvents, such as chloroform, and add the solute to a great excess of water which will
dissolve the chloroform and precipitate the lipid. Sonication of
such lipid precipitates may lead to the formation of spherical
bodies. Another procedure which has been used in the author's
laboratory consists in drying a very thin layer of lipid to the
wall of a large flask. The thin film of lipid is first mechani-
cally agitated in water, then sonicated. The lipid film will
come off the wall of the flask and many of the peeled-off fragments will close again, forming a spherical liposome. The method
described here has been used for the preparation of immune adjuvant liposomes from synthetic glycolipids (Behling et al.,
1976). The preparation of N-palmitoyl-D-glucosamine, a synthetic
glycolipid, has been carried out according to the procedure of
Fieser et al. (1956).

Materials and Equipment

   N-palmitoyl-D-glucosamine (NPG). Preparation is described below
   50% Propanol
   Ethanol, purest available quality
   Tetrahydrofuran
   Palmitoyl chloride
   10% Na_2CO_3 in water
   1.8% NaCl in water
   50 ml round flask, 24/40 ℈
   Reflux condenser with 24/40 ℈
   Büchi rotating vacuum evaporation equipment (see Figure 12)
   Sonicating instrument
   5 cm diameter Büchner funnel with 1000 ml filter flask
   Melting point apparatus

Procedure

   1. Dissolve 1.8 g D-glucosamine · HCl in 20 ml 10% Na_2CO_3.
Take 2.8 g palmitoyl chloride and dissolve it in 10 ml tetrahydrofuran. Add the palmitoyl chloride dropwise to the D-glucosamine solution under constant stirring at room temperature.
Continue stirring for 60 min. Now add the above mixture to 300 ml
water under constant stirring at room temperature. Transfer the

Fig. 12. Vacuum
distillation ap-
paratus, Büchi type

suspension to the cold room for 30 min and filter the white
precipitate on a Büchner funnel. Recrystallize it twice from
hot 50% propanol. This white substance is N-palmitoyl-D-glucosamine (NPG).

   2. Dissolve 2 mg NPG in 30 ml ethanol in a 500 ml round flask
on a hot water bath. Use a reflux condenser to prevent evaporation. After complete dissolution, cool it to room temperature
and attach it to a Büchi rotating evaporator. Apply vacuum, and
immerse the flask in a lukewarm water bath. Remove the solvent
while slowly rotating the flask. The NPG forms an invisible thin
film on the inside of the flask.

   3. Remove the flask from the vacuum distillation unit and add
20 ml double-distilled cold water. Close the flask, immerse it
for a short while in ice-cold water and then shake vigorously
for several minutes. This shaking will remove the fine lipid
film from the inside of the flask. Pour the contents of the
flask into a 50 ml beaker and sonicate the suspension for 2 min
at 1.7 A.

Evaluation

Determine the melting point of the recrystallized NPG. It should
be 202°-204°C. Determine the dry weight content of the liposome
suspension by using the procedure described in Exercise No. 47.
Place a droplet of this on a microscope slide and observe the wet
suspension by darkfield microscopy. You will see amorphous fragments as well as round spherical bodies. At least 50% of the
suspension should consist of spherical bodies, which are the
liposomes.

   To test the adjuvant effect of NPG liposomes for sheep red
blood cells, the following procedure is recommended. Adjust the
washed sheep red blood cells to a final concentration of 5×10^7
SRBC/ml in 1.8% saline. Adjust the liposome suspension to 50 µg/ml
using distilled water. Add equal volume of liposome suspension
to the SRBC suspension to obtain an isotonic condition. Allow
this combined suspension to stand at room temperature for 60 min

with occasional gentle agitation. Inject a total of 0.4 ml
intraperitoneally into each mouse. This suspension contains
1×10^7 SRBC + 10 µg liposomes. The animals can be killed four
days after the injection and their immune response to SRBC can
be quantitated by the PFC assay, as described in Exercise No. 66.
As control, inject only 1×10^7 SRBC in one group of mice. Another
group of mice should receive 10 µg liposomes alone.

Use and Limitations

Lecithin, cholesterol, and a number of other lipids can be used
to prepare liposomes. Sessa and Weissman (1968) prepared liposomes with male or female steroids. One can mix several lipids,
in different ratios, and form a mixed liposome. Furthermore, one
may mix lipids with proteins or other macromolecules. Kinsky
(1972) incorporated an antigen in the lipid model membrane and
studied the antigen-antibody-complement interaction. With this
model system, lysis of liposomes could be achieved. Similar model
systems were used for the study of immune lysis by Humphries and
McConnel (1974). Gregoriadis and associates (1974) entrapped
drugs in liposomes and used this system in cancer chemotherapy.
For the same purpose Rahman et al. (1974) encapsulated actinomycin
C into liposomes.
   N-palmitoyl-D-glucosamine was found earlier by author and
collaborators to have a mitogenic effect (Rosenstreich et al.,
1974).

References

Bangham, A.D.: Progr. Biophys. Mol. Biol. *18*, 31 (1968)
Behling, U.H., Campbell, B., Chang, C., Rumpf, C., Nowotny, A.: J. Immunol.
   117, 847 (1976)
Fieser, M., Fieser, L., Toromanoff, E., Hirata, Y., Heyman, H., Teft, M.,
   Bhattacharya, S.: J. Am. Chem. Soc. *78*, 2825 (1956)
Georgoriadis, G., Wills, E.J., Suwain, C.P., Tavill, A.S.: The Lancet, p. 1313
   June, 1974
Humphries, G.K., McConnel, H.M.: Proc. Natl. Acad. Sci. USA *71*, 1691, 1974
Kinsky, S.C.: Biochim. Biophys. Acta *165*, 1 (1972)
Rahman, Y.E., Cerny, E.A., Tollaksen, S.L., Wright, B.J., Nance, S.L.,
   Thomson, J.F.: Proc. Soc. Exp. Biol. Med. *146*, 1173 (1974)
Rosenstreich, D.L., Asselineau, J., Mergenhagen, S.E., Nowotny, A.: J. Exp.
   Med. *140*, 1404 (1974)
Sessa, G., Weissman, G.: J. Lipid Res. *9*, 310 (1968)

Exercise No. 19

Isolation of Murine Transplantation Antigen Preparation

The products of H-2 genes are among the most potent transplantation antigens in mice. Much attention has been devoted to the
isolation and chemical characterization of these antigens, particularly because of their importance in immunogenetic studies.

Several methods were applied to their isolation, some of the
first procedures used nonionic detergents to solubilize membranes
and liberate H-2 antigens from them, as developed by Kandutsch
and Stimpfling (1963). Schwartz and Nathanson (1971a) used NP-40
nonionic detergent to solubilize and liberate the H-2 antigens.
The crude extracts obtained were further purified by elaborate
procedures. The Exercise described here is based on this publication as well as on a later report by Schwartz et al. (1973), but
it does not include any further steps of purification.

Materials and Equipment

Spleen cells from 20 A/J mice (H-2^a) and from 20 C57Bl/6J
  mice (H-2^b)
Nonionic detergent NP-40. This is available from Shell Chemical
  Co., New York, NY. The detergent is an alkyl-aryl-ethyl-
  enedioxide derivative.
Gauze, 10 × 10 cm cuts
Glass funnel, 5 cm diam
Tris buffered saline, pH 7.5. Mix 40.6 ml 0.1 N HCl and 25 ml
  0.2 *M* Tris (24.2 g Tris in 1000 ml). Add water to 500 ml.
  Dissolve in the same solution 4.5 g NaCl and 70 mg MgCl$_2$
Tweezers and scissors
Tissue grinder (Kontes Glass Company, Vineland, NJ. 08360)
2% Na dodecylsulfate (SDS) in water, also containing 1%
  2-mercaptoethanol
Hemocytometer and microscope
0.22 µ Millipore membrane and filtration assembly (Millipore
  Corp., Bedford, MA 07130)
Anti-H-2a and anti-H-2b allocenic antisera

Procedure

1. Tease the spleens apart with tweezers or with a tissue
homogenizer in Tris-buffered saline as described in Exercise
No. 66 or No. 89. Filter the cells through two layers of gauze
placed in a 5 cm diam glass funnel. Centrifuge the cells and discard the supernatant. Wash the sediment twice with buffered saline
Discard the washings. Resuspend the cells in buffered saline so
that the cell concentration will be 10^9 cells/ml. Use a hemocytometer to determine the cell count, as in Exercise No. 82.

2. Add to the cell suspension an equal volume of 1% NP-40
dissolved in the above Tris-buffered saline. Mix gently but
thoroughly and let it stand at 4°C for 15 min. Sediment the
nuclei and the unbroken cells at 27000 g for 10 min. The supernatant contains the H-2 antigens. Force this through an 0.22 µ
pore-size Millipore or other similar quality filter using vacuum
or preferably pressure.

Evaluation

The homogeneity of the preparation can be analyzed by SDS gel
electrophoresis, staining the bands for proteins and for glycoproteins as in Exercise No. 10 and 45. Many components will be
visible.

To establish the specificity of the isolated H-2 preparations, one useful assay is the inhibition of monospecific alloantisera by the H-2 containing extracts. Sanderson (1965) elaborated such a procedure by inhibiting the cytotoxicity of alloantisera by similar preparations. The activity of the isolated H-2 can be measured by quantitative inhibition of cytotoxicity. Exercise No. 90 in this Manual can be adapted for such determinations. Manson and Simmons (1969) used isolated transplantation antigens to induce lymphoblast formation, which was measured by the enhanced incorporation of tritiated thymidine into sensitized lymphocytes. Colombani and Colombani (1976) used a complement fixation method for HLA typing on various substrates such as platelets, lymphocytes, and other cell types. Quantitative complement absorption was used for studying varations in antigenic content.

Use and Limitations

Schwartz and Nathanson (1973) used strain-specific tumor cells grown in ascites form for the isolation of H-2 antigens. They used meth-A, a methylcholanthrene fibrosarcoma induced and maintained in BALB/cJ (H-2^b) mice, or EL-4 lymphosarcoma induced and maintained in C57BL/6J, or MTC mastocytoma P815 cells grown in DBA/2J mice. Ascites tumors can give a very high yield in cell number and this single cell suspension does not require further treatment but washing with buffered saline or Hank's balanced salt solution. Schwartz and Nathanson (1971a) also labeled these viable tumor cells in vitro with ^{3}H-fucose. Since this will be incorporated into the glycoprotein H-2 antigens, they followed up this marker during the purification of the H-2 antigens, such as column chromatography and gel electrophoresis.

Further purification of such labeled extract has been achieved by reacting the preparation with monospecific alloantisera, such as anti-H-2 serum produced in allogenic mouse strain and precipitating the complex with an antimouse immunoglobulin goat antiserum. The precipitate can be washed repeatedly and dissociated with 2% sodium dodecylsulfate containing 1% 2-mercaptoethanol. Separation of this mixture has been carried out by SDS polyacrylamide gel electrophoresis by Schwartz et al. (1973).

Several other methods have been elaborated for the isolation of H-2 antigens of mice and HL-A histocompatibility antigens of man. Papain was used by Schwartz and Nathanson (1971b) to release H-2 antigens. 70% of the H-2 receptors were removed from the cell surface, but only 15% could be recovered in the supernatant in active form, indicating the sensitivity of these antigens to proteolytic enzymes. Cunningham-Rundles and Good isolated HL-A histocompatibility antigens from human peripheral lymphocytes by the papain as well as by the NP-40 detergent procedures (1977). Papermaster et al. (1972) employed the extraction procedure of Reisfeld et al. (1971) using 3 molar KCl. The extract was further purified by centrifugation and preparative electrophoresis.

References

Colombani, J., Colombani, M.: In: Manual of clinical immunology. Rose, Friedman (eds.) ASM publication, p. 805, 1976

Cunningham-Rundles, C., Good, R.A.: Transpl. Proc. *9*, 587 (1977)
Kandutsch, A.A., Stimpfling, J.J.: Transpl. *1*, 201 (1963)
Mason, L.A., Simmons, T.: Transpl. Rev. *6*, 81 (1971)
Papermaster, B.W., Papermaster, V.M., Reisfeld, R.A., Pellegrini, M.A.,
    Ferrone, S., Kahan, G.D., Terasaki, P.I., Takasugi, M., Albert, E.E.:
    In: Cellular antigens, Nowotny, A. (ed.), Berlin-Heidelberg-New York:
    Springer 1972
Reisfeld, R.A., Kahan, B.D.: Transpl. Rev. *6*, 81, 1971
Sanderson, A.R.: Immunology *9*, 286 (1965)
Schwartz, B.D., Kato, K., Cullen, S.E., Nathanson, S.G.: Biochem. *12*, 2157
    (1973)
Schwartz, B.D., Nathanson, S.G.: J. Immunol. *107*, 1363 (1971a)
Schwartz, B.D., Nathanson, S.G.: Transpl. Proc. 3, 180 (1971b)

Exercise No. 20

Isolation of Bacterial H-Antigens

The cells of certain bacterial species are motile; they are able
to penetrate semisolid agar layers. This motility of cells frequently occurs in Gram-negative species and is due to the existence of flagella on the bacterial cells. These hair-like flagella
are highly antigenic and, since the thorough studies of Craigie
(1931), it is generally accepted that H-antigenic properties of
certain bacteria are associated with the flagella. The nature of
flagellar antigens was reviewed by Koffler and Smith (1971).
Weibull (1949a, 1949b) showed that the flagella consist of
fibrous proteins and can be purified by fractional sedimentation
in a high-speed centrifuge and also by ammonium sulfate precipitation. In the method described here, motile bacteria will be selected according to the procedure of Craigie. The flagella will
be separated from the cells and their fibrous protein content,
called flagellin, will be partially purified. The method described
here is a modification of the descriptions of Weibull and also
of Koffler and Kobayashi (1957).

Materials and Equipment

> *Proteus vulgaris* X 19, or *S.typhi* O901, or *Serratia marcescens* strains
> Beef extract
> Beef infusion broth
> Peptone
> Gelatin
> Agar
> Sodium chloride
> 1 *N* Sodium hydroxide
> 1 *N* Hydrochloric acid
> Formaldehyde
> Ammonium sulfate solution (saturated)
> High-speed centrifuge
> Stirring motor or Waring Blendor
> Bacteriologic loop
> Cultivating flasks, 1000 and 2000 ml Erlenmeyers

    18 × 150 mm test tubes
    Glass tubing, 8 mm diam
    Vacuum distilling apparatus (Fig. 12)
    Magnetic stirrer and bars
    pH Meter

Procedure

Proteus vulgaris or *S.typhi* O901 or *Serratia marcescens* strain may be used in these assays, but other Gram-negative bacteria may also be tried.

1. First prepare the "motility agar" according to the following description: Dissolve 5g NaCl, 3 g beef extract, 10 g peptone in 300 ml distilled water. Stir until completely dissolved. Mix 80 g gelatin and 4 g agar with 700 ml distilled water and agitate at room temperature for 30 min. Combine the two solutions, pour into a 2000 ml Erlenmeyer flask and sterilize by autoclaving.

2. Cut 8 mm diameter glass tubing into approximately 50 mm lengths. Leave both ends of the small tubes open. Place one such tube into an 18 × 150 mm test tube and add enough of the above medium to it so that the upper half of the 50 mm-long glass tubing will remain out of the liquid. Prepare approximately six test tubes in the same way, plug them with cotton, and sterilize again, being careful to reduce autoclave pressure slowly.

3. Take a bacteriologic loop and inoculate the *Proteus* or other strain into the 8 × 50 mm small glass tubing by touching the top of the medium in this small tube with the loop. Do not inoculate other parts inside the test tube. Incubate this culture tube for 24 h at 37°C. Those cells which are motile will descend to the bottom of the small tubes and will stark growing outside of the tubing as well. These cells will come up to the surface of the medium in the 18 × 150 mm test tubes.

4. Transfer these motile cells into another sterile tube, again inoculating only the inside of the small glass tubing. Repeat the passage of the motile cells about four times. The strain thus obtained will be rich in flagellar organelles.

5. To isolate these organelles, larger volumes of beef infusion broth cultures must be inoculated. Take 500 ml medium and inoculate it with the actively motile strain. Incubate the culture for 24 h at room temperature; then add 50 ml saline containing 4% formaldehyde. Mix and let stand at room temperature for approximately one half hour.

6. It is sufficient to shake the culture to detach the long, filamentous flagella from the cells. Therefore, immerse a motor driven stirring rod, or a stainless steel stirring blade into the culture and agitate the culture vigorously for 60 min. The lowest speed of a Waring Blendor will have the same effect in 10 min. Too intensive stirring may result in the release of O-antigens from the surface of the bacterial body and this will contaminate the H-antigen preparations. Centrifuge the filtrate at 1000 g for 30 min to sediment the cells. The supernate will contain the detached flagella.

7. Further methods of purification of the flagellin have been described. The procedure given here is a simplified version of these methods. The flagella must be sedimented at 25,000 g for 2 h. Most laboratory centrifuges do not have sufficient capacity

to take 550 ml supernate and reach the above speed. Therefore,
it is recommended that the solution be concentrated by vacuum
distillation to approximately 100 ml volume prior to centrifugation.

8. After centrifugation, discard the supernate and resuspend
the sediment in 100 ml water. Be sure that no aggregates remain
undispersed. Adjust the pH of the viscous solution to pH 2 by
adding 1 *N* HCl under constant stirring. Let the solution stand
at room temperature for 30 min, then centrifuge at 35,000 g for
1 h. Discard the sediment. Neutralize the supernate with 1 *N* NaOH
to pH 7.4. This solution will contain the flagellin liberated by
acid disintegration from the flagella. The molecular weight of
this protein is approximately 40,000. Further purification of the
preparation can be achieved by salt fractionation. Ammonium sulfate saturation between 50% and 55% gives a purified flagellin
preparation.

Evaluation

Antigenicity of the flagellar preparations can be compared with
each other or with other proteins or polysaccharide antigens on
a weight basis. Polymerized flagellin is highly immunogenic in
rats. A single injection causes high IgM and later high IgG levels
Flagellin is also immunogenic but to a lesser degree and it
initiates only IgG response. If the flagellin is split by CNBr
and the breakdown products are isolated, one of them (Fragment
A) is not only poorly immunogenic, but may induce immune tolerance
if given in a certain dose or according to a specific schedule
(Nossal and Ada, 1971).

Homogeneity of the preparation can be investigated in geldiffusion systems by analyzing the precipitin lines formed between
the above preparations and anti-H and anti-O rabbit serum,
respectively.

Limitations

The procedure described here for the selection of motile bacteria
may not be fully effective in eliminating nonflagellar cells.
This is the main reason why the passage of cells through the
motility agar has to be repeated several times to reduce this
possibility.

References

Craigie, J.: J. Immunol. *21,* 417 (1931)
Koffler, H., Kobayashi, T.: Arch. Biochem. Biophys. *67,* 246 (1957)
Koffler, H., Smith, R.W.: In: Cellular Antigens, ed. by Nowotny, A., Heidel-
   berg-New York: Springer 1971
Nossal, G.J.V., Ada, G.L.: In: Antigens, Lymphoid Cells, and the Immune System.
   Acad. Press (1971)
Weibull, C.: Biochim. Biophys. Acta *3,* 378 (1949a)
Weibull, C.: Arkiv Kemi *1,* 573 (1949b)

Exercise Nr. 21

Extraction of Bacterial O Antigens (Endotoxins or lipopolysaccharides)

Extraction of O-antigens (or endotoxins) from bacterial cells
was first described by Boivin, Mesrobeanu and Mesrobeanu (1933).
The trichloroacetic acid extraction procedure used by these
workers dissolves a peptide-containing glycolipid from the cell
walls. In some cases, depending on the bacterial strain used,
the TCA extract also contains nucleic acids derived from the
cells. Numerous other procedures for the isolation of endotoxic
O-antigens from bacteria have also been published. This Exercise
in *Part A* presents a slightly modified version of the original
Boivin, Mesrobeanu and Mesrobeanu procedure.

The use of 90% phenol for the dissociation of proteins from
polysaccharides was introduced into immunochemical procedures
by Palmer and Gerlough (1940). The procedure was applied to
prepare antigenic substances from *Salmonella typhi* bacteria. A more
convenient modification of the above procedure was introduced by
Westphal and Lüderitz (1954), who applied 45% phenol and elevated
temperature to extract endotoxins *(Part B)*. The method can be
applied for the isolation of polysaccharides from tissues, cells,
and other natural products. The phenol-water procedure has also
been applied to the isolation of nucleic acid preparations from
mammalian cells and viruses. The basic principle of the method
is that proteins and lipoproteins will be dissolved in the
phenol phase while the upper water phase will contain the water
soluble and liberated lipopolysaccharides and polysaccharides
together with nucleic acids.

Rough mutant strains do not synthesize the entire polysaccharide
moiety of the lipopolysaccharide (LPS). Such products were called
endotoxic glycolipids (EGL), (Kasai and Nowotny, 1967) and were
found to be directly extractable from lyophilized bacterial cells
using chloroform and methanol (Chen et al., 1973). This procedure
has been found to be superior in author's laboratory as compared
to any other extraction methods used for similar purposes. It is
very simple, gives a good reproducible yield of highly active EGL
and the preparations obtained can be further purified by relatively
easy procedures of lipid chemistry. Preparative TLC has been used
in author's laboratory to obtain the first chromatographically
homogenous EGL preparations, ready for meaningful analysis of its
chemical structure (Chen et al., 1975, Ng et al., 1976). *Part C*
of this Exercise describes the isolation and purification of the
endotoxic glycolipid from S minnesota R 595 heptoseless rough
(Re) mutant.

Part A. Trichloroacetic Acid Method

Materials and Equipment

> 5 g Washed and lyophilized *Serratia marcescens* or any other
>   Gram-negative cells
> 0.1 *N* Sodium hydroxide
> 5% Trichloroacetic acid (TCA) solution in water

Methanol containing 0.2% magnesium chloride
Servall Omnimixer or other similar high-speed homogenizer
  with 400 ml capacity
Centrifuge with 1000 ml capacity
250 ml Plastic centrifuge flasks
Dialysis bags, 2 or 5 cm inflated diameter
3 or 5 gal capacity glass jar for dialysis, with stirrer
Filter flask
Büchner funnel
Vacuum distillation equipment (Fig. 12)

Procedure

1. Add 150 ml 5% cold TCA to the 5 g dried bacteria and homogenize the mixture in a Servall Omnimixer at top speed under ice-
water cooling for 2 min, followed by medium-speed stirring for
8 min. Centrifugation at 6000 g for 30 min separates the O-antigens from the sedimented cells and cell debris.

2. Siphon off the supernate of the centrifuge tubes. If the
sediment is hard enough, the clear supernate may be simply
poured off. Transfer the supernate into dialysis bags and dialyze
against distilled water. Changing the water twice daily for three
days will remove the TCA and any other diffusible components
from the extract. For better yield, the sedimented residue may
be extracted twice more using 100 ml 5% TCA in each step.

3. Filter the contents of the bags through two layers of analyti
filter paper using a Büchner funnel with vacuum. This crude extract is ready for concentration by vacuum distillation and freeze
drying. Adjust the pH to 7.4 using 0.1 N NaOH. Concentrate the
filtered extract to approximately 200 ml in vacuum and slowly
add 2 vol cold methanol which contains 0.2% $MgCl_2$, in the cold
room, under constant stirring. A precipitate is formed which can
be sedimented at 5000 g for 30 min in a refrigerated centrifuge.
Remove the supernate and dissolve the sediment in 200 ml water,
and add 400 ml of the above methanol. Centrifuge as above and
redissolve the sediment in 200 ml water, dialyze for three days
against distilled water, changing the outer fluid twice daily.
Concentrate the extract down to approximately 100 ml in vacuum
distillation equipment, and freeze-dry it (see Exercise No. 32
for vacuum distillation and lyophilization).

Evaluation

The material obtained is an endotoxic O-antigen preparation,
showing all the characteristic biologic activities of endotoxins.
Some of these assays are described in this Manual. Chemically,
the preparation is a lipopolysaccharide containing covalently
bound peptides. Several exercises given in this manual may be
applied for the chemical analysis of the constituents.

Use and Limitations

As has been observed, this material as well as all other endotoxin
preparations obtained by other extraction procedures from different
bacteria are heterogeneous. This has been revealed by column

chromatography using Amberlite XE 220 anion exchanger (Nowotny, 1966, and Nowotny et al.,1966).

Part B. Phenol-Water Procedure

Materials and Equipment

In addition to materials and equipment listed in *Part A:*
90% phenol. (Prepared from freshly distilled crystalline
 phenol by adding to it 10% glass distilled water and heated
 in a boiling water bath)
Stirrer motor with glass rod
70°C water bath

Procedure

1. Take 10 g lyophilized bacteria and mix with 300 ml distilled water, immersing the container in a 70°C water bath. After a homogeneous suspension has been obtained, add 300 ml 90% phenol. Continue the stirring until the temperature inside the container reaches 70°C. At that point the water and phenol will form a homogeneous system. Continue stirring at that temperature for approximately 10 min, then cool the material and centrifuge it a 3000 g for 30 min. The slightly milky appearing supernate will contain the crude lipopolysaccharide and nucleic acid fractions.

2. Separate the upper phase and repeat the extraction of the phenol phase with freshly added 300 ml distilled water. Again raise the temperature of the emulsion to 70°C and proceed as before. Add the new supernate to the first one. Repeat the extraction for the third time with 300 ml distilled water. Combine all water extracts.

3. Pour the water phases into a dialysis bag and dialyze in the cold room against distilled water for three days, changing the outer fluid twice daily.

4. Filter the extracts and concentrate them down to 100 ml using a flash evaporator or any conventional vacuum distillation equipment. Take 300 ml cold methanol containing 3 ml of the 20% $MgCl_2$ in ethanol, and under constant stirring add it to the concentrated crude extract in the cold room. A rapidly sedimenting precipitate will be obtained. Let the preparation stand at cold room temperature for 60 min, then centrifuge it at 3000 g for 30 min. Discard the supernate. Dissolve the sediment in 100 ml water and add it to 200 ml cold methanol (without $MgCl_2$). Centrifuge as above after 60 min and discard the supernate. Repeat the dissolution of the sediment in 100 ml water and its precipitation with 200 ml cold methanol.

Dissolve the final sediment in 100 ml distilled water and concentrate it down to approximately one-half using vacuum distillation. Reconstitute the extract to 100 ml with distilled water and repeat the vacuum distillation in order to remove the methanol from the solution. The material is now ready for lyophilization.

Evaluation

The biologic activity and chemical constituents of the material
can be analyzed in the same manner as the TCA-extracted preparation.

Use and Limitations

Further purification of this endotoxin can be obtained by ultra-
centrifugation. The endotoxic lipopolysaccharide has a very high
molecular weight and will sediment at 100,000-110,000 g in 2 h.
The nucleic acids which are present in the water extract will
remain in the supernate. The endotoxic lipopolysaccharide obtained contains, in addition to bound fatty acids, a few percent
of bound amino acids. As earlier results showed, column chromatography on ion exchanger resins of similarly purified preparations
revealed the presence of several carbohydrate-containing materials
therefore it must be considered as still heterogeneous.

   While the phenol-water procedure has a very wide application
in the extraction of bacterial O-antigens, it can be shown that
not all bacteria will release all their endotoxic constituents
during phenol extraction. Different procedures have sometimes
been found more advantageous for the best yield of a highly
active material from other strains.

Part C. Chloroform-Methanol Extraction of Endotoxic Glycolipid (EGL)

From Rough Mutants

Material and Equipment

     Lyophilized *S. minnesota* R 595 cells
     Chloroform (freshly distilled)
     Methanol (purest quality available)
     Glass centrifuge tubes, 200 or 250 ml capacity
     Medium fine glass filter

Procedure

   1. Mix 800 ml chloroform with 200 ml methanol to obtain a
chloroform: methanol = 4:1 (CM41) mixture.
   2. Add 10 gm lyophilized bacterial powder to 200 ml vigorously
stirred CM41 at room temperature. Continue stirring for 3 h.
   3. Filter the suspension on a medium fine glass filter and
save the extract. The cell residue has to be extracted twice more
for 3 h at room temperature using 200 ml CM41 for each extraction.
Pool the filtrates.
   4. Dry the combined extracts in a rotating Büchi apparatus
(Fig. 12), remove the organic solvents completely by vacuum
distillation. Dissolve the residue in 50 ml CM41. Transfer it
into a 200 (or 250) ml glass centrifuge tube. Add to it under
constant stirring 70 ml pure methanol, to precipitate the endotoxic
glycolipid. Close the centrifuge tube tightly with a rubber stopper
or rubber cap and centrifuge at 3000 g for 30 min in a refrigerated

centrifuge. Discard the supernatant. Dissolve the sediment in
45 ml CM41 and precipitate it again with 70 ml methanol. Centrifuge as above, discard the supernatant. Repeat the dissolution
and precipitation of the endotoxic glycolipid once more.

5. The preparation can be stored dissolved in CM41 at cold
room temperature indefinitely. If lyophilization is preferred,
the precipitate obtained in step 4 may be dispersed in 20-30 vol
water, sonicated for a few seconds and lyophilized.

Evaluation

To dissolve the endotoxic glycolipids, a dry aliquot of it has
to be sonicated in distilled water (not saline) until a fine
colloidal solution is obtained. 0.1 mg/ml dry weight concentration can be obtained with ease and such solution will be stable
in cold room temperature for several years without noticeable
change in the biologic activity. If physiologic NaCl concentration is required, 9 vol glycolipid solution should be mixed with
1 vol 9% NaCl shortly before use. The activity of such dispersed
endotoxic glycolipids can be determined in the same assays as
used for TCA- or phenol-water-extracted endotoxin preparations.

Thin-layer chromatographic separation as well as chemical and
biologic analyses of endotoxic glycolipids is described in Exercise 58, based on the method of Chen et al. (1973 and 1975).

References

Boivin, A., Mesrobeanu, I., Mesrobeanu, L.: C.R. Soc. Biol. *113,* 490 (1933)
Chen, C.H., Johnson, A.G., Kasai, N., Key, B.A., Levin, J., Nowotny, A.:
    J. Inf. Dis. *128S,* 43 (1973)
Chen, C.H., Chang, C.M., Nowotny, A.M., Nowotny, A.: Anal. Biochem. *63,* 183 (1975)
Kasai, N., Nowotny, A.: J. Bact. *94,* 1824 (1967)
Ng, A.K., Chang, C.M., Chen, C.H., Nowotny, A.: Inf. Immun. *10,* 938 (1974)
Nowotny, A.: Nature London, *210,* 278 (1966)
Nowotny, A., Cundy, K.R., Neale, N.L., Nowotny, A.M., Radvany, R., Thomas,
    S.P., Tripodi, D.J.: Ann. N.Y. Acad. Sci. *133,* 586 (1966)
Palmer, J.W., Gerlough, T.D.: Science *92,* 155 (1940)
Westphal, O., Luderitz, O.: Angew. Chem. *66,* 407 (1954)

Exercise No. 22

Preparation of ^{14}C-Labeled Endotoxin and Endotoxic Glycolipid (EGL)

LPS can be labeled by ^{51}Cr which forms an undefined complex with
the toxic components of the structure (Braude et al., 1955).
Such compounds were used extensively in studying the fate of
[51]Cr-LPS, under normal as well as pathologic conditions. A short-
coming of this otherwise excellent procedure is that the [51]Cr
label is detected by it with or without the carrier LPS molecule,
which means that the presence of the label not necessarily indicates the presence of the entire complex. The procedure described here was used in our laboratory by Dr. Chen-lo H. Chen

and it applies covalently bound ^{14}C which is incorporated into
the entire bacterial cell through biosynthesis. Extraction and
purification of the labeled endotoxic LPS or EGL is carried out
as described in Exercise 21.

Warning: Since radioisotopes are used in this exercise, the
radiation safety regulations must be followed. Absorption of
^{14}CO$_2$ gas developed during fermentation is an absolute must.
Consultation with the radiation safety office is imperative
before the Exercise can be carried out.

Materials and Equipment

Sodium acetate-UL ^{14}C, 100 µCi (specific activity 55 mCi/mM,
  ICN Isotope and Nuclear Division, Cleveland, OH 44128)
Salmonella minnesota 1114 and *Salmonella minnesota* R 595 bacteria
Growth medium, ph 7.2:
A. 1.5% tryptone (Difco)
   0.5% beef extract (Difco)
   0.3% NaCl
   0.23% disodium phosphate
   0.5% yeast extract (Difco)
B. 20% glucose
   0.4% magnesium sulfate
A and B should be autoclaved separately. After this is done,
 mix 19 vol A with 1 vol B
T-soy agar slants for the growth of bacteria
1 *N* NaOH
90% phenol
Sidearm culture flask, 500 ml, made by Bellco Glass Co.,
  Vineland, NJ 08360) Fig. 13 A
Gas washing flask containing 10 *N* NaOH as ^{14}CO$_2$ trap. Another
  similar flask is needed filled up half way with conc H$_2$SO$_4$.
  Fig. 13 B
Spectronic 20, Bausch & Lomb Spectrophotometer for test tube
  readings
Water bath at 37°C
Filtertube, Fig. 13 C, filled with glass wool

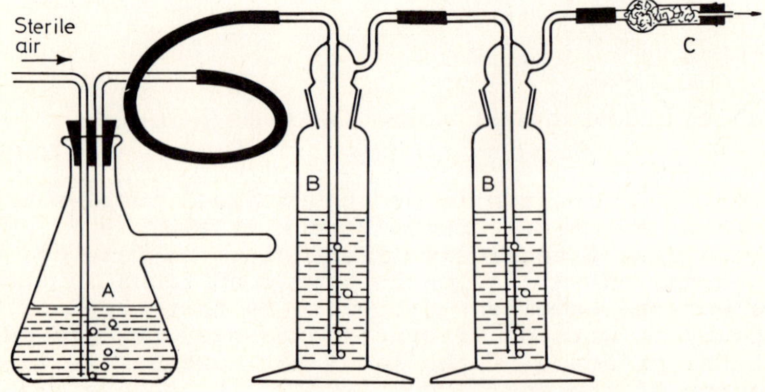

Fig. 13. Assembly for small scale cultivation of ^{14}C labeled bacteria

Procedure

1. Transfer the bacteria from stock culture to new T-soy agar slants and incubate at 37°C. Use 24 h culture as inoculum for liquid growth medium.

2. Inoculate two test tubes, each containing 10 ml liquid medium, and incubate them on a shaker at 37°C overnight.

3. Add 140 ml sterile medium to the culture flask. Place the flask into the 37°C water bath. Add the two tubes of overnight culture to the flask. Mix well by aeration as shown in Figure 13. Follow the growth by reading at 660 nm with the sidearm of the culture flask, using Spectronic 20. Plot the growth curve.

4. As optical density approaches 0.5, add 100 µCi ^{14}C sodium acetate to the culture, wash the contents of the vial with a small amount of sterile liquid medium, and add this to the culture. Now check the pH of the culture. A drop of the medium should be taken with a sterile Pasteur pipette and applied to pH paper. Adjust the pH of the culture to 7-7.5 with several drops of N NaOH. Mix well by aeration. Check the pH again.

5. Continue the growth until early stationary phase is reached. Stop the growth by adding 1.0 ml 90% phenol. Aerate for about 10 min and cool the culture in an ice bath.

6. Centrifuge at 10,000 g for 15 min in a closed container and wash the cell pellet three times with water.

7. LPS extraction can be carried out directly from the centrifuged and washed pellet. The cells may also be stored after lyophilization. For the extraction of EGL by the chloroform-methanol procedure of Chen et al. (1973), lyophilized cells should be used (see Exercise 21).

8. The extracted LPS and EGL preparations should be tested for endotoxicity and for radioactivity.

Evaluation

The specific activity of the isolated LPS or EGL preparation can be measured. For the measurement of biologic activities of the labeled endotoxins use Exercise No. 91 or No. 94 or No. 95.

Baserga (1967) developed the autoradiographic method for the detection of cellular or subcellular distribution of isotope-labeled compounds. Detection of injected endotoxin in tissues or in cells can be done by these procedures.

Use and Limitations

The same procedure can be used to label other components or subcellular organelles of the bacterial cells, however for optimal labeling of such preparations compounds other than ^{14}C acetate can be more suitable.

A disadvantage of the biosynthetic ^{14}C labeling procedure is that the label will be distributed throughout the cell. This means that not only the compound intended to be labeled will have ^{14}C incorporated, but also other cell constituents which may contaminate the preparations. In the case of LPS, frequent contaminants are nonendotoxic polysaccharides, nucleic acids, and sometimes proteins. Detection of the ^{14}C label in a labeled

endotoxin-injected host may come from such contaminants, unless
very careful purification of ^{14}C endotoxin has been carried out.
This can be achieved relatively easily in the case of EGL by using
thin layer chromatographic procedures such as described in Exercise No. 58, based on the method of Chen et al. (1973). Complete
purification of LPS is much more circumstantial. For these purposes, ion exchange chromatography has been found to be acceptable (Nowotny et al., 1966).

One should also emphasize here the obvious, that degraded and
no longer endotoxic products of ^{14}C endotoxin preparations will
still carry the label. The presence of ^{14}C, therefore, may not
mean the presence of toxic endotoxin, but only of its breakdown
products.

Ulevitch reported (1978) a new method to introduce ^{125}I into
LPS. According to this author the specific activity of the prepara
tion is very high (5μCi/μg LPS) and in spite of profound chemical
alterations of some functional groups in the LPS structure, the
labeled preparations are fully active biologically.

References

Baserga, R.: Methods in Cancer Research *1*, 45 (1967)
Braude, A.I., Carey, F.J., Sutherland, D., Zalesky, M.J.: Clin. Invest. *34*, 850
    (1955)
Chen, C.H., Johnson, A.G., Kasai, N., Key, B.A., Levin, J., Nowotny, A.:
    J. Infec. Dis. *128S*, 43 (1973)
Nowotny, A., Cundy, K.R., Neale, N.L., Nowotny, A.M., Radvany, R., Thomas,
    S.P., Tripodi, D.J.: Ann. N. Y. Acad. Sci. *133*, 586 (1966)
Ulevitch, R.J.: Immunochem. *15*, 157 (1978)

Exercise No. 23

Isolation of Bacterial Capsular Polysaccharides, K Antigens

Some bacterial strains are known to form a capsule around the
cells which consist of acidic polysaccharides. This capsule is
the outermost anatomic layer of the cells; its thickness is considerable, but its consistency is gel-like. This loose structure
can be removed from the cells relatively easily and may be purifie

Extracellular polysaccharides which originate from the capsule
of certain bacteria are serologically specific, but often cross
reactivity can be observed between capsular polysaccharides of
different strains.

The modified method described here for the isolation of capsula
bacterial polysaccharides is based on the publication of Henriksen
and Eriksen (1961) and Eriksen (1965).

Materials and Equipment

Klebsiella pneumoniae Type C cell culture (American Type Culture
    Collection No. 10273)
1% Phenol in water

2% Solution of cetylpyridinium chloride
2% Sodium chlordie solution
95% Ethanol
20% Magnesium chloride in ethanol
Celite (reagent grade)
Centrifuge
Plastic centrifuge tubes
Lyophilizer
Dialysis bags and jars
Petri dishes with 2% lactose agar
Stirrer or shaker
Waring Blendor
Berkefeld N. filter

Procedure

1. Inoculate 50 lactose-agar plates with *Klebsiella pneumoniae*
Type C. After 48 h growth at 37°C, wash the surface of each agar
plate with 5 ml 1% phenol. Repeat the washing with another 5 ml
phenol. Pool the washings and agitate the collected cells very
vigorously for 60 min at room temperature with a stirrer or in
a closed container on a shaker.

2. Add to the suspension 50 g reagent grade Celite, under
continuous stirring. After a homogeneous suspension has been
obtained, centrifuge it in a nonrefrigerated centrifuge at
10,000 g for 2 h. Lift the clear supernate and wash the sediment
with 50 ml 1% phenol. Centrifuge as above and pool the supernates.
If the combined supernates are not water clear, filter it through
a Berkefeld N filter.

3. The viscous filtrate contains the extracted polysaccharides.
This can be fractionated by the addition of 2% cetylpyridinium
chloride (CPC). Add the CPC solution dropwise, under constant
stirring, to the polysaccharide extract until further addition
of CPC does not result in additional precipitation. Centrifuge
at 1000 g for 15 min, wash the sediment twice with distilled water.
Add the washings to the supernate, which contains the neutral
polysaccharides. Save this fraction for further experiments.

4. The sediment contains the complex formed between acidic
polysaccharides and the CPC detergent. This can be dissociated
by adding 300 ml 2% NaCl to the sediment. Homogenize it by stirring or by high-speed mixing in a Waring Blendor for 10 min. The
acidic polysaccharides will be dissolved and separated from the
CPC by centrifugation at 1000 g for 15 min. Discard the sediment.
Precipitate the acidic polysaccharide from its solution by adding
3 vol of 95% ethanol. Carry out this precipitation at cold room
temperature. Centrifuge the precipitate formed after 60 min at
1000 g for 30 min, and wash the sediment twice with cold 95%
ethanol. Discard the washings. Dissolve the precipitated polysaccharides in 100 ml water, dialyze for three days in the cold
room and lyophilize.

5. To isolate the neutral polysaccharides, 3 vol cold 95%
ethanol must be added to the supernate obtained after the removal
of the CPC precipitated acidic polysaccharides. It forms a precipitate overnight in the cold room. If it fails to do so, addition of magnesium chloride (from the 20% $MgCl_2$ solution in
ethanol) to a final 0.2% $MgCl_2$ content, will form a white pre-

cipitate. This can be sedimented by centrifugation at 1000 g for
30 min and washing twice with 95% ethanol. Discard the supernate
and washings. Dissolve the sediment in 50 ml water, dialyze in
the cold room as above, and lyophilize.

Use and Limitations

It must be noted that similar procedures are generally applicable
for the isolation of capsular polysaccharides from a wide variety
of microorganisms.
   The sediment obtained after the centrifugation of the cells
with Celite may be used to extract cell-wall polysaccharides from
the sediment by other procedures, as described in Exercise No. 21.

References

Eriksen, J.: Acta Path. Microbiol. Scand. *64*, 347 (1965)
Henriksen, S.D., Eriksen, J.: Acta Path. Microbiol. Scand. *51*, 259 (1961)

Exercise No. 24

Extraction of Teichoic Acid

The so-called teichoic acids, discovered by Baddiley and associates, can be found in Gram-positive bacteria. This material
is one of the components of the bacterial cell wall and its
composition varies. The backone of these polymers is either
polyribitol or polycerol joined together through phosphodiester
bridges. Certain carbohydrates may be attached to the polyols,
and D-alanine may also be esterified to the hydroxyl groups of
the backbone. The presence of the intracellular glycerol teichoic
acids in the Gram-positive bacteria seems to be uniform. Although
the cell walls frequently contain glycerol or ribitiol teichoic
acids, they are not universally present in each Gram-positive
strain. The extraction procedure used in this experiment follows
the outlines given in the paper by Armstrong et al. (1958) with
some modifications (Baddiley, 1964).

Materials and Equipment

    Dried cells of *Staphylococcus aureus H*
    Physiologic saline
    Trichloroacetic acid, 10% solution in water (TCA)
    Ethanol
    Acetone
    Diethyl ether
    Gelman filter
    Dialysis bags
    Servall homogenizer (or any other corresponding high-speed
     mixer)

Magnetic stirrer
Centrifuge and tubes with rubber caps

Procedure

1. Suspend 20 g bacterial cells in 2000 ml physiologic saline
overnight in the cold room on a magnetic stirrer. Wash the centrifuged sediment twice with 500 ml physiologic saline, and extract the wet sediment three times with cold 10% TCA (100 ml in
each run). Homogenize with the high-speed mixer at top speed for
1 min. Centrifuge the homogenate and combine the extracts. These
extracts contain mostly polyglycerol phosphate and nucleic acids.
These can be precipitated by adding 2 vol cold ethanol to the
filtered, pooled extracts. The precipitate has a tendency to
adhere to the wall of the glass container as a gum. This can be
reduced to an easily workable powder by rubbing with pure ethanol.
2. After 12 h, the precipitate can be centrifuged, washed
with acetone, and then with diethyl ether. The yield should be
approximately 60-70 mg.
3. The polyribitol teichoic acids may be extracted from the
residue which was obtained after the previous extraction. In
order to obtain this preparation, mix the sedimented cell debris
with 300 ml 10% cold TCA and homogenize the mixture at top speed
for 5 min in the cold room. Let the homogenate stand in the cold
room overnight, then homogenize again at top speed for 2 min.
Centrifuge the cells, filter the supernate first through filter
paper, then through a Gelman 0.2 µ membrane filter (or any other
comparable synthetic filter). The extract contains nucleic acids
and teichoic acids, in addition to large numbers of other TCA-
soluble cell constituents.
4. Add 1 vol cold ethanol to this extract. Centrifuge, add
40 ml acetone to the precipitate, centrifuge again, discard the
supernate, and add 40 ml diethyl ether. Mix the precipitate with
the ether by using a glass rod, close the centrifuge tube with a
rubber cap, and centrifuge it in an explosion-proof refrigerated
centrifuge. Discard the supernate. Add 10 ml water to the sediment
and try to homogenize the mixture with a glass rod for 10 min.
The teichoic acid will go into solution.
5. Centrifuge the suspension on the following day, and precipitate the teichoic acid from the supernate by adding 2 vol
acetone. Leave the precipitate in the cold room until the next
day, then centrifuge it, and wash it once with acetone and dry
the sediment in a vacuum desiccator. The resin-like material is
partially purified teichoic acid. The yield of the procedure
from 20 g *Staphylococcus aureus H* is approximately 20 mg.

Use and Limitations

The products obtained are not homogeneous, but these preparations
are still not as heterogeneous as similar TCA-extracted products
of Gram-negative bacteria. Further purification of the intracellular teichoic acid can be achieved by DEAE cellulose column
chromatographic process, developed by Wicken and Baddiley (1963).
During such attempts of purification lipoteichoic acid has been
isolated from several Gram-positive bacterial strains. In these

molecules long chain carboxylic acids are ester-bound to the
glycerol. These substances are active in most of the assays
characteristic for the Gram-negative lipopolysaccharides (Wicken
and Knox, 1977).

The very low yield (approximately 0.1% of the cell dry weight)
may indicate that not all the teichoic acids were extracted from
the cells. The yield may be improved by prolonged TCA extraction,
but the products obtained will contain higher amounts of non-
teichoic acid components.

According to investigations of Baddiley and associates, the
teichoic acids determine the serologic group specificity of
Staphylococcus aureus H and some other Gram-positive bacteria.

Use of purified cell walls increases the yield and homogeneity
of the preparation.

References

Armstrong, J.J., Baddiley, J., Buchanan, J.G., Carss, B., Greenberg, G.R.:
   J. Chem. Soc. 4344 (1958)
Baddiley, J.: Endeavor *23*, 33 (1964)
Wicken, A.J., Baddiley, J.: Biochem. J. *87*, 54 (1963)
Wicken, A.J. and K.W. Knox: Microbiology 1977, p. 60. ASM publication.

Exercise No. 25

Extraction of *Pneumococcus* Type XIV Polysaccharide

The polysaccharide is isolated from the autolysate of Type XIV
Pneumococcus cells and purified by deproteinization and by repeated
precipitation using sodium acetate containing ethyl alcohol. This
isolation procedure is based on the publication of Goebel, Beeson
and Hoagland (1939).

Materials and Equipment

    Beef infusion broth
    Bacterial culture *(Diplococcus pneumoniae* Type XIV)
    Glucose
    Witte's peptone
    Sterile 0.3% sodium chloride solution
    Sodium acetate
    Ethanol
    Acetone
    Fermentor
    Centrifuge with continuous flow attachment
    Vacuum distillation apparatus
    Freeze drying equipment
    Büchner funnel
    Dialysis bags and jars

Procedure

1. Prepare 50 l beef infusion broth for the cultivation of
the cells by adding 0.1% of glucose and 1.0% of Witte's peptone.
Inoculate 200 ml of this broth with bacteria and cultivate them
overnight at 37°C. The next morning, centrifuge the culture under
sterile conditions at 3000 g for 30 min. Discard the supernate.
Suspend the sediment in approximately 100 ml fresh broth and add
it to the fermentor containing the 50 l of medium. Let the growth
proceed for 6 h. Centrifuge the cells in a continuous flow system
under sterile conditions, and discard the supernate. Suspend the
cells in 2000 ml sterile 0.3% NaCl solution, and let them autolyze
at 37°C for about ten days.

2. Centrifuge the cells and the cell debris at 6000 g for
60 min and discard the sediment. Concentrate the supernate by
vacuum distillation to approximately 500 ml. Proteins must now
be removed from the clear supernate. Large amounts will be precipitated by lowering the pH to 3.5. Centrifuge at 3000 g for
30 min and discard the sediment. Further deproteinization can be
achieved by the method of Sevag (1934).

3. The precipitation of the partially purified polysaccharide
can be primed by adding 10% (w/v) of sodium acetate to the solution. Filter if necessary after the salt has been dissolved.
Add cold ethyl alcohol to the solution until a final concentration
of 75% is reached. Sediment the polysaccharide by centrifugation
at 3000 g for 30 min. Discard the supernate and dissolve the
sediment in 500 ml distilled water. Adjust the pH to 7.4 if necessary with 0.1 N NaOH. Dialyze it against approximately 50 vol
distilled water overnight, continuously agitating the outer
fluid. Filter the contents of the dialysis bag through two layers
of paper in a Büchner funnel, applying suction. Repeat the alco-
holic precipitation by dissolving 10% of sodium acetate in the
solution and add 1.5 vol cold ethanol. After 60 min at cold room
temperature the precipitate formed must be separated by centrifugation. Discard the supernate and dissolve the sediment in 500 ml
water. Adjust the pH to 7.4. Dialyze the material as before
overnight.

4. Final precipitation of the polysaccharide is carried out
by pouring the polysaccharide solution into 10 vol acetone.
Centrifuge the precipitate; discard the supernate. Dissolve the
sediment in 100 ml water, reduce its volume to half, replace the
water, reduce it again. This removes the remnants of acetone.
The material is ready for freeze drying.

Austrian studied the cross reactivity of pneumococcal and
streptococcal polysaccharides and developed new approaches for
inducing immunity to pneumoniacausing microorganisms. The materials
used by Austrian and associates to prepare vaccines were capsular
polysaccharides (1977).

The usual yield of this procedure in the laboratories of
Goebel and co-workers was approximately 0.5 g from a 50 l culture.
This preparation may be used as the starting material for further
studies described in this manual.

References

Austrian, R.: J. Inf. Dis. *136* Suppl, S. 38-42 (1977)
Goebel, W.F., Beeson, P.B., Hoagland, C.L.: J. Biol. Chem. *129,* 455 (1939)
Sevag, M.G.: Biochem. Z. *273,* 419 (1934)

Exercise No. 26

Isolation of Phytohemagglutinin

Some proteins, present in plants, are able to agglutinate red
blood cells. These phytohemagglutinins (or lectins) of certain
plant seeds may have a rather high specificity, which makes them
useful in the differentiation of human blood subgroups such as
A_1 and A_2 (Boyd and Reguera, 1949). Some lectins behave as incomplete antibodies, which means that they do not agglutinate
red blood cells in saline, but do cause agglutination if the
cells are suspended in plasma proteins. The majority of lectins,
on the other hand, are nonspecific; they agglutinate all human
red blood cells and a large variety of animal cells (Boyd, 1963).
An interesting property of these lectins was discovered by
Nowell (1960) who reported that human leucocyte cultures obtained
through agglutination of red blood cells with lectin will undergo
mitosis, while leucocytes not exposed to lectin do not show
mitotic activity.
The Exercise described here yields phytohemagglutinin, according to the procedure of Rigas and Osgood (1955).

Materials and Equipment

    100 g Phaseolus vulgaris (red kidney beans)
    Heparinized human blood
    Group A red cells
    Group B red cells
    Group O red cells
    Difco TC-199 tissue culture medium
    Penicillin
    Streptomycin
    95% Ethanol
    Sodium chloride
    2000 ml beaker
    1% Sodium chloride solution
    Ammonium sulfate, saturated solution
    0.1 M Phosphate buffer, pH 8
    Dialysis tubing and jars
    8 × 150 mm Test tubes
    Centrifuge and tubes
    Waring Blendor
    Lyophilizing equipment

Procedure

1. Take 100 g beans in a 2000 ml beaker and add 500 ml 1% NaCl.
Let stand in the cold room for 24 h. Transfer the swollen beans
and the saline to a Waring Blendor and homogenize for 1 min at
top speed. Cool the container by immersing it in ice water; repeat
the homogenization for 1 min at the same speed. Let stand overnight in the cold room.
2. Centrifuge the homogenate; discard the sediment. Adjust
the pH of the supernate to 5.7, cool in ice and salt to approximately -2°C, add 1.5 vol 95% precooled ethanol, stir a few minutes

then sediment in a refrigerated centrifuge at 1000 g for 30 min.
Discard the sediment.

   3. Measure the volume of the supernate, cool it again as
above, and add 2 vol cold 95% ethanol. Stir and centrifuge as
above. Discard the supernate.

   4. Dissolve the sediment in 70 ml 1% NaCl, add 20 ml cold
95% ethanol. If any precipitate is formed, centrifuge and discard
the sediment. Add 180 ml 95% ethanol in the cold, as above. Centrifuge the sediment and discard the supernate.

   5. Dissolve the sediment and make up to 150 ml with pH 8.0,
0.1 M cold phosphate buffer. Add 100 ml saturated ammonium sulfate
(to reach 40% saturation), stir with a glass rod, then centrifuge
the precipitate formed at 1000 g for 30 min and discard the sediment. Measure the volume of the supernate and add 1 vol saturated
ammonium sulfate, stir in the cold, centrifuge as above, and
discard the supernate. Dissolve the sediment in 100 ml water;
dialyze against distilled water for at least three days in the
cold room, changing the water twice daily. Lyophilize the product.

Evaluation

Dissolve 25.6 mg of the above product in 10 ml saline. Prepare
ten twofold serial dilutions of this material with saline. The
last dilution will contain approximately 2.5 µg/ml. Pipette
0.2 ml of each dilution into a small test tube and add 0.2 ml
2% vol/vol suspension of human Group A red blood cells. Prepare
two other series of twofold dilutions of the same lectin solution.
Add to one of these a 2% suspension of Group B and to the third
series a 2% suspension of Group O human erythrocytes. Shake all
the tubes and incubate them for 30 min at 37°C, then for 30 min
at room temperature. Read the last dilution which shows hemagglutination. Compare the hemagglutinating effect of the lectin
on the three different human blood groups. Exercise No. 89 in
this Manual describes the lymphoblast assay. In this procedure
the enhanced mitosis caused by PHA, LPS or other mitogens is
measured by determining the incorporation rate of ^{3}H thymidine
into the rapidly dividing cells.

Use and Limitations

This mucoprotein lectin contains two components, as shown by
electrophoresis. Another procedure also developed by Rigas and
Osgood separates an electrophoretically homogeneous protein
with higher hemagglutinating activity and with enhanced effect
on mitosis. This protein is not soluble in water but easily
soluble in saline. It contains between 3% and 5% of bound carbohydrates.

   Several other methods have also been described for the isolation of phytohemagglutinins. They vary slightly according to the
source of the starting material. The procedure described here
may be used for a variety of different plant seeds. The isolation
and chemical studies of a hemagglutinating glycoprotein obtained
from soybean flour was described by Lis, Sharon, and Katchalski
(1966). For review on Lectins see Sharon (1977).

It should be mentioned that not only human, but also dog, cat, rabbit, chicken, frog, and a number of other erythrocytes can be agglutinated by phytohemagglutinins of *Phaseolus vulgaris* obtained by the above procedure.

References

Boyd, W.C., Reguera, R.M.: J. Immunol. *62,* 333 (1949)
Boyd, W.C.: Vox Sang. *8,* 1 (1963)
Lis, H., Sharon, N., Katchalski, E.: J. Biol. Chem. *241,* 684 (1966)
Nowell, P.C.: Cancer Res. *20,* 462 (1960)
Rigas, D.A., Osgood, E.E.: J. Biol. Chem. *212,* 607 (1955)
Sharon, N.: Sci. American, June 1977, p. 108

Exercise No. 27

A and B Blood Group Antigen Preparations From Erythrocytes

The ABO group antigens of human red bloods cells (RBC) are localized in the stroma (synonyms are membrane or ghost) of the cells. Hemolysis of the RBC is carried out in distilled water and subsequent repeated washing of the insoluble residue leads to a stroma preparation which is still light red from the presence of remnants of hemoglobin. It was observed that dilute alkali treatment appears to solubilize the prepared stroma without changing the reactivity of A or B or AB antigens with isoagglutinins. Subsequent careful neutralization of this alkaline preparation keeps the stroma solubilized (Nowotny, 1955). This is described in *Part A* of this exercise.

Another procedure for group antigen isolation is also given here in *Part B,* which utilizes cetyltrimethylammonium bromide (CTB) for the mild dissolution of the stroma, followed by ethanol fractionation of the extract (Nowotny and O'Neill, 1964, and Abdelnoor and Nowotny 1971).

Materials and Equipment

    Group A or B blood (containing 2% sodium citrate)
    Physiologic saline, pH 7.4
    0.01 *N* Acetic acid or dry ice
    0.1 *N* Sodium hydroxide
    0.1 *N* Hydrochloric acid
    0.5% Cetyltrimethylammonium bromide solution
    Methanol, anhydrous
    250 and 50 ml capacity centrifuge tubes
    Centrifuge, refrigerated

Part A. Solubilization With Dilute NaOH

    1. Mix 2 ml citrated whole blood and 40 ml saline in 50 ml tubes and centrifuge at 500 g for 10 min. Siphon off and discard the supernate.

2. To the packed cell sediment add 40 ml physiologic saline,
stir the contents of the tube gently with a glass rod, and centrifuge again under the above conditions. The saline wash should
be discarded, and washing with fresh saline repeated twice more.

3. To the sedimented and washed RBC add 40 ml distilled water
and centrifuge again at 5000 g for 10 min. Discard the supernatants. This procedure must be repeated at least four times or
until the supernatant does not show significant amounts of hemoglobin. For final washing of the residue, use either 0.01 N acetic
acid or CO_2 saturated water (dry ice dissolved in water). This
will release additional hemoglobin.

4. Suspend the washed sediment, consisting of the red blood
cell stroma, in physiologic saline to a total of 2 ml. Add 0.2 ml
0.1 N NaOH and thoroughly mix with a glass rod. Observe the clear-
ing of the suspension. After 30 min at room temperature add 8 ml
saline and exactly 0.2 ml 0.1 N HCl under constant and vigorous
stirring. Adjust the pH to 7.5. This preparation can be stored
in the cold room.

Part B. The Use of Detergent

1. Take 2 ml blood, either A or B, wash the cells with saline
and prepare the stroma as before. Do not add NaOH.

2. Add 4 ml 0.5% CTB to the centrifuge tube which contains
the washed stroma, and stir with a glass rod for 30 min at room
temperature. Centrifuge at 5000 g for 30 min. Lift the yellow
supernate and wash the insoluble sediment twice with 4 ml CTB.
Pool the supernates.

3. Cool the extract in salted ice; then add 1 vol anhydrous
methanol. Leave the turbid solution in the cold room. A precipitate will form in approximately 1 h. Spin down at 5000 g
for 30 min in a refrigerated centrifuge. Wash the sediment with
10 ml 50% cold methanol and discard the washings.

4. Dissolve the sediment in 5 ml distilled water and dialyze
in the cold room against distilled water for three days, changing
the outer fluid twice daily. The material can be lyophilized
or stored as is in the cold. The addition of 1:10,000 merthiolate
will preserve the solution.

Evaluation

These two stroma preparations can be tested for group antigen
activity in a hemagglutination inhibition assay, which is described in Exercise No. 68. Group B stroma will yield more active
preparations than A. The same preparations also show activity
in inhibiting viral hemagglutination.

The heterogeneity of the obtained crude extracts can be best
studied by using the SDS gel electrophoretic procedure as described in Exercise No. 10.

Use and Limitations

The materials can be used as immunizing agents for the production
of hemolysins or other antibodies. They serve well as crude starting materials for further purification of erythrocyte antigens.

Several authors described different methods of extracting blood
group antigens from human erythrocytes. For the most part, alcohols at different concentrations, alone or combined with other
solvents, were used in these experiments. The various isolation
procedures to fractionate antigenic components of erythrocytes
have been reviewed by Nowotny (1971). The most outstanding contributions to this field were published by Yamakawa and co-workers
(1952), Koscielak (1963), and Hakomori and Jeanloz (1961), as
well as Hakomori and Stycharz (1968), and Hakomori and Kobata
(1974).

References

Abdelnoor, A., Higgins, M., Nowotny, A.: In: Cellular Antigens. p. 153
    Nowotny, A. (ed.) Berlin-Heidelberg-New York: Springer 1971
Hakomori, S.-I., Jeanloz, R.W.: J. Biol. Chem. *236*, 2827 (1961)
Hakomori, S.-I., Kobata, A.: In: The Antigens, Sela, M. (ed.) London,
    New York: Acadmic Press, vol. II., p. 79, 1974
Hakomori, S.-I., Stycharz, G.D.: Biochemistry, *7*, 1279 (1968)
Koscielak, J.: Biochim. Biophys. Acta *78*, 313 (1963)
Nowotny, A.: Acta Physiol. Acad. Sci. Hung. *7*, 31 (1955)
Nowotny, A. (ed.): In: Cellular Antigens. p. 161, Berlin-Heidelberg-New York:
    Springer 1971
Nowotny, A., O'Neill, G.J.: Lecture, Xth Congress Int. Soc. Hematol. Stockholm,
    1964
Yamakawa, T., Suzuki, S.: J. Biochem. *39*, 393 (1952)

Exercise No. 28

Isolation of NN Blood Group Antigen From Human Erythrocytes

The human NN blood group system was discovered by Landsteiner and
Levine in 1927. Analyses of the isolated antigens showed that
these materials are glycoproteins (Springer and Ansell, 1958;
Klenk and Uhlenbruck, 1960). A relatively simple procedure was
described by Springer, Nagai and Tegtmeyer (1966) for the isolation of the NN antigen from human erythrocytes. This Exercise
uses their method.

Materials and Equipment

    NN type human blood, 1000 ml, containing citrate as anticoagulan
    Acetic acid, glacial
    Ethanol, containing 0.1% sodium acetate
    Phenol
    Physiologic saline
    Mechanical shaker
    Centrifuge, 3000 ml capacity
    Beckman L or comparable preparative ultracentrifuge

Procedure

1. Centrifuge 1000 ml NN blood for 40 min at 2000 g, discard plasma and buffy coat and wash the packed red cells three times with an equal volume 0.85% aqueous NaCl. Lyse the erythrocytes in 10 vol distilled water at 5°C. Adjust the pH to 5.3 with acetic acid and add phenol to a final concentration of 0.2%.

2. Allow the preparation to stand overnight, then siphon off the supernatant fluid, add 10 vol water and readjust the pH. Repeat this procedure six times with phenol addition at every second water change. Then collect the sediment by centrifugation and store at -20°C until use. The yield should be 5-8 g per liter of blood.

3. Extract the NN active material from the stroma using the following modifications of the phenol-water extraction procedure (Exercise No. 21): (a) The aqueous phase should contain 0.85% NaCl (Stalder and Springer, 1960) since little active material will be obtained otherwise; (b) Carry out the extraction for 2 h on a mechanical shaker at 23-25°C (Hotta and Springer, 1964) in order to avoid degradation of the NN antigen, which was demonstrable biologically and chemically after extraction at elevated temperatures.

4. Extract the cells three times and purify the aqueous phase by fractional centrifugation in a Beckman L preparative ultracentrifuge. Material sedimenting between 32,000 and 96,000 g is most active.

5. Dissolve the active fraction in water and fractionate with ethanol at 23-25°C. The most active materials precipitate between 50% and 55% ethanol. Dissolve this ethanol precipitated material in 100 ml pH adjusted saline and dialyze for three days against 10 vol of the same saline solution. The material may now be dried from the frozen state.

Evaluation

Extracts of *Vicia graminea* seeds prepared according to the method of Ottensooser and Silberschmidt (1953) are useful hemagglutinins for NN erythrocytes. This reaction can be inhibited by NN substances (see Exercise No. 68).

Use and Limitations

The incomplete procedure described here yields a nonhomogeneous material. The complete method as elaborated by Springer and associates (1966) has been claimed to yield a homogeneous substance with a molecular weight of 595,000. Immunodeterminant groups are sialic acid derivatives and β-galactopyranosyl structures. For the quantitative determination of sialic acid see Exercise No. 57 in this manual.

References

Hotta, K., Springer, G.F.: Sangre *9,* 183 (1964)
Klenk, E., Uhlenbruck, G.: Z. Physiol. Chem. *319,* 151 (1960)
Landsteiner, K., Levine, P.: Proc. Soc. Exp. Biol. Med. *24,* 941 (1927)

Ottensooser, F., Silberschmidt, K.: Nature London *172*, 914 (1953)
Springer, G.F., Ansell, N.: Proc. Natl. Acad. Sci. (USA) *44*, 182 (1958)
Springer, G.F., Nagai, Y., Tegtmeyer, H.: Biochemistry 5, 3254 (1966)
Stalder, K., Springer, G.F.: Fed. Proc. *191*, 7 (1960)

Exercise No. 29

Isolation of Forssman Hapten From Erythrocytes

Forssman (1911) showed that guinea pig organs contain an antigen
which produces sheep hemolysins if injected into rabbits. It
has been shown since that time that similar heterophile antigens are present not only in a rather wide variety of different animal species, but also in some bacteria. Those animals
which do not have these heterophile antigens will produce antibodies which react with Forssman antigens of all different sources
Although these antigens show a wide cross-reactivity, it is not
known whether their chemical structure is identical in all species
Attempts to isolate and purify the antigens from these tissues
showed that a hapten can be dissociated easily from a nonspecific
carrier (Sordelli et al., 1921). This hapten reacted with Forss-
mann antibodies, but did not induce antibody production. It was
first thoroughly investigated by Landsteiner and Levene (1925).
Papirmeister and Mallette (1955) isolated this material from
sheep red blood cells. The isolation of a water-soluble Forssman
hapten was published by Rapp and Borsos (1966).

Materials and Equipment

    500 ml Sheep blood containing 2% sodium citrate to prevent
     clotting
    Physiologic saline
    Acetone
    Methanol
    Ethyl acetate
    Dry ice
    Desiccator
    Glass jar, 12 l capacity
    Freezer
    Lyophilizer (Fig. 14 or 15)
    Laboratory shaker
    Centrifuge tubes with rubber caps or stoppers
    Refrigerated centrifuge, 4 × 250 ml capacity

Procedure

    1. Centrifuge the 500 ml citrate blood at 500 g for 30 min
and discard the plasma. Wash the erythrocytes three times using
a total of 2000 ml physiologic saline. Discard the washings.
    2. Pour the washed red blood cells into approximately 10 l
distilled water, rinse the centrifuge tubes. Mix and keep the
hemolysates in the cold. In approximately 2 h the hemolytic

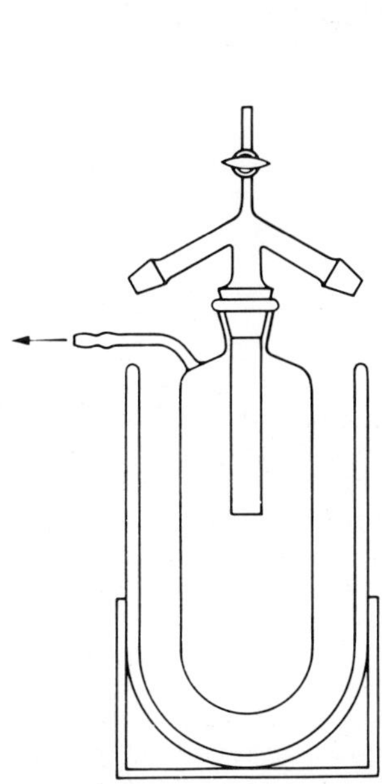

Fig. 14. Freeze drying
equipment, capacity up
to 500 ml

Fig. 15. Freeze drying equipment, Vir-Tis type
(courtesy of the VirTis Company, Inc. Gardiner,
NY) 12525)

residue of the erythrocytes (stroma) will sediment and the supernatant fluid may be lifted by slow siphoning. Remove and discard
as much of the supernate as you can without losing stroma. Fill
the jar with distilled water, mix and let stand in the cold room
for another 2 h. This time throw two or three nut-sized pieces
of dry ice into the jar. Stir it slowly then let it stand. The
dry ice will lower the pH of the water and hasten the sedimentation of the stroma by simultaneously enhancing the release of
hemoglobins. Remove the supernate as above. Repeat the washing
with distilled water and dry ice as above once more. After de-
canting the supernate, centrifuge the sediment at 1000 g for
30 min. Finally, suspend the stroma in approximately 100 ml water
and lyophilize.

3. Take all the stroma obtained and suspend it in 200 ml anhydrous methanol. Transfer it to centrifuge tubes, stir well
with a glass rod, close the opening of the tubes with an airtight
rubber cap or stopper and agitate them on a shaker for 24 h. Centrifuge the suspension and save the supernate. Replace the
methanol and extract the residue again for 24 h as before. Repeat
the extraction twice more and pool all four methanol extracts.

4. Place the extracts in the freezer and keep at approximately
-10°C for 24 h. A white precipitate will settle out. Siphon the
supernate carefully without letting the temperature rise above
-5°C. Centrifuge the residue at -10°C for 20 min at 1000 g. Approximately 70% of the active material is in the precipitate.

5. Dissolve the precipitate in 20 ml warm ethyl acetate at approximately 50° to 60°C. Pool the material and cool the solution to 0°C. The precipitate formed may by dried in a vacuum desiccator.

Evaluation

This product is partially soluble in water and shows considerable activity in the hemolysis inhibition assay which can be measured by the method of Papirmeister and Mallette (1955).

Use and Limitations

A basically similar procedure may be used to extract Forssman antigen from other tissues. The procedure given here is incomplete and the purification of the materials should be continued if the chemical composition of the Forssman hapten is to be investigated.

The heterogeneity of the methanol extract as well as of the final product described here may be well demonstrated by thin layer chromatography (see Exercise No. 42), using a chloroform: methanol:water = 100:40:10 mixture.

Springer studied extensively blood group and Forssman antigenic determinants shared by microbial cell walls and mammalian cell membrane surfaces (1971). The most recent studies on the structure of the Forssman antigens, and their appearance on normal and malignant human cells come from the laboratories of Hakomori and associates (Siddiqui and Hakomori, 1971, Stellner et al., 1973, and Hakomori et al., 1977).

References

Forssman, J.: Biochem. Z. *37*, 78 (1911)
Hakomori, S.-I., Wang, S.-M., Young, W.W., Jr.: Proc. Soc. Natl. Acad. Sci. USA *74*, 3023, 1977
Landsteiner, K., Levene, P.A.: J. Immunol. *10*, 731 (1925)
Papirmeister, B., Mallette, M.F.: Arch. Biochem. Biophys. *57*, 94 (1955)
Rapp, H.J., Borsos, T.: J. Immunol. *96*, 913 (1966)
Siddiqui, B., Hakomori, S.-I.: J. Biol. Chem. *246*, 5766, 1971
Sordelli, A., Fischer, H., Wernicke, R., Pico, C.: C. R. Soc. Biol. *84*, 173 (1921)
Springer, G.F.: Progr. Allergy *15*, 9, 1971
Stellner, K., Saito, H., Hakomori, S.-I.: Arch. Biochem. Biophys. *155*, 464, 1973

Exercise No. 30

Isolation of Ragweed Pollen Allergens

The most common seasonal hayfever on the American continent is caused by the pollens of ragweed *(Ambrosia elatior* and *Ambrosia trifida)*. Several extraction procedures are described in the litera-

ture, most of them yielding crude but potent allergenic preparations. King and associates used a combination of various isolation methods, including several column chromatographic steps,
and isolated two active allergens from ragweed pollens: Antigens
K and E (1962, 1964, 1967). The method given here is a simplified
version of their procedure and the crude product obtained here
can be used for further purification.

Materials and Equipment

    20 g low ragweed pollen (available from Greer Laboratories,
     P. O. Box 800, Lenoir, NC 28645)
    1000 ml diethyl ether
    1000 and 2000 ml Erlenmeyer flasks with $ 24/40
    Büchner funnel, 15 cm diam, and 1000 ml filter flask
    Approximately 1 N NH_4OH (dilute 10 ml concentrated NH_4OH to
     170 ml)
    Tris (hydroxymethyl) aminomethane
    0.05 M pH 7.9 Tris-HCl buffer. Dissolve 24.2 g Tris in 1000 ml
     water (0.2 M). Take 197 ml 0.2 N HCl, and 303 ml 0.2 M Tris
     and make up with water to 2000 ml. Several liters of this
     buffer solution will be necessary for this Exercise
    0.025 M pH 7.9 Tris-HCl buffer. Dilute the above buffer with
     water 1:1
    Sephadex column (see Fig. 1) 50 × 4 cm
    $(NH_4)_2SO_4$, reagent grade
    DEAE cellulose, 30 g. Activate it as described in Exercise
     No. 3
    Sephadex G 25
    Chromatographic column for DEAE cellulose, 20 × 4 cm

Procedure

    1. Extract fatty and oily materials from the pollen by mixing
20 g with 250 ml diethyl ether at room temperature in a 1000 ml
Erlenmeyer flask. Work under a well ventilated and explosion-
proof chemical hood. The ether is extremely flammable! Extinguish
all cigarettes and flames and turn off all spark producing motors
or other instruments in the entire laboratory where the extraction is carried out. Close the flask with the $ stopper, shake
the contents carefully but constantly for about 30 min at room
temperature. Loosen the stopper from time to time. Filter the
contents on a Büchner funnel, using vacuum produced by a water
aspirator. Transfer the pollen back to the same Erlenmeyer flask
and add another 250 ml diethyl ether, close the flask and extract
it for 30 min again. Filter as above. Discard the filtrates.
Finally, add 500 ml ether to the pollen and let it stand overnight, well closed at room temperature under the hood. Filter it
again the next day and let the pollen dry on the Büchner funnel
while air is sucked through it by vacuum. Discard all the filtrates. After 1 or 2 h, disconnect the vacuum, remove the pollen
from the filter, spread it under the hood on a large sheet of
aluminum foil to let the last traces of ether evaporate.
    Instructors: It is recommend that step 1 above be carried out
by experienced personnel and students be provided with already
defatted pollen.

2. Transfer the defatted pollen to a 2000 ml Erlenmeyer flask
and add 1500 ml water. Stir the suspension with a motor-driven
glass rod at room temperature for 5 h. Filter it on a Büchner
funnel, using good vacuum, and discard the pollen residue.

3. Adjust the pH of the extract to 7.0 using 1 N NH_4OH.
Measure the volume of the extract and add $(NH_4)_2SO_4$ crystals to
0.9 saturation. This will require 0.68 g of $(NH_4)_2SO_4$ crystals
per ml of solution at room temperature. Let it stand for 2-3 h,
after which a precipitate will form.

4. Centrifuge the brown precipitate at 1000 g for 30 min.
Discard the supernatant and dissolve the sediment in 70 ml 0.05 M
pH 7.9 Tris buffer. Stir it well with a glass rod and dialyze the
contents against 0.05 M Tris buffer for three days. Change the
outer fluid daily. Centrifuge the contents of the dialysis bag
at 5000 g for 60 min. Use the clear supernatant for chromatographic purification.

5. The next step is chromatography of the supernatant on a
Sephadex G 25 column. The Sephadex has to be previously equilibrated with 0.05 M Tris buffer. Fill the column as described
in Exercise No. 2. Apply the whole extract you obtained in step
4 to this column. Develop the column with approximately two void
volumes of the same Tris buffer. In the case of the column size
described here the void volume will be approximately 300 ml.
This step will remove the brownish color of the extract as well
as other small molecular weight components which did not pass the
dialysis membrane. According to some authors this step may be
omitted.

6. Precipitate the collected eluate with $(NH_4)_2SO_4$ at 0.9
saturation, as described above. Collect the precipitate by centrifugation and dissolve the sediment in a total of 60 ml 0.025 M
Tris buffer. If necessary, adjust the pH of the solution to 7.9
with NH_4OH.

7. Take the 20 × 4 column, fill it up bubble-free with freshly
activated DEAE in OH form. The DEAE cellulose should be suspended in 0.025 M pH 7.9 Tris buffer. Now wash the column with approximately 200 ml of the same buffer. Apply the preparation
obtained in step 6 and first elute the column with 600 ml 0.025 M
Tris buffer. The flow rate should be approximately 300 ml/h. This
eluate will contain several fractions, but these are weaker in
inducing allergic reactions than those remaining on the column.

8. To elute these more active components, use 0.05 M Tris
buffer containing 0.2 M NaCl, pH 7.9. 500 ml of this solution
will bring down a mixture of components, some of them highly
active. Approximately 1 g allergenic preparation can be obtained
in this fraction from 200 g pollens.

Evaluation

Direct skin testing of the allergens can be carried out by injecting 0.05 ml solutions dissolved in pH 7.4 saline intracutaneously into ragweed pollen sensitive patients. To compare the
potencies of the preparation, 10 X dilutions are made. The lowest
dose which still elicits erythema and wheal is recorded. Start
with the lowest concentrations since sensitive individuals may
react very strongly, and such testings should be carried out by
allergologist clinicians only. Injections can be substituted by

scratching the skin surface, approximately 1 cm long, with a
hypodermic needle, and applying one drop of the various dilutions
to the scratch. Inflammation can be observed after 15 min in both
assays. The method of the end-point titration by skin testing
for allergens has been described by Gleich and Yunginger (1976).
    Lichtenstein and Osler (1964) developed an in vitro assay in
which lymphocytes of sensitized human patients are exposed to
allergens in tissue cultures and the amount of histamine released
into the supernatant is measured. This assay can also be used
to compare activities of various allergen preparations. Antibodies to antigen K or E inhibited the histamine release from
sensitized leukocytes in the above assay, according to King
et al. (1967).

Use and Limitations

For further purification, King and associates used Sephadex
100 column chromatography (1972) and obtained at least six fractions. Rechromatography of these revealed the existence of numerous subfractions. Most of these were active in the skin test on
sensitive patients, although to a different degree. Final purification by King et al. resulted in two highly active and apparently
homogeneous preparations which were named Antigen E and K (King
et al., 1967, and King, 1972).
    Needless to say, individuals allergic to pollen should not
be permitted to be in the laboratory during the preparation of
the allergen. It is also recommended that workers on such projects
should always wear a face mask to minimize their exposure to
pollen.

References

Gleich, G.J., Yunginger, J.W.: In: Manual of clinical immunology. Rose, Fried-
    man (eds.) ASM publication, p. 575, 1976
King, T.P.: Biochemistry *11*, 367 (1972)
King, T.P., Norman, P.S.: Biochemistry *1*, 709 (1962)
King, T.P., Norman, P.S., Connell, J.T.: Biochemistry *3*, 458 (1964)
King, T.P., Norman, P.S., Lichtenstein, L.M.: Biochemistry *6*, 1992 (1967)
Lichtenstein, L.M., Osler, A.G.: J. Exp. Med. *120*, 507 (1964)

Exercise No. 31

Ultrasonic Disruption of Bacteria and Density-Gradient Centrifugation of the Products

In addition to those mechanical procedures which disrupt cells
by grinding them with solid particles, such as sand or small
glass beads, the most widely applied is the destruction of cells
by ultrasonication.
    To explain the theory of ultrasonic disruption we have to
consider the fact that a sonic or ultrasonic wave traveling
through a medium consists of alternate compressions and rarefac-

tions. If this wave, or extension-compression cycle, is applied
to a liquid and is strong enough, microscopic bubbles are formed.
Gaseous bubbles are formed at the compression region of the sonic
wave and travel at high speed towards the rarefaction region.
The resulting flow of bubbles and liquid causes rings of high
intensity, and the friction between these rings and the suspended
cells disrupts the cells. Summarizing, present theories hold
that bubbles are formed by gas dissolved in the medium and these
bubbles are ejected at high speed from the sonic generator like
bullets. When such streams or bubbles or swiftly moving liquid
hit the cells, the resulting friction will disrupt them.

The result of such a disruption is that a certain percentage of
the cells will be ruptured and their intracellular content will
be released. Some cells remain intact; some will leave behind
only cell walls and protoplasm membranes as insoluble residues.
For further information regarding the action of ultrasonic waves,
see Britten and Roberts (1960), Hughes and Nyborg (1962), or
Lindstroem (1955).

The separation of these products can be carried out by density-
gradient centrifugation as described by Meselson, Stahl and
Vinograd (1957). A continuous density gradient is produced in a
centrifuge tube, using, for example, caesium chloride or Ficoll
or sucrose. The greatest density is on the bottom of the tube
and the suspended mixture of the breakdown products will be
layered on the top. Centrifugal force will cause the components
to descend through the solution until they reach a density grade
which has the same specific gravity as the component itself.
Heavier components will be found in the denser layers towards
the bottom of the tube, light components stay closer to the surface. Controlled temperature in the centrifuge is necessary because thermal convections may disturb the formation of equilibrated sharp bands. A properly selected density-gradient range
may result in several bands, each corresponding to one of the
components of the distupted cell. This is the so-called isopycnic
gradient centrifugation or equilibrium centrifugation, where the
components of a mixture will be separated based on the differences
in their density. For accurate details of equilibrium sedimentation in density gradients, see the work of Meselson, Stahl and
Vinograd (1957).

This exercise gives the so-called moving zone method which is
also called band centrifugation procedure. The density gradient
here serves mostly as stabilizer. The method is used to separate
components which differ from each other in their sedimentation
velocity. The separation stops before equilibrium is reached.
To determine the density of the components, the isopycnic centrifugation in a carefully prepared linear density gradient has to
be used.

Materials and Equipment
<u></u>

    Lyophilized *E. coli* O8 bacteria
    Ten sucrose solutions from 5% to 50% in water with 5% increases
    Beaker, 50 ml
    Ultrasonication instrument (MSE or other model)
    Plastic centrifuge tubes, 20 ml
    Preparative ultracentrifuge with swinging buckets

<u>Procedure</u>

1. Suspend 0.2 g packed sediment of E. coli or other Gram-negative bacteria in 1.8 ml distilled water.
2. Sonicate this suspension in a small beaker for 15 min under cooling in crushed ice.
3. Pipette 1.7 ml 50% sucrose directly into the bottom of a 20 ml plastic centrifuge tube. Carefully layer on the top of it 1.7 ml 45% sucrose solution, followed by 1.7 ml of the 40% concentration. After all sucrose concentrations have been used, finally pipette 1.7 ml water as the top layer. Place the centrifuge tube in the cold room overnight, but do not disturb the layers. Through spontaneous diffusion, a rather smooth sucrose gradient will be obtained in the tube.
4. Layer 0.2 ml of the sonicated *E. coli* 08 cells in water on the top of the density gradient in the centrifuge tube, put on the cap and place the tube in the refrigerated ultracentrifuge head.
5. Start the centrifuge according to the manufacturer's instructions, raise its speed to 35,000 g. Continue centrifugation for 120 min.

<u>Evaluation</u>

Carefully remove the centrifuge tubes from the centrifuge and observe the formation of layers. Record the results of sedimentation.

Several procedures may be used to isolate the layers formed in the density gradient. One may attempt to lift the visible layers from the centrifuge tube using a syringe, or one may punch a hole in the bottom of the plastic centrifuge tube and collect the drops manually or with the help of a fraction collector provided with a drop counting device. A good, commercially available instrument for the preparative isolation and simultaneous analysis of the layers produced in a density gradient is available from ISCO (Instrumentation Specialties Co., Inc., Lincoln, NE). The ISCO density gradient fractionator injects the dense solution into the bottom of a density-gradient centrifuge tube, floating and forcing the contents of the tube upward through a flow cell of a recording ultraviolet absorbance monitor. The flow cell is mounted directly above the tube to prevent mixing and turbulence of the separated zones. The effluent from the flow cell may be collected in any automatic fraction collector.

<u>Use and Limitations</u>

Not all bacterial species can be disrupted by this treatment. Some of them require more drastic disintegration. Mammalian cells will be ruptured easily, but several sensitive constituents will undergo marked inactivation during sonication. For the isolation of such substances (e.g., transplantation antigens of cells), the N_2 gas decompression method is much more suitable. Repeated freezing and thawing can also be applied to disrupt certain cells.

Density-gradient centrifugation is used not only for isolation of subcellular components, but also for determination of the density of macromolecules, such as single or double stranded DNA and RNA. Labeled substances with known density may serve as internal standards (markers) for such studies.

Density of proteins averages 1.3, nucleic acids are above 1.7. For equilibrium sedimentation of these materials, cesium chloride must be used, where a high density can be achieved. In sucrose gradient, where the specific gravity of the 50% solution at 20°C is 1.2296, the same substances will settle out after extensively long centrifugation. On the other hand, separation of certain cellular organelles (e.g., ribosomes) requires a narrower sucrose density gradient interval, such as 5% to 20%. "Ficoll", a syntheti polymerized sucrose produced by Pharmacia, Uppsala, Sweden, is gaining in popularity because it was found to give better separation of subcellular components than sucrose, due to high density, lower viscosity and lower osmotic activity.

The procedure as described here is eminently suitable to purify viruses from concentrated virus suspensions. It can be also used to separate 7 S and 19 S immunoglobulins. In that case the centrifuge has to run 18-20 h. The 19 S components will be near the bottom, while the 7 S globulins will be in the upper half of the tube.

References

Britten, R.J., Roberts, R.B.: Science *131*, 32 (1960)
Hughes, D.E., Nyborg, W.L.: Science *138*, 108 (1962)
Lindstroem, O.: J. acoust. Soc. Amer. *27*, 654 (1955)
Meselson, M., Stahl, F.W., Vinograd, J.: Proc. Nat. Acad. Sci (USA) *43*, 581 (1957)

Exercise No. 32

Vacuum Distillation and Freeze Drying

Vacuum distillation is based on the fact that under vacuum the boiling point of the materials is lower than at normal atmospheric pressure. The use of vacuum distillation for distilling off solvents is required where elevated temperatures would destroy the resolved components. Working with some biologically active materials which are dissolved in water, the solution can be concentrated using vacuum distillation at 20° or 30°C without inactivating the preparation if the pressure is reduced to 14 or 16 mm Hg. The complete removal of water by vacuum distillation, however, damages the biologic potency of many preparations. Some protein-containing solutions foam considerably, and in these cases vacuum distillation should be avoided. Frequently, the freeze drying (lyophilization) procedure gives much better results.

If the pressure is lower (increased vacuum), even lower boiling points can be reached. If water is frozen, and a very high vacuum (equal to or less than 50 µ Hg) is applied to it, the ice will sublime from the preparation. After all the ice has sublimed from the sample, only the solid material remains as a light and porous dry substance. The biologic activity of the preparation is usually hardly affected.

In this exercise both procedures will be performed. Vacuum distillation will be applied to reduce the volume of the crude O-antigen extract obtained in Exercise No. 21. The concentrated extracts of the O-antigen will be dried finally by lyophilization.

Materials and Equipment

    Dry ice
    Acetone, technical grade
    Octanol
    Round flask, with ℥ 24/40 joints
    McLeod vacuum gauge
    Adjustable water bath
    All glass vacuum distillation equipment (Büchi type), see
     Fig. 12
    Water aspirator
    Freeze drying equipment, Fig. 14 or Fig. 15
    Desiccator with phosphorus pentoxide
    Vacuum pump
    Face shield

Procedure

1. It is important that the vacuum distillation unit be kept clean. Quite often some drops of the solution will splash, through foaming or sudden boiling, into the condenser inlet, dry there, and eventually contaminate the next solution if they are not removed by careful cleaning.

2. Transfer 200 ml of the dialyzed solution into a 500 ml flask; a few drops of octanol should be added to prevent excessive foaming.

3. The condenser of the vacuum distillation unit should be attached to the cold water faucet, and the air outlet of the system connected to an efficient water aspirator pump. Turn the water on. Check that all the joints are properly greased.

4. Attach the 500 ml 24/40 flask with 200 ml filtrate to the distillation unit; hold it in place for a few seconds until sufficient vacuum is built up to prevent it from falling off. Turn on the rotating knob, adjusting the speed to approximately 1-2 rps. Immerse one-third of the rotating flask into the water bath adjusted to approximately 35°C. Watch the solvent in the flask; if too intensive foaming occurs before boiling starts, air should be allowed to enter the system for a split second, to break the foam. Repeat this several times if foaming persists. Eventually, renewed addition of a few drops of octanol may be necessary. With bacterial O-antigen solutions, such as this one, there is usually no difficulty.

If the crude extract is reduced to approximately 50 ml, it may be prepared for freeze drying in the same flask.

5. The lyophilization apparatus must be prepared first. The trap should be immersed in a Dewar thermos flask half filled with acetone, and the thermos should be filled slowly with crushed dry ice. Insure that all the joints are properly greased. Attach the system to an efficient mechanical vacuum pump and turn on the motor.

6. Cool approximately 200 ml acetone in a pot with a few
pieces of crushed dry ice. After a minute or so, the temperature
reaches -30° to -50°C. Immerse the flask containing the concentrated O-antigen and shell-freeze its contents by slowly rotating the flask in the cooled acetone.

7. Wearing a face shield, attach the flask containing the
shell-frozen material to the proper joint of the lyophilization
equipment. Rotate the flask until the grease forms a continuous
transparent film in the joint. Be sure that no air enters the
system.

The heat required for the sublimation is taken from the frozen
solvent, holding its temperature at a low level. Therefore, the
formation of a snowy ice crust on the outside of the frozen
flask shows that the system is working properly. This crust will
remain on the flask until all the ice sublimes away from the
contents. Thereafter, the temperature of the flask reaches room
temperature. Do not disconnect the system if you can feel any
cold spots by touching the flask.

8. After the lyophilization is complete, allow air to enter
the system slowly and disconnect the flask from the equipment.
Wipe the greased joints carefully, remove the dried material
from the flask and determine its weight.

Evaluation

Measure the yield of crude endotoxic O-antigen preparation.
Determine the moisture content of the lyophilized material by
weighing around 20 mg material into a small beaker. Cover it
with aluminum foil, punch small holes in it with a pencil. On
an analytical balance, determine the exact weight of the beaker
with the material in it, covered with the foil. Place the material in a vacuum desiccator and dry it over P_2O_5 under lower
than 10 μ Hg pressure for 48 h. Reweigh the beaker and the substance in it. Determine the loss of weight. Good lyophilization
should not leave more than a few percent moisture in the materials.

Use and Limitations

Preparations containing higher than 1% salt concentration may
melt during lyophilization due to lowered melting point (eutectic
mixtures). Alcohols will also significantly lower the freezing
point of the solutions. All materials which have similar effect
must be removed by dialysis before freeze drying.

The lyophilization should not be considered as a perfectly
harmless method for drying preparations. Several biologically
active macromolecules will undergo irreversible structural
changes, and some of them, such as antihemophylic globulins,
will lose activity.

2 Structural Studies

2.1 Qualitative Methods

2.1.1 Hydrolytic Procedures

Exercise No. 33

Determination of Optimal Hydrolytic Conditions

During hydrolysis, not only degradation of large molecules, but also destruction of liberated units such as monosaccharides or amino acids takes place. Both the liberation and destruction of these components depend upon the strength of the hydrolysis, i.e. the acid or alkali used, its concentration, temperature, and the length of the reaction. Obviously, 100% recovery of the liberated components in the hydrolysate is possible only if the hydrolysis is complete, while liberated units are fully resistant to further degradation. No such hydrolytic procedure exists. Therefore, quantitative determination of the monosaccharide or amino acid constituents of a biopolymer always results in a "minimum value" which is significantly less than the actual percentage.

In order to obtain the closest value in the determination of the composition of natural products, the optimal hydrolysis conditions must be determined for each unknown preparation. Some information regarding the optimal conditions of hydrolysis for certain groups of biopolymers can be obtained from the literature. It is known that polysaccharides can be split more easily into monosaccharides than proteins can be cleaved into their amino acid units, and the glycosidic linkages of aminohexoses are more resistant to acid hydrolysis than similar bonds of aldohexoses. Determination of certain amino acids requires alkaline hydrolysis because of the acid sensitivity of the free amino acids. Some monosaccharides are also much more sensitive to acidic hydrolysis than others, and for their quantitative determination individual hydrolytic procedures have to be elaborated in preliminary experiments. For more accurate analysis, the time curve of hydrolysis must be measured, as described here. For this Exercise a glycoprotein will be hydrolyzed and the optimum conditions will be determined for the liberation of (a) aldoses and (b) amino acids or amino sugars.

Materials and Equipment

    Phytohemagglutinin (PHA) from Exercise No. 26
    2 *N* Sulfuric acid
    0.1 *N* Sodium hydroxide
    10 *N* and 6 *N* Hydrochloric acid
    Microburet
    Acetone
    Dry ice
    Potassium hydroxide pellets

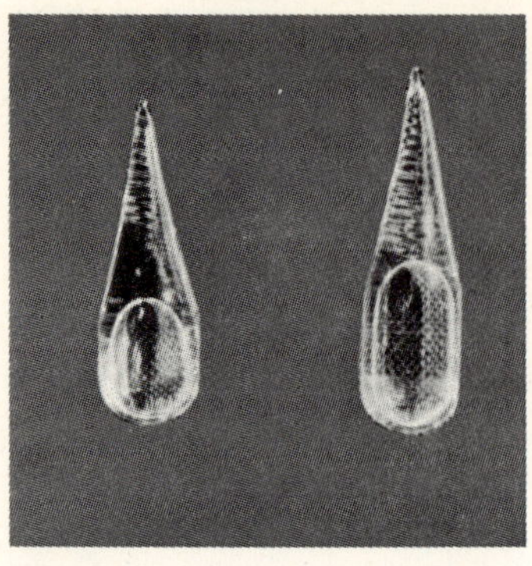

Fig. 16. "Cold fingers"

Oil bath, adjustable to 100°C or boil-
  ing water bath with water level regu-
  lator
16 × 150 mm test tubes, borosilicate
Glass marbles larger than 16 mm in diam,
  or better, "cold fingers" shown in
  Fig. 16 which are made from the same
  size test tubes
Wire test tube rack
Oxygen gas torch
Vacuum pump
Glass cutter knife
Desiccator

Procedure

Part A. Carbohydrate Liberation

1. Dissolve 10 mg PHA in 2.5 ml water in a test tube and add
to it 2.5 ml 2 N sulfuric acid. Cover the tube with a marble or
"cold finger" and place it in a boiling water bath. Accurately
withdraw 0.1 ml samples after 1, 2, 4, 6, and 8 h of hydrolysis.
Pipette these 0.1 ml samples into an electrometric titration
vessel (see Fig. 17). Add 5.0 ml water and titrate the samples
with 0.1 N NaOH to pH 7. Use a microburet. Add water to a final
volume of 10 ml.

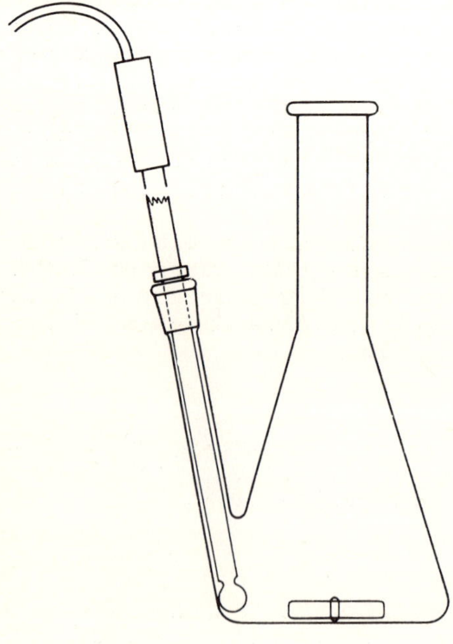

Fig. 17. Electrometric titration vessel
using a combination glass electrode of
5 mm diameter. The electrode is available
from Arthur H. Thomas, Philadelphia, PA
19/07.

2. Take 1.0 ml of these neutralized dilutions (containing
20 µg PHA hydrolysate per ml) and measure the reducing carbohydrate content with the micromethod of Park and Johnson as described in Exercise No. 53.

Part B. Liberation of Amino Acids and Amino Sugars

To determine the hydrolysis time which liberates the maximum
amount of amino acids, 5 N hydrochloric acid final concentration
is used. If the liberation of hexosamines is to be investigated,
the same procedure may be used, but a lower hydrochloric acid
concentration, such as 3 N is recommended.

    1. Pipette 0.5 ml samples of the above PHA solution into
16 × 150 mm borosilicate tubes, and add 0.5 ml 10 N HCl for amino
acid (or 0.5 ml 6 N HCl for amino sugar) hydrolysis. Seal the
tubes in an oxygen flame by heating the rim of the test tube to
a red glow, attach a glass rod to this, and after melting the
upper 20 mm of the tube, pull it out to form a capillary. This
can be easily sealed in the flame. Do not leave a sharp needle-
like tip on the sealed tube which may break off, but keep the
sealed end of the tube in the flame for a few seconds to permit
it to round off by melting. After the tubes have been sealed,
wrap them in several layers of a strong fabric which may prevent
the showering of glass splinters in case the tube should explode.
Never fill the tubes more than one-fifth full. Attach a small
metal number plate to them and place them in a boiling water
bath or 100°C oil bath.

    2. Remove the tubes one by one after 1, 2, 4, 6, and 8 h of
hydrolysis. Freeze their contents in acetone and dry ice to stop
hydrolysis. Scratch a line with a glass cutter at the upper
one-fifth of the tube, and press a red-hot glass rod to this
point. The tube will crack and it may now be opened by striking
it gently at the split with a glass rod. Discard the upper part
and place the bottom part of the tube containing the frozen
acidic hydrolysate into a vacuum desiccator which contains potassium hydroxide pellets, and evacuate it. Do the same with the
other samples after the hydrolysis time has expired. Under good
vacuum, the tubes will be dry and free of hydrochloric acid
overnight.

    3. Dissolve the yellow-brown residue in accurately measured
2 ml distilled water (which corresponds to 1 mg hydrolyzed PHA
per ml). Determine the amino acid or amino sugar content of the
samples as described in Exercise No. 51 and Exercise No. 55.

Evaluation

Plot the values obtained against time of hydrolysis and determine
from the curve the optimal hydrolysis time for (a) reducing
carbohydrates, and (b) amino acids or amino sugars.

Use and Limitations

Practically the same procedure may be applied to most of the
carbohydrate-containing materials such as oligosaccharides,
glycolipids, mucoproteins, etc. The use of sealed tubes is re-

commended for peptides and proteins, but also may be used for
carbohydrate hydrolysis. It increases the rate of reaction and
often reduces extensive destruction of liberated components due
to prolonged hydrolysis. The sealed tube technique must be
applied where volatile components are liberated from the sample
during hydrolysis, as we will see in a later experiment for
determining the shortchain carboxylic acids of natural products
(see Exercise No. 43 or No. 64.)

If the aim of the experiment is to obtain partially hydrolyzed
products, the above hydrolysis time curve will serve as a guide
to estimate how long the hydrolysis should go on in order to
achieve the desired degree of degradation.

The main limitation of these procedures has been discussed in
the introduction. It must be reemphasized that the values obtained
for certain components are usually lower than their actual percentages in the unhydrolyzed materials and corrections must be
made to obtain an approximately valid result. The losses in certain known constituents may be eliminated if the percent destruction of this material has been determined under identical
conditions in separate experiments.

Another procedure to determine the amount of destroyed material
during the hydrolysis can be applied if the hydrolysis curve
shows a rather linear decline after the maximum has been reached.
This decline can be very steep, as, for example, in the case of
2-keto-3-deoxy carbohydrate derivatives. In similar instances
the reading points after the maximum can be connected with a
straight line and the concentration of the analyzed constituent
in the hydrolyzed natural product can be approximated by extrapolating this line to zero time hydrolysis. An example of this
method can be found in Kasai and Nowotny (1967).

It is self-evident that the course of hydrolysis will not be
identical under pressure or under vacuum. Which one of these
will be the most suitable for the material to be analyzed must
be carefully investigated, together with other variants (such as
acid, concentration, temperature, time, etc.) in preliminary
experiments, similar to those outlined above.

Reference

Kasai, N., Nowotny, A.: J. Bacterol. *94,* 1824 (1967)

Exercise No. 34

Hydrolysis of Lipopolysaccharide With Acid

Hydrolysis of polysaccharides does not require such strong acidic
treatment as does hydrolysis of proteins. However, although the
conditions are relatively mild, destruction of the liberated
carbohydrates takes place easily. It is important, therefore, to
keep the time of hydrolysis as short as possible while trying to
achieve a complete hydrolysis of the polysaccharide molecule.

In this Exercise endotoxic O-antigen obtained from Exercise
No. 21 will be hydrolyzed. For these lipopolysaccharides, hydroly-

sis with 1 N H_2SO_4 at boiling water bath temperature for 16 h proved to be satisfactory, liberating the maximum amounts of monosaccharides (Fromme et al., 1958). Hydrolysis with the same acid and at the same temperature for 3 h liberates approximately 60%-70% of the total monosaccharides from bacterial lipopolysaccharides. The qualitative composition of the hydrolysate obtained after 3 h of hydrolysis is identical to that obtained after total hydrolysis.

In this Exercise the more convenient three-hour hydrolysis will be carried out.

Materials and Equipment

    Endotoxic O-antigen (from Exercise No. 21)
    1 N Sulfuric acid
    Barium carbonate
    Octanol
    Centrifuge with 50 ml capacity tubes
    Test tube, 16 mm × 150 mm
    Glass marbles or "cold fingers"
    Water bath with water level regulator (or oil bath)
    Funnel, 6 cm in diam
    Round flask ℭ 24/40, 100 ml capacity
    Büchi type flash evaporator with ℭ 24/40 adapter (Fig. 12)
    Vacuum desiccator

Procedure

1. Prepare a boiling water bath. Adjust the water level regulator.

2. Weigh 20 mg endotoxic O-antigen into a test tube and add 4 ml 1 N H_2SO_4. Place the tube in the water bath with a marble or "cold finger" on it.

3. After 3 h of hydrolysis, remove the sample, pour the content into 50 ml centrifuge, tube and rinse the tube twice with 4-5 ml water. With a spatula, add approximately 0.5 g solid $BaCO_3$ in small amounts to the hydrolysate as in Exercise No. 43. Remove the precipitated $BaSO_4$ and the excess $BaCO_3$ by centrifugation at 3000 g for 15 min. The supernatant usually contains non-sedimented precipitate, and therefore should be filtered through analytic paper. The filtrate is collected directly into a round, 100 ml capacity flask. To this filtrate add two drops of octanol and evaporate to complete dryness in vacuum, using a Büchi-type flash evaporator. The dried sample should be placed in a vacuum desiccator over Drierite and kept for further use.

Several authors prefer to use anion exchangers in HCO_3^- form to remove the sulfuric acid. Freshly activated Dowex 1 or Amberlite IRA 400 in bicarbonate cycle may be used. Keep adding the proper ion-exchanger resin to the hydrolysate and after each addition check the pH of the supernate with indicator paper. When it becomes neutral, filter it through a coarse grade glass filter, applying suction. Wash the resin with a few ml water, concentrate the de-ionized hydrolysate to dryness in vacuum.

Evaluation

The hydrolysate will be analyzed by paper chromatography, as
described in Exercise No. 36. The amount of liberated reducing
carbohydrates may be measured by the method given in Exercise
No. 53. Qualitative analysis of the other constituents, such as
amino acids, peptides, phosphoric acid esters, carboxylic acids,
etc. may be carried out using the procedures outlined in Exercise
No. 43.

Use and Limitations

The optimal conditions for polysaccharide hydrolysis must be
determined individually for each polysaccharide in preliminary
experiments, and can be established by measuring the amount of
liberated monosaccharides after different lengths of hydrolysis
carried out with different acid concentrations. This has been
described in Exercise No. 33.

Reference

Fromme, I., Luderitz, O., Nowotny, A., Westphal, O.: Pharm. Acta Helv. *33*,
391 (1958)

Exercise No. 35

Partial Hydrolysis of Lipopolysaccharide With Ion Exchanger

Strongly acidic cation exchangers, such as sulfonated resins, act
as insoluble acids on dissolved substances which come into prolonged contact with them. The use of ion exchangers for hydrolysis
of natural polymers has been described, but the practical advantages of this procedure have not been fully exploited. The
method given in this exercise has been elaborated and used for
the partial hydrolysis of endotoxic bacterial lipopolysaccharide
preparations (Nowotny, 1957, unpublished).

In *Part A* of this Exercise partial hydrolysis will be carried
out in the presence of chloroform in order to separate simultaneously those split products of the lipopolysaccharide which are
soluble in chloroform.

In *Part B* of this Exercise partial hydrolysis will be carried
out with ion exchangers using a continuous dialysis apparatus.
In partial hydrolytic procedures, samples taken at a certain
time contain all the products of the hydrolysis, from liberated
monomers to large, barely affected macromolecules. Oligomers
(such as oligosaccharides or peptides) which are desired mostly
for the elucidation of the structure of the polymer, are not
always formed in sufficient amounts if the hydrolysis is of
short duration, and they will be further degraded during prolonged hydrolysis. Hydrolysis with ion exchangers results in a
good yield of oligosaccharides and the time of the reaction is
easily adjustable to desired experimental conditions. The liberation of certain split products from a polymer usually occurs in

sequential steps, cleaving first those constituents which are
linked to the structure through especially acid-sensitive linkages. Such constituents are the carbohydrates at the end of the
polysaccharide chains. Naturally, other acid labile linkages in
the polysaccharide will also be split almost simultaneously,
producing degradation products with new additional nonreducing
and reducing end groups. Also, in most cases, there is a measurable time difference between the liberation of certain constituents, which fact provides useful information regarding the possible linkage and position of the split products in the biopolymer.

The experiment described in *Part B* has been developed to observe the sequential release of the carbohydrates from the
lipopolysaccharide with simultaneous protection of the split
products from further degradation (Nowotny, 1965, unpublished).
Ion exchanger Amberlite 120 in hydrogen form and *Serratia marcescens*
lipopolysaccharide are placed in a dialysis bag and stirred continuously while kept at elevated temperature. The outer fluid
of the dialysis bag is continuously changed at a very slow rate,
and collected in a fraction collector. As soon as the ion exchanger inside the bag splits off a particle of the molecule which
is small enough to pass the dialysis membrane, such a molecule
will be separated from the resin through dialysis, and in this
way will be protected from further hydrolysis. The procedure is
described here as it has been used for the studies of serologic
determinant groups of *Serratia marcescens* O8 lipopolysaccharides.

Part A. Discontinuous Method

Materials and Equipment

> *Serratia marcescens* endotoxin, obtained as described in Exercise
>   No. 21, or any other similar natural product
> Amberlite IR 120 resin in hydrogen form, mesh size 200 or
>   higher, freshly activated and washed
> Chloroform, reagent grade
> 9% Sodium chloride
> Three reflux condensers with T joints
> Test tubes, 18 × 150 mm, with above T joints
> Sintered glass filter funnel, coarse grade
> Filter flask, 50 ml capacity
> Centrifuge tubes, 15 ml capacity
> Water bath
> Alundum crystals

Procedure

1. Dissolve 40 mg lipopolysaccharide in 20 ml water. Prepare
three tubes as follows: Pipette 5 ml of the above solution into
an 18 × 150 mm test tube and add 5 ml chloroform. Measure 2 g
freshly activated and washed Amberlite resin in hydrogen form
into the tubes. The addition of a few alundum crystals is advisable. Attach the tubes to the reflux condensers and immerse them
halfway into a water bath at 75°C. The chloroform will boil in
the tubes, agitating the three phase system and continuously
extracting the liberated lipid components.

Remove the first tube after 24 h. Filter its contents through
the glass filter into a flask, transfer the filtrate into a 15 ml
capacity centrifuge tube. Spin it for 10 min at 3000 g to facilitate the separation. Lift the supernate with a pipette and save
it for further experiments. Add 5 ml distilled water to the
chloroform, close the tube, shake it vigorously, repeat the centrifugation and separation as before. Combine the water extracts.
Repeat the extraction once more.

2. The second and third tubes are prepared at the same time
and in exactly the same way as the first. Hydrolyze the second
tube for two days, and the third sample for three days. Proceed
as described above.

3. For quantitative studies, it is advisable to extract the
combined water phases twice with 5 ml chloroform. The pooled
water phases and the chloroform extracts may be made up to 20 ml
each for further studies.

Evaluation

If chromatography is to be used for the separation of the split
products, it is better to concentrate the extracts and take them
up in a small volume such as 0.5 ml. If quantitative chemical
analytic procedures are to be used to determine the degree of
degradation, such as measurement of milliequivalents reducing
aldehyde groups (Exercise No. 53 or No. 61), or number of milliequivalent ester groups (Exercise No. 63), the dilute solutions
can be used. For serologic studies, such as inhibition of passive
hemagglutination (Exercise No. 69), the water-dissolved samples
must be adjusted to physiologic sodium chloride concentration
by adding 1/10 vol of 9% sodium chloride.

Use and Limitations

Although the method has been elaborated for bacterial lipopolysaccharides, it may be applied to any natural products which
can be hydrolyzed by acids. For alkaline hydrolysis of biopolymers, anion exchangers can be used in hydroxyl form in the same
system. One of the advantages of the application of ion exchangers
for the cleavage of biopolymers is that the procedure is relatively mild, and therefore the degree of hydrolysis can be more
easily regulated. Another advantage is that the insoluble resin
particles can be removed from the reaction mixture by simple
filtration, which at the same time stops the hydrolysis instant-
ly. The appearance of undesirable side products of acidic hydrolysis, such as the dark-colored humin precipitates, were not
observed in using ion exchangers. Some of the limitations of
the procedure are obvious from the above remarks. For example,
complete hydrolysis of certain products can hardly be achieved
by ion exchangers, even after several weeks of reaction time.
The ion exchanger may also bind split products from the hydrolysates. However, this binding may be dissociated easily and this
selective property of resins may be utilized for the purposes
of the investigation. The solubility of the ion exchanger resins
is negligible in water, but if they are to be used for the hydrolysis of substances dissolved in organic solvents, the solubility

chart of the resins used should be consulted. Such information
is available from the manufacturer.

Part B. Continuous System

Materials and Equipment

> *Serratia marcescens* endotoxin, as in *Part A*
> Amberlite IR 120 resin in hydrogen form, as in *Part A*
> 9% Sodium chloride
> N_2 gas or micro "Vibromixer" (available from Chemapec, Inc.,
>   Hoboken, NJ 07030)
> Dialysis bag, 18-20 mm inflated diam
> Water reservoir for adjustable steady flow rate (Fig. 18 or 19)
> Equipment for continuous dialysis (Fig. 20)
> Automatic fraction collector with 10 ml volumetric siphon
>   (Fig. 21)
> Circulating constant temperature bath
> Nylon stocking

Procedure

1. Dissolve 40 mg lipopolysaccharide in 10 ml distilled water
on a magnetic stirrer. While this is being prepared, assemble
the equipment. The volumetric siphon (Fig.21) of the fraction
collector should be in position above the collector. The out-
flow tube 10 should be right above No. 19, and sidearm funnel
9 of Fig. 20 should be under glass tube 8.

2. Assemble the water reservior unit as shown in Fig. 19.
Put the 1000 ml capacity container 1 on the table, fill it completely with distilled water. Put in stoppers 3 and 4. Note that
stopper 4 has an aerating slit cut in its side, while stopper 3
does not. The stoppers are connected with a 10 cm long, 15 mm
diam glass tube. On the top of the tube place a 200 ml separatory
funnel 5, and connect container 1 with funnel 5 with strong
springs 2. Keep the stopcock closed and turn the assembled unit
upside down, placing it in a holder ring as shown in Fig. 19.
Connect funnel 5 with rubber tubing 7 to capillary bore class
tube 8, which has an internal diameter of 1 mm and an outside
diameter of 5 mm. The end of the tube is ground flat. Open
stopcock 6 and fill the tube with water. The relative position
of tube 8 to the water level in funnel 5 will determine the flow
rate. Do not use the stopcock for regulation, but keep it open
during the operation. Adjust the flow rate by changing the position of the reservoir. In this experiment a flow rate of approximately 5 ml/h is necessary, which means that the flow rate must
be about one drop/min. The size of capillary tube 8 given above
will deliver approximately 0.1 ml water/drop. Fill tube 8 above
the small sidearm funnel 9 of Fig.20 and regulate the flow rate
to approximately 1 drop/min. Turn off the stopcock of funnel 5.
Instead of the constant flow assembly described here (Fig. 19),
a commercially available, easily adjustable dropping funnel may
be used (Kontes Glass Co., Vineland, NJ 08360, cat. No K-63465)
which delivers any liquid at a constant rate. This funnel is
shown in Fig. 18.

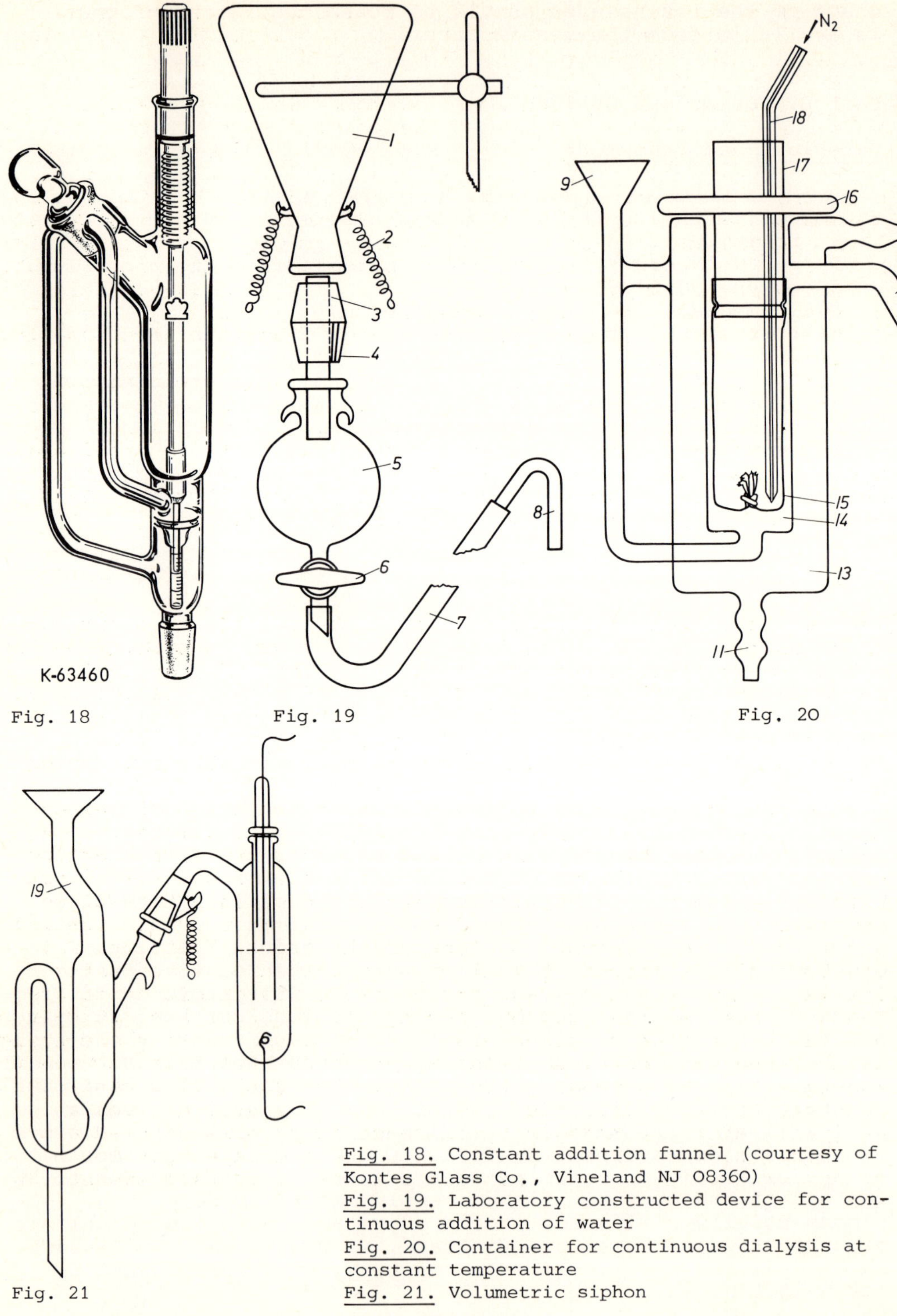

Fig. 18

Fig. 19

Fig. 20

Fig. 21

Fig. 18. Constant addition funnel (courtesy of Kontes Glass Co., Vineland NJ 08360)
Fig. 19. Laboratory constructed device for continuous addition of water
Fig. 20. Container for continuous dialysis at constant temperature
Fig. 21. Volumetric siphon

3. Connect heating jacket 13 of Fig.20 with a circulating
water thermostat through ports 11 and 12. Start the circulation
and adjust the temperature to 75°C.

4. Prepare the dialysis bag 15. Cut a 20 cm long piece of
dialysis tubing of approximately 18-20 mm inflated diameter.
Make a tight knot at one end of the tube, fill the tube with
water and, by squeezing it with the hands, make sure that it
does not leak. Put glass tube 17 into the bag with one end of
the glass approximately 6-7 mm from the knot in the dialysis bag.
Secure the bag in this position with a rubber band. Pipette 10 ml
lipopolysaccharide into it, and add 2 g freshly activated Amberlite IR 120 resin. Place the bag attached to the glass tube into
compartment 14. Adjust tightly fitting cork or stopper collar 16
so that it holds the bag in the position shown in Fig. 20.
Carefully place the capillary 18 into the bag and start bubbling
N_2 slowly, making sure that the capillary does not rub against
the wall of the bag. Sometimes the material has a tendency to
foam. In such cases a micro Vibromixer blade should be used
instead of N_2 bubbles to stir the resin within the bag.

5. Open stopcock 6, turn on the fraction collector, and place
an empty tube under the 10 ml volumetric siphon 19. Let the
hydrolysis proceed for 48 h uninterrupted. Calculate the time
which was required to fill one siphon, and record it.

Evaluation

Take aliquots from the tubes, measure the quantitative carbohydrate content of each fraction by phenol-sulfuric acid. Plot the
values found against the tube number (or time of hydrolysis).
Measure the hemagglutination inhibitory effect of the collected
samples as described in Exercise No. 69. In order to obtain
physiologic salt concentration for passive hemagglutination,
mix 1 vol of the collected fluid with 1/10 vol 9% sodium chloride
solution.

For the analysis of the carbohydrate composition of the content of the tubes by paper chromatography, approximately 50 µg
carbohydrate-containing sample must be applied on the starting
spot of the paper chromatogram. The collected samples will not
have the proper concentration, therefore the contents must be
concentrated either by vacuum distillation or by lyophilization
and taken up in one-tenth or less of the original volume. This
can be easily achieved if 1 ml of each sample is transferred
into an 18 mm test tube, and its contents frozen either in dry
ice and acetone or simply by placing the tubes into a -20°C freezer. Put the frozen samples into a freeze dryer, with the opening
of the tube covered with a fine Nylon screen which is secured to
the test tube by a tight rubber ring. The lyophilized contents of
the test tubes will be dissolved in 0.1 ml distilled water.
Analytic paper chromatography of these samples is carried out
as described in Exercise No. 36. Preparative paper chromatography or paper electrophoresis is described in Exercise No. 36
and No. 37.

Use and Limitations

The same experiment can be carried out with basic ion exchangers, and not only polysaccharides but other polymers can be hydrolyzed in this way. The use of soluble acids or alkalis instead of ion exchangers for the preparation of partially hydrolyzed split products gives a faster hydrolysis, but in such case the outer fluid of the dialysis must contain the same concentration as the fluid inside the bag to maintain a continuous acidity or alkalinity throughout the experiment. Because in such case the split products remain in contact with the hydrolyzing agent, further degradation cannot be avoided. If a refrigerated fraction collector is available, the continued hydrolysis may be considerably slowed down, although the removal of acid or alkali from the collected samples must still be carried out. The same equipment described here may be used.

Soluble ion exchangers have been developed. Their average molecular weight is much higher than the pore size of the dialysis membrane. The hydrolysis with such a dissolved ion exchanger takes place in a homogeneous system, and therefore the rate of reaction is higher. In some cases the separation of the soluble ion exchangers from the residual hydrolytic products may present difficulties. Soluble ion exchangers were used for the hydrolysis of blood group antigens by Painter (1960). Enzymes may also be used for the sequential cleavage of proteins or polysaccharides instead of acid or alkali, using the same apparatus. A similar approach has been described by Fasold, Linhart and Turba (1962).

A limitation of the procedure described here is the difficulty of avoiding microbial growth inside the dialysis bag and in the collected tubes. Due to the long reaction time, the reaction mixture may be contaminated unless the temperature of the hydrolysis is high enough. To avoid significant growth of microorganisms in the collected samples, one must use a refrigerated fraction collector, or bactericidal or bacteriostatic agents must be added to the receiving tubes. Transfer of the receiving tubes with their contents to a refrigerator as soon as possible often satisfies these requirements.

References

Fasold, H., Linhart, P., Turba, F.: Biochem. Z, *336*, 191 (1962)
Painter, T.J.: Chem. and Indust. 1214 (1960)

2.1.2   Separation and Isolation of the Split Products

Exercise No. 36

The Methods of Analytic and Preparative Paper Chromatography

Part A of this exercise gives instructions on how to find the proper chromatographic solvent in orienting analysis. The method which is described in *Part B* applies a thick filter paper for the preparative chromatographic separation of oligosaccharides.

Basically, the same procedure can be applied for the separation
of peptides or other split products. Obviously, in these latter
instances different spray reagents and different chromatographic
solvents must be used.

Part A. Preliminary Runs

Materials and Equipment

    Hog gastric mucin (crude blood group A substance), or any
      similar natural product
    2 N Sulfuric acid
    Amberlite IRA 400 resin in HCO_3^- form
    Ninhydrin reagent (see Exercise No. 43)
    Phenol
    Ethyl acetate
    n-Butanol
    Pyridine
    Alkaline silver nitrate reagent (see Exercise No. 43)
    Regular test tubes
    Test tubes, 10 × 120 mm
    Marbles or "cold fingers"
    Whatman No. 1 or any other filter paper of comparable quality
    Test tubes, 18 × 150 mm or larger, with corks or rubber stoppers
    Boiling water bath
    Hot air fan (hair dryer)
    Pipette, 0.1 ml capacity
    Hot air oven
    5 mm diameter glass tube

Procedure

1. First prepare a partial hydrolysate of the hog gastric
mucin which is used as a model substance in this exercise. Any
other polysaccharides, glycoprotein or proteins may be used in
similar setups. In this description 0.5 N H_2SO_4 will be used
for the partial hydrolysis. Take 10 mg mucin, dissolve it in
a test tube in 1 ml water at room temperature. Add 1 ml 2 N H_2SO_4,
cover the tube with a marble and place it in a boiling water
bath for 120 min. Cool the tube and add to it approximately 2 g
Amberlite IRA 400 resin in HCO_3^- form. Filter the contents of
the tube and check the pH, which should be between 5 and 7. The
material is now ready for separation.
2. For the paper chromatographic fractionation, the proper
solvent system has to be found. This can be easily achieved by
trying several solvent combinations in preliminary experiments
using Whatman No. 1 paper. If this has to be done, first find
the proper size large test tubes, and corks or rubber stoppers
which will fit tightly into these tubes. Cut the stoppers in half
as shown in Fig. 22. Pipette 1 ml experimental chromatographic
solvent into the bottom of the tubes, do not wet the inside walls.
Solvent systems which may be considered should include different
n-butanol:pyridine:water ratios or ethyl acetate:pyridine:water
mixtures for oligosaccharides. Phenol saturated with water may
also be tried. This is especially useful in the separation of

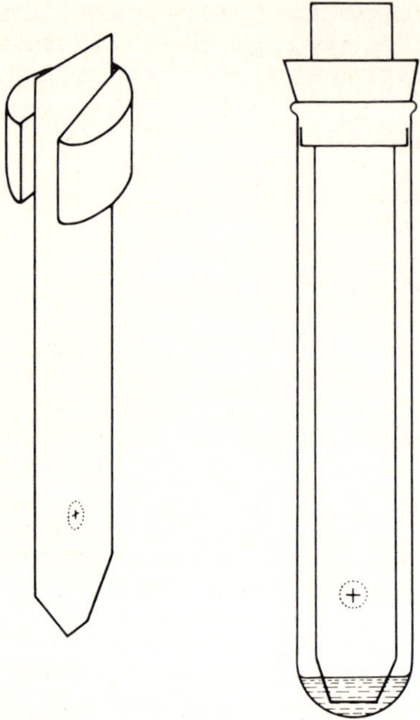

Fig. 22. Paper chromatography in test tubes,
useful for preliminary runs

amino acids and peptides. Close the tubes with a stopper and let
them stand at room temperature while the paper strips are prepared. Cut strips 14 mm wide and 200 mm long, shape one end as
shown in Fig. 22. Apply 0.01 ml of the partial hydrolysate to
the strips; dry them with warm air. Be sure that the applied
samples do not form a spot larger than 5 mm in diam. Place the
upper part of the strips between the two halves of the stopper
so that when the stopper is positioned firmly in the tube, the
end of the paper strip will reach into the chromatographic solvent. The strip should hang free without touching the walls or
bottom of the test tube. It is important to observe these per-
cautions carefully. If the ascending solvent front is uneven,
the components which will be chromatographed will not move up
in the center of the paper strip, but will move along the edge
with considerable tailing. If the solvent front is uneven and it
has not yet passed the applied spot, the strip may be taken out,
dried quickly with warm air, and a proper chromatographic start
may be tried again. To have an even solvent front, it is imperative to cut the end of the paper strip as shown in Fig. 22.

   3. After the solvent has reached the stopper, take the chromatogram out, dry it, and develop it with alkaline $AgNO_3$, as
described in Exercise No. 43, or with any other chromatographic
spray reagent for carbohydrates. If components other than carbohydrates are also to be detected, use the corresponding reagents. Some of them are described in Exercise No. 43.

   4. If the aim is to find the serologically active components
on the paper chromatogram, run another strip under exactly
identical conditions, but do not develop it with spray reagent.
Dry it at 90°C for 5 min, then for at least 1 h under a well-
ventilated fume hood to remove all the chromatographic solvents

from the paper. With a pencil, mark those parts of the unstained
strip which may have carbohydrates as indicated on the stained
strips. Cut out the area of the starting point also; it may contain serologically active carbohydrates which are just faintly
visible with this method of detection. Cut the zones into small
pieces with a scissors, and put each zone into a 10 × 120 mm test
tube. Add 0.5 ml saline and agitate the contents with a glass
rod. Let stand for 30 min at room temperature with occasional
mixing. Centrifuge the tubes at 500 g for 30 min and lift the
supernate for inhibition of hemagglutination as described in
Exercise No. 69.

If a sufficient amount is available, it is feasible to use
0.5 ml 1:100 diluted antiserum instead of saline. In such case
the centrifuged supernate will contain the absorbed serum. Those
zones cut out of the paper chromatogram which contain serologically active oligosaccharides will reduce the titer of the serum.
This can be measured by passive hemagglutination.

5. Oligosaccharides have a very low R_F value in most solvents.
In order to enhance the migration of these, paper chromatography
may be repeated on the same strip. Dry the strip as above, and
place it in the same tube again. After two or more repeated
chromatographies, the monosaccharides will be piled up on each
other near the solvent front, but the oligosaccharides will be
better separated from each other. This procedure is often used
in preparative paper chromatography, which is described in the
next paragraph.

Evaluation see *Part B.*

Use and Limitations see *Part B.*

Part B. Preparative Procedure

Materials and Equipment

Large paper chromatographic jar, descending type (Fig. 23)
Whatman No. 3 or any other filter paper of comparable thickness
 and quality
Other materials and equipment as listed in *Part A*

Procedure

1. If the proper solvent mixture is known, the preliminary
experiment described above may be omitted. In this Exercise the
use of the upper phase of ethyl acetate:pyridine:water = 2.5:1:3.5
mixture is recommended, which is a good paper chromatographic
solvent for the separation of oligosaccharides and monosaccharides.

2. Take a 15 × 40 cm strip, cut from Whatman No. 3 paper. Draw
a pencil line parallel to and 10 cm distant from the shorter
edge of the paper. This will be the starting line. Measure 0.2 ml
of the partial hydrolysate into a small test tube. Pull out a
5 mm diameter glass tube into a fine capillary which will be
used to apply the sample on the paper strip. Pull up an aliquot
into the capillary and by touching the starting line with the
tip, move it rather quickly along the pencil line. The sample
will be taken up and adsorbed by the paper. Do not come closer
than 10 mm to the edge of the paper. When one streak is applied,

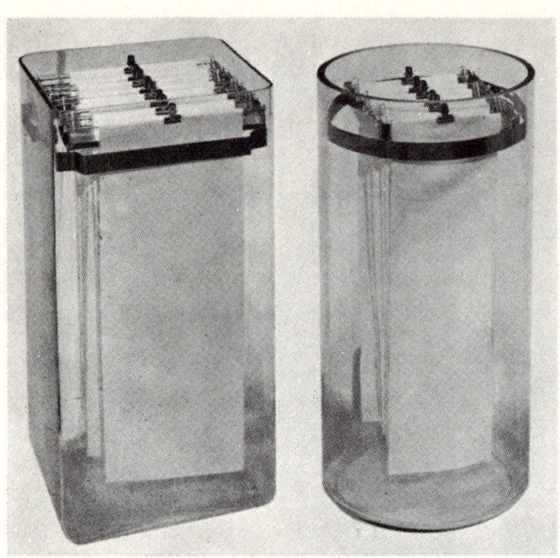

Fig. 23. Descending paper chromatographic
jars (courtesy of Precision Scientific
Company, Chicago, Ill.)

Fig. 24. Ascending paper
chromatographic jar

immediately take a fan and dry the sample with warm air. Repeat
the application, try to distribute the 0.2 ml evenly on the starting line. The streak should never be wider than 10 mm.

   3. Place the shorter end of the paper strip in the dry, empty
tray in the chromatography jar as shown in Fig. 25. Pipette 30 ml
chromatographic solvent into the tray, cover the jar properly.
As the solvent front comes close to the other end of the paper,
remove it carefully from the chamber and dry it for 5 min at 90°C
in a hot air oven.

   4. Cut 2 cm wide strips from both longer edges of the chromatogram. Develop them with alkaline $AgNO_3$ reagent. After this
is done and the strips are dry, lay them back in their original
position along the edges of the unstained paper chromatogram.
Dark brown to black spots on the two narrow indicator strips
will indicate the position of carbohydrates on the entire chromatogram. Now cut out those areas of the unstained strip which
contain carbohydrates according to the indicator strips. Number
them with pencil and elute the carbohydrates as follows.

   5. With a needle and thread, sew additional clean paper strips
to the cut-out portions of the chromatogram, as shown in Fig. 26.
Take another similar chromatography jar as above. Place the
upper edge of the attached paper in the empty well of the jar,
secure its position with a heavy glass rod, and place a test
tube or small Erlenmeyer flask under the pointed end of the
other attached paper. It is feasible to elute all other chromatographic cut-outs in the same jar simultaneously. Fill the trough

Fig. 25. Stepwise illustration of descending paper chromatography

with distilled water. This will slowly descend along the paper, washing out the carbohydrates. The first 10 ml collected from the eluate will contain practically all the carbohydrates.

The elution can be carried out also by cutting the paper into small pieces and soaking them in an acid-washed 50 ml centrifuge tube in 20 ml glass-distilled water. Sonication will greatly facilitate the release of components from the cellulose fibers. Centrifuge the suspension at 3000 g for 30 min and lift the supernate. Repeat the extraction twice more. Pool the extracts and concentrate it by vacuum distillation or freeze drying.

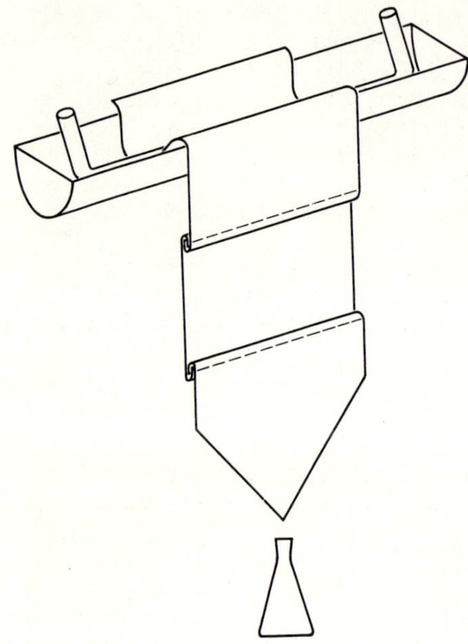

Fig. 26. Elution of isolated components cut out of a chromatogram

Evaluation

The serologic activity of the eluates may be determined as described above, without concentrating the eluate, if the eluted volume is not more than approximately 10 ml.

For quantitative determination of carbohydrates present in the obtained fractions, either the simple phenol-sulfuric procedure (Exercise No. 52) or the much more sensitive ferri-ferro cyanide reaction (Exercise No. 53) may be used.

For structural studies of the obtained oligosaccharides, several methods described in this manual can be used (Exercises Nos. 34, 40, 46 and 61).

Use and Limitations

For preliminary paper chromatographic experiments in test tubes, larger tubes or graduate cylinders, with correspondingly larger paper strips may be used. It is important to have a stopper which will close the container perfectly so that a saturated atmosphere may be achieved.

Naturally, peptides can be separated by the here-described procedure also, but the solvent system applied should be changed.

If those components which contain amino acids or amino sugars
are to be eluted, ninhydrin spray of the narrow indicator strips
is the best detection procedure. Any other sprays which are described for paper chromatography, such as Elson-Morgan reagent
for hexosamines, Hanes-Isherwood reagent for phosphorus, etc.,
can also be used here (see Exercise No. 43).

A serious limitation of this procedure is that the higher the
oligosaccharides, the lower their R_F values. Some of them do not
move from the starting point at all. This means that a clear
separation is very difficult if several oligosaccharides are
present in a partial hydrolysate, which is usually the case. It
is recommended to use the above described repeated chromatography or one of the more demanding column chromatographic processes in such cases (see Heftmann 1963 or Whistler and Wolfrom
1962). Paper electrophoresis sometimes helps in the separation
of oligosaccharides where paper chromatography fails. Often
several steps must be applied. Preliminary purification may be
achieved by paper chromatography or column chromatography, and
paper electrophoresis is recommended for additional fractionation. In any case, before structural studies of the obtained
oligosaccharides can be started, the homogeneity of the fractions
obtained must be proven beyond doubt.

References

Heftmann, E.: Chromatography, New York: Reinhold Publ. Corp. 1963
Whistler, R.L., Wolfrom, M.L.: Methods in carbohydrate chemistry, Vol. 1.
  New York: Academic Press 1962

Exercise No. 37

Separation and Isolation of Peptides or Oligosaccharides by High-Voltage Paper
Electrophoresis

Sometimes the use of high-voltage paper electrophoresis (HPE),
which was developed by Michl (1951), is more advantageous for
the isolation of partially hydrolyzed natural products than chromatographic procedures. In this Exercise the separation of peptides and oligosaccharides will be demonstrated. A similar procedure has been used by Simmons, Lüderitz and Westphal (1965)
in the studies of bacterial O-antigens.

Materials and Equipment

    Partially hydrolyzed preparation (obtained in Exercise No. 34
      or No. 35)
    Buffer, pH 5.2, pyridine:acetic acid:water = 70:30:900
    n-Heptane (practical grade)
    Ninhydrin spray reagent (see Exercise No. 43)
    Alkaline silver nitrate spray reagent (see Exercise No. 43)
    High-voltage electrophoresis equipment, either commercial
      instrument or homemade chamber as shown in Fig. 27

   High-voltage power supply, with 0-5000 V output and 500 mA
    capacity. This is commercially available
  Whatman No. 3 or any similar thick filter paper
  Drying oven

Procedure

   1. Take a filter paper strip (A) long enough to reach the two
ends of the rack (B). The width of the paper and its thickness
will determine the resistance of the system. A 10 cm wide Whatman
No. 3 strip impregnated with the above buffer will let approximately 200 mA through if 2000 V is given to the two electrodes
(C). This requires effective cooling which can be achieved through
glass coils (D) which are immersed in the heptane phase (E).
Good cooling can be achieved by running tap water, but ice water
should be circulated in the coils if a temperature lower than
20°C is desirable. The temperature can be read by the thermometer
(F).

   2. Prepare the chamber. Fill it with buffer (G) over the height
of the top of the dividers (H) but not to cover dividers (I).
These serve to keep the pH changes in the buffer due to electro-
lytic processes away from the two ends of the paper strip.

   Now fill the chamber with practical grade heptane as indicated
by the upper broken line in Fig. 27. Cover the chamber. The cooling coils should be immersed in the heptane. Start the circulation of cold water in the coils.

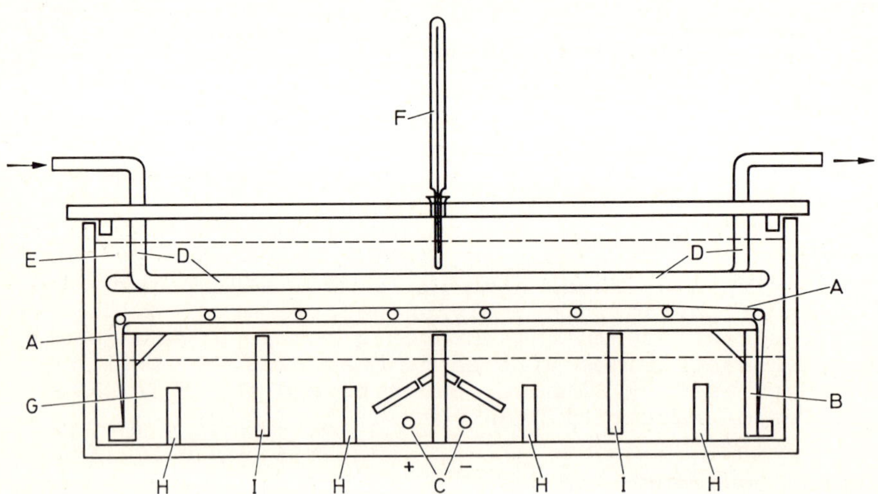

Fig. 27. High-voltage paper electrophoresis chamber constructed in the
laboratory

   3. Stretch a dry paper (A) on the clean and dry rack (D). Draw
a pencil line across the center of the strip; this will be the
starting line. Also mark the positive end of the paper. Apply
0.2 ml hydrolysate on the center line.

4. Draw buffer up into a large pipette, and, by touching the paper with its smooth tip, moisten the paper starting from the two ends and moving the pipette evenly across the paper strip. After the two ends are wet, apply buffer closer to the center line with the pipette. Work alternately on both ends so that the two buffer fronts will meet exactly at the point where the samples have been applied. This may sound difficult, but it can be accomplished with some practice. Do not make the paper dripping wet, just moisten it and let the buffer be absorbed by the paper. Do not come closer than 2-3 cm from the center line with your pipette. Let the buffer fronts meet at the center line. Instead of a pipette, a clean fine paint brush may also be used for moistening the paper strip.

5. When the paper on the rack is evenly moist, the strip is properly prepared. Now place the rack in the chamber without delay, submerging the paper in the heptane phase, while the two ends of the paper strip attached to the rack are immersed in the buffer. Cover the chamber and connect it to the power supply. Apply 2000 V to the two electrodes. Approximately 200-300 mA current will go through the system if everything has been well prepared and the paper strip connects the two electrode chambers. Let the electrophoresis proceed for 60 min. Turn off the power supply, wait a few seconds, then disconnect the electrophoretic chamber. Lift out the rack, let it dry at room temperature, or remove the wet paper and dry it in a hot-air oven at 90°C for 5 min.

6. Cut off margins and develop them with ninhydrin for peptides or with $AgNO_3$ for carbohydrates. After the position of these constituents is found in the marginal indicator strip, cut the unstained paper and elute the zones as described for preparative paper chromatography (Exercise No. 36).

Evaluation

For determination of serologic reactivity, the same assays can be used as in Exercises No. 69 or 73 or 78.

For the determination of constituents in the split products, total hydrolysis may be applied followed by various qualitative and quantitative analytic methods (see Exercises Nos. 36, 40, 41, 51, 53).

Use and Limitations

High-voltage paper electrophoresis has many advantages when compared with paper chromatographic procedures (Michl, 1958). First, up to 1-2% salts present in the hydrolysis mixture do not disturb the separation. They will be removed from the starting line by the electric field, their migration usually being at a higher rate than the migration of other components which will be separated. Another great advantage is that from the electrophoretic mobility we can draw conclusions regarding the net electric charge of the separated components, which is often very helpful in structural studies. The rapid separation achieved by HPE gives a sharp differentiation between the components. A detailed study of the analysis of amino acids by paper electrophoresis has been published by Blackburn (1965).

A disadvantage is that HPE separates only those constituents
which have a net electric charge. Electrically neutral components
such as the large majority of aldohexoses, oligosaccharides, or
neutral amino acids, will not migrate from the starting line.
For their separation, paper chromatography is more suitable.
Neutral carbohydrates may also be separated in an electric field
in the form of their borate complexes.

Another disadvantage of HPE is that a more expensive apparatus
and somewhat more complicated procedure must be used. One must
also be aware of the fire hazard involved in the use of relatively large volumes of heptane, and similarly of the danger of
high voltage. Students may not carry out this experiment without
close supervision by instructors.

References

Blackburn, S.: Meth. Biochem. Anal. *13*, 1 (1965)
Michl, H.: Monatsh. Chem. *82*, 489 (1951)
Michl, H.: J. Chromat. *1*, 93 (1958)
Simmons, D.A.R., Lüderitz, O., Westphal, O.: Biochem. J. *97*, 807 (1965)

Exercise No. 38

Demonstration of Ion Exchange Column Chromatography of Amino Acids

Ion exchange may take place between the ionic components of a
solvent and the counter ions of a solid phase. Such materials
occur in nature, but synthetic ion exchange resins with far
greater capacity have been developed and made commercially available. Cross-linked high polymers, having $-OH$, $-SO_3H$, $-COOH$,
$-N = (CH_3)_2$, etc. groups are able to bind cations. One of the
best known groups of cation exchangers comprises the sulfonated
polystyrene derivatives. Other polymers having amino groups in
their structure will bind anions. These resins, coming into contact with a solution containing another cation or anion, may
release their loosely held counter ions and pick up another
from the solution. For example, Dowex 50 cationic exchanger
having bound H^+ ions will release hydrogen and pick up sodium
if it makes contact with sodium chloride solution. The released
hydrogen forms hydrochloric acid with the chlorine anions. If
such a HCl-containing filtrate is mixed with an anion exchange
resin, such as Dowex 1 in OH^- form, the chlorine will be ex-
changed with OH^- and the filtrate of this mixture is salt-free
neutral water.

The strength of the binding forces depends upon several factors. The law of mass action, adsorption and distribution coeffi-
cients play the most important role in this phenomenon. If a mixture of different compounds such as nucleotide bases or amino aci
comes into contact with an ion exchanger, their binding to the
resin will be different and characteristic of the individual compound. By slowly washing an ion exchange column with distilled
water, those components which were hardly or not at all bound to
the resin will leave the bottom of the column. To elute the stron

bound components from the resin, different electrolyte concentrations must be used.

In this Exercise an artificial mixture of three amino acids will be separated by ion exchange chromatography in order to acquaint the student with the column chromatography procedure and to demonstrate separation by it.

Materials and Equipment

Aspartic acid, valine, and leucine
1.5 *N* and 4 *N* Hydrochloric acid
4 *N* Sodium hydroxide
Concentrated ammonium hydroxide
Ninhydrin solution (see Exercise No. 43)
Dowex 50, 200 mesh resin
Chromatographic column, size 1 cm × 25 cm (Fig. 2)
Separatory funnel, 500 ml capacity
Büchner funnel, 10 cm diam
Filter flask, 1000 ml capacity
Automatic fraction collector with drop counting device (Fig. 3)
Regular test tubes
Bacteriologic loop, approximately 2-3 mm in diam
One 20 × 20 cm sheet of Schleicher and Schuell No. 2043, or
 any other good quality filter paper
Tray with glass cover
Hair dryer
Chromatographic drying oven

Procedure

1. Activation of Dowex 50 resin is usually done with a strong acid. In order to remove any impurities and to obtain maximal capacity of the resin, it is recommended to transfer the resin from hydrogen form to sodium form and back again to hydrogen form.

To approximately 100 g resin, add 200 ml 4 *N* HCl. The resin must be agitated at room temperature for 5 min, filtered through a Büchner funnel, washed with approximately 500 ml distilled water. Suction should be applied after filtration has been completed. Transfer the resin to a beaker and mix with approximately 200 ml 4 *N* NaOH for 5 min, and filter again through a Büchner funnel. Wash with 500 ml distilled water as before. Allow the resin to stand in 200 ml 4 *N* HCl for 30 min, filter again as above, and wash with distilled water until the pH of the filtrate is identical with the pH of the distilled water used for washing. This resin is activated and ready for use.

2. Make a thick slurry of the resin in a beaker, using approximately 100 ml 1.5 *N* HCl. Mount the chromatographic column on a stand above a sink. Pour the resin slurry into the column along a glass rod, trying to avoid air bubble formation in the column. The resin will sediment in the column; supernatant HCl may be removed with a pipette to speed the preparation of the column. Add further slurry until the sediment filling the column reaches approximately 20 cm in height.

3. Fix the column to the stand of the automatic fraction collector. The outlet of the column must be connected with the
volumetric siphon or with the drop counter. By driving 1.5 N HCl
through the column, check and adjust the functioning of the
siphon or drop counter and the automatic turntable. For details
of this operation, the instruction manual of the automatic fraction collector should be read carefully. In this Exercise, approximately 1 ml samples will be collected in properly marked,
clean test tubes which are placed in the automatic turntable.

4. The column should be drained until the fluid surface reaches the level of the resin. Discard the HCl collected up to
this point. Carefully layer 0.5 ml of the amino acid mixture
dissolved in 1.5 N HCl onto the top of the column. This solution
contains 2.5% aspartic acid, 2.5% valine and 2.5% leucine. Drain
the column until the 0.5 ml mixture enters the resin bed. Immediately after that, carefully add approximately 2 ml activated
resin slurry in 1.5 N HCl to the column.

5. Place a separatory funnel above the column, and with a
flow rate of approximately 40-50 drops per minute, drive 50 ml
1.5 N HCl through the system, collecting approximately 1 ml samples
in each test tube of the turntable.

Evaluation

The three components leave the column separately and will be
distributed in three groups of test tubes. In order to find these
tubes, their contents must be analyzed. For quantitative studies,
samples are taken from each tube and the amino acid content is
measured by a proper quantitative analytic chemical procedure.
However, the separated components can be detected by means of
a rapid spot test. Divide an approximately 20 cm × 20 cm sheet
of filter paper into squares, 2 cm × 2 cm in size. Place a loop
of solution from every tube onto the center of these squares and
write the number of the tube on the square with a pencil. The
loop must be rinsed and flamed after each application. The sam-
pling from each test tube can be done during the chromatography
immediately after collection.

When the spotting is completed, dry the sheet at room temperature. A hair dryer will facilitate the drying. Place the sheet
over a tray containing concentrated ammonium hydroxide, and replace the glass cover. The paper must not touch the solution,
but the NH_3 gas will neutralize the acid in the spots. After
5 min, remove the paper and hang it in a fume hood for another
5 min to remove the excess ammonia. Spray the paper evenly under
a fume hood with 0.3% ninhydrin solution, avoiding excess use of
the spray. The paper should appear translucent but should not be
saturated with the reagent. Transfer the sheet to a chromatography drying oven which has been previously adjusted to 90°C.
The main amino acid spots will become visible on the sheet within
5 to 10 min. Do not overheat. Examination of the paper will
reveal those test tubes which contain amino acids.

For complete separation, a certain length of column is necessary. If the column is not long enough, the components will over-
lap. To check the homogeneity of the fractions, a one-dimensional
paper chromatogram may be made from those tubes which gave positiv
spot tests.

Use and Limitations

Three amino acids have been selected as models to demonstrate ion
exchange column chromatography of amino acids. These three are
easily separable on a short column such as the one used here.
Many amino acids occurring in natural products are much more
difficult to separate. Basic amino acids, for example, require
a different chromatographic system if a clear separation is
desired.

The separation of all amino acids present in the total hydrolysate of a protein require a well-standardized column chromatographic setup. In these systems, first of all, much longer columns filled with different ion exchangers are applied. The temperature of the column is constant, the column being supplied
with a water jacket in which thermostatically controlled water
is circulated. The flow rate of the elution is one of the most
important factors which should be standardized if reproducible
experiments are to be obtained. Obviously, the concentration
and pH of the eluants are equally important. To avoid the occur-
rence of artifact peaks, continuous gradient changes in the pH
or concentration of the eluant have to be applied whenever possible. To obtain this, several devices have been constructed,
many of them commercially available. Fig. 28 shows a "Varigrad"
multichamber; Fig. 4 shows a homemade device which gives a linear
gradient between two different concentrations.

For reproducible and comparative analyses of amino acids on
a certain ion exchanger column, the following three conditions
have to be kept strictly standardized: (1) temperature, (2) flow
rate, and (3) reproducible gradient changes of the concentration

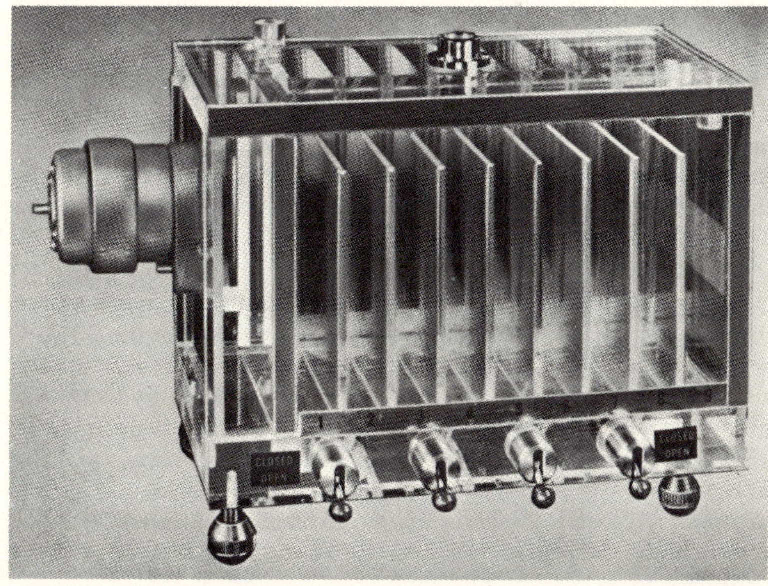

<u>Fig. 28.</u> "Varigrad" for gradient elution of chromatographic columns

of the eluant. Several automatic amino acid analyzers are manfactured, all based on the above principles.

Research efforts on improvement of amino-acid analyzer systems lead to one of the latest modifications of chromatographic methods which is the high-pressure liquid chromatography (HPLC). The HPLC achieves the speed and reproducibility of the gas-liquid chromatographic procedures. Chromatographic time is reduced to minutes due to the applied high pressure, while excellent resolution is obtained. Advantage of the system is its applicability to non-volatile solutes. Disadvantage of HPLC versus GLC is in the detection systems, since the best detectors for HPLC are UV flow cells, which somewhat limits its applicability. HPLC is extensively used in food, drug and cosmetics industries. The adsorbents are ion exchangers or a variety of synthetic and semisynthetic substances. Several reviews were written on this method, such as by Brown (1973), or by Rajcsanyi and Rajcsanyi (1975).

The references listed here are only a few from the most relevant literature which can be consulted for further details.

References

Brown, P.R.: High pressure liquid chromatography. Biochemical and biomedical
    application, London-New York: Academic Press, 1973
Calmon, C., Kressman, T.R.E.: Ion exchangers in organic and biochemistry.
    New York: Interscience 1957
Heftmann, E.: Chromatography, New York: Reinhold Publ. Corp. 1963
Lederer, E., Lederer, M.: Chromatography. Amsterdam: Elsevier 1957
Moore, S., Stein, W.H.: J. Biol. Chem. *211*, 907 (1954)
Rajcsanyi, P.M., Rajcsanyi, E.: In Chromatographic science series vol, VI,
    New York: Dekker, 1975

Exercise No. 39

Gas-Liquid Chromatography for Fatty Acid Separation

In the gas-liquid chromatographic system (GLC) the mobile phase is a carrier gas and the stable phase is a liquid, which impregnates an inert porous material.

Volatile components can be analyzed at low temperature; some liquids and solids only at elevated temperatures. The material being analyzed is injected into a column which is filled with the porous material impregnated with the stable phase. The temperature is carefully controlled, as is the flow rate of the carrier gas. The gas passes through the column and carries the components of the injected mixture. The adsorbent in the column will hold back those components which are easily soluble in the stable liquid phase; less soluble components will be somewhat retarded, and insoluble material will leave the column without any delay. The sequence of the components leaving the column is determined by their partition coefficient between the stable and mobile phases. GLC is especially useful for the separation of those compounds which are volatile at room or at elevated temperatures, such as carboxylic acid esters. For details, stu-

dents are referred to the textbooks of Pecsok (1959) and Burchfield and Storrs (1962).

In this Exercise, students will get acquainted with the method
and with the instrument by separating some esters. It is essential that students read beforehand the manufacturer's instruction
manual for the instrument they use.

Materials and Equipment

Mixture of acetic, butyric, isobutyric and hexanoic acid
  methyl esters. Dissolve 0.1 ml of each in a total of 10 ml
  chromatoquality hexane. Preparation of fatty acid methyl
  esters is described in Exercises No. 43 and No. 64
Chromatoquality hexane
Hamilton lambda syringe, 10 μl capacity
Conical centrifuge tube, 5 ml capacity
Centrifuge
Gas chromatograph equipped with flame ionization detector,
  temperature programmer, and an automatic recorder
Chromatographic column for fatty acid methyl esters. A 5 mm
  diameter and 200-250 cm long coiled column is recommended.
  One of the best adsorbents is 15% polyethylene glycol suc-
  cinate on acid-washed 100 mesh Chromosorb-W. Such columns
  are commercially available

Procedure

1. The gas chromatograph must be stabilized before the Exercise. This cannot be achieved in a few minutes or hours; it
usually requires a day or more. Some of the parts of the GLC are
never turned off unless it is not to be used for several weeks.
Therefore, the chromatograph used in this Exercise has to be set
up, stabilized, and adjusted to the proper sensitivity in ad-
vance, by the instructor. The setting and the controls of the
instrument will be explained to the students.

2. First, injection of the ester sample (5 μl) will be performed at 75°C. Record the column temperature, the flow rate of
gases, the injection port temperature, and the sensitivity of
the electrometer.

3. Because of the small retention volume of these methyl esters,
in approximately 20 min a new injection can be made. Raise the
temperature of the column to 100°C. Introduce the same amount
into the column and record the settings again.

4. After the previous exercise has been completed, raise the
temperature to 125°C. Introduce the next injection, record the
new settings and observations concerning the changes in retention
volume.

5. Now cool the column to 75°C; wait until the temperature is
stabilized. Increase the flow rate of nitrogen carrier gas by
50%. Inject the same amount of methyl esters as before. Record
settings and measure the retention time.

6. Compare the effect of temperature on the retention volume
and compare the effect of flow rate in step No. 2 and No. 5 on
the retention time.

Evaluation

For the identification of the observed peaks, their retention
time gives useful data. This is the time elapsed between the injection and the appearance of the highest point of the peaks. It
must be mentioned that retention time values may be used for
identification only if the flow rate of the gases, the temperature of the column, and the rate of temperature programming are
well established and reproducible. The use of retention volume
at a certain temperature gives more reliable information regarding the identity of the observed peaks. The retention volume is
the amount of gas which went through the column between the time
of injection and the appearance of the component in the chromatogram.

Obviously, authentic samples made from chromatographically
homogeneous materials will give the most reliable clue to the
identification. These carboxylic acid or carbohydrate samples
have to be treated in a manner identical to that described above.
It is strongly advised to include so-called internal standards
in the unknown mixtures. In the selection of these, one has to
consider those substances which hardly ever occur in similar
products. Inositol is suitable for bacterial lipopolysaccharides.
Aliphatic hydrocarbons may be used in the analysis of unknown
carboxylic acid methyl esters. Authentic samples, together with
the internal standards, must be analyzed individually on the
same column, under conditions identical to those used for the
analysis of the unknown samples. (See also Wehrli and Kovats,
1959).

By the use of authentic carboxylic acid derivatives, the
students have to collect data about the retention time or volume
of the peaks of the different compounds. It is necessary to
repeat the run of the authentic samples at least five times,
under most carefully controlled conditions, in order to measure
the relative standard deviations.

If, in an unknown sample, the position of the peaks does not
allow an unquestionable identification of the components, the
authentic sample which is closest to the observed pattern must
be rechromatographed with the unknown. If the height of the
questionable peaks increases but their shape does not change
(such as formation of asymmetric Gauss curves or "shoulders"),
the results can be accepted as proof of identity. A similar
procedure is widely used for identification of components in
column or paper chromatographic analysis as well as in electrophoretic investigations.

For the calculation of the relative amounts of different components in the hydrolysate, the following procedure may be used.
If the exact amount of added inositol in the unknown sample is
known, fairly reliable information can be obtained with regard
to the amount of other constituents in the unknown sample. To
achieve this, the areas under the peaks must be compared.

Use and Limitations

Gas-liquid chromatography is as widely used today as any other
chromatographic procedure. Further developments of the method
have made possible the preparative isolation of fractions suit-

able for spectroscopic analyses. Capillary columns give a superior resolution of fatty acid methylesters. Inside diameter is
usually 0.25 mm while the length varies between 25 and 100 m.
This latter length can have a column efficiency expressed as
theoretical plate number >300,000. The inside of the capillary
wall is coated with the liquid stationary phase. One of the
best capillary columns on the market is produced by Applied
Science Laboratories, Inc., State College, PA 16801.

The series *Chromatographic Science* is an outstanding collection
of diverse articles dealing mostly with theoretical aspects of
chromatography but several chapters are also quite useful for
practical application. Vol. 5 of this series by Novak (1975)
discusses in detail quantitative analyses by GLC. Vol. 7 by
Gudzinowicz et al. (1976) deals with GLC, Mass Spectrometry and
the integration of these two powerful procedures of structural
studies.

References

Burchfield, H.P., Storrs, E.E.: Biochemical applications of gas chromato-
   graphy. New York: Academic Press 1962
Gudzinowicz, B.J., Gudzinowicz, M.J., Martin, H.F.: In: Chromatographic
   science series, vol. VII Parts 1-3. New York: Dekker, 1975
Novak, J.: In: Chromatographic science series, vol. V. New York: Dekker, 1975
Pecsok, R.L.: Principles and practice of gas chromatography. New York:
   Wiley & Sons 1959
Wehrli, A., Kovats, E.: Helv. Chim. Acta *42*, 2709 (1959)

Exercise No. 40

Carbohydrate Analysis by Gas-Liquid Chromatography

Monosaccharides may be rendered volatile by various procedures.
Analysis of methylated monosaccharides was described first by
McInnes and associates (1958). This procedure was applied to a
number of fully methylated aldohexoses by Bishop and Cooper (1960),
and Kircher (1960). Trimethylsilylation was developed by Sweeley
et al. (1963) and was used extensively for the study of the
carbohydrate composition of various natural products. This procedure clearly distinguishes between α and β anomers of carbohydrates, which became one of its disadvantages. In studying the
carbohydrate components of heteropolysaccharides, frequently so
many peaks were found in a gas chromatography pattern that it
became difficult to identify the individual peaks.

The most often used and probably the most satisfactory procedure is the conversion of monosaccharides into their alditol
acetates. The procedure was developed by Sawardeker et al. (1965).
Several modifications of this procedure were published.

Oligo and polysaccharides must be hydrolyzed first in order
to liberate the monosaccharides. This must be carried out under
optimal hydrolytic conditions, where maximal liberation of the
monosaccharides is achieved with minimal degradation. Exercise
No. 33 in this Manual gives the procedure for finding such op-

timal hydrolytic conditions. The next step in the procedure is
neutralization of the hydrolyzed and liberated monosaccharide
mixture, and conversion of their C-1 carbon atoms to a primary
alcohol using $NaBH_4$. The method described in this Exercise is
based on the description by Kannan et al. (1974), as elaborated
in author's laboratory by Dr. Dezsö Gaál in 1976.
  In this Exercise commercial carbohydrate samples and two
hydrolyzed lipopolysaccharide aliquots will be used, with inositol
added to the hydrolyzed mixtures as an internal standard.

Materials and Equipment

   Gas chromatograph (any model with programmable temperature
    can be used)
   Of the various columns suitable for carbohydrate analysis,
    the 3% OV-225 80/100 Supelcoport column, 6 ft. long and 1/8"
    diameter, was found to be the most suitable since this column
    can be heated up to 250°C without noticeable disintegration
    of the stationary phase. OV-225 is a silicon polymer which
    contains methylphenyl and cyanopropyl functional groups.
    This column was developed by Lonngren and Pilotti (1971) and
    is available from Applied Sci. Labs. Inc., State College,
    PA 16801
   Vacuum distillation equipment. (Büchi type shown in Fig. 12
    or any other models are suitable)
   Endotoxic lipopolysaccharide extracted from *Serratia marcescens* O8
    and from *Serratia marcescens* Bizio ATCC *174. These LPS prepara-
    tions can be obtained by trichloroacetic acid extraction or
    by other procedures described for their isolation in Exercise
    No. 21 of this Manual
   1 *N* H_2SO_4 (or 1 *N* HCl)
   0.8 *N* $NaBH_4$
   Acetic acid (glacial)
   Methanol
   Phosphorus pentoxide
   Acetic anhydride (freshly distilled before use)
   Toluene
   Chloroform
   Acetone
   Amberlite IR120 resin in OH form (or other anion exchanger,
    freshly activated and washed)
   Büchner funnel (6-8 cm diam) with suction flask and Whatman
    No. 1 filter paper
   Clinical tabletop centrifuge
   18 × 150 mm test tubes provided with a 12/30 ‡ and glass stop-
    pers
   Oil bath adjusted to 100°C (±1°)
   Round flask with 24/40 ‡, 125 ml capacity
   Relux condenser with 12/20 ‡ joints
   Magnetic stirrer and bar
   pH meter
   Vacuum desiccator containing fresh P_2O_5
   The following carbohydrates will be used as standards: D-
    mannose, D-glucose, D-glucoheptose and D-glucosamine. Obtain
    the purest available preparations.

<u>Procedure</u>

1. Weigh 10 mg LPS into a 18 × 150 ℔ 12/30 test tube and add
1 ml 1 *N* H_2SO_4 (or 1 *N* HCl). Connect the tube to the reflux condenser with ℔ 12/30 joint and immerse it in the oil bath at 100°C.
The hydrolysis of LPS samples is virtually complete after 8 h.
Cool the tube and transfer the contents into a 100 ml beaker.
Rinse the tube three times with 5 ml distilled water to complete
the transfer. Place a small magnetic bar in the beaker and place
the beaker on a magnetic stirrer.

2. Carefully immerse the electrode of a pH meter in the liquid
and start adding to it slowly anion exchange resin in OH cycle,
until the pH reaches approximately 6. For 1 ml *N* H_2SO_4 approximately 0.5-0.7 g resin will be required. Filter the neutralized
mixture on Whatman No. 1 paper into a filter flask through a
Büchner funnel. Wash the resin three times with approximately
10 ml distilled water to remove hydrolysate remaining in the
resin.

3. Evaporate the filtrate and the combined washes to complete
dryness using a Büchi type rotating vacuum flash evaporator, in
a 125 ml ℔ 24/40 round flask. Place the dried flask in a vacuum
desiccator over P_2O_5, evacuate it with a vacuum pump and let it
stand overnight.

4. Carry out the reduction of the liberated monosaccharides
by dissolving the dry hydrolysate in 2.2 ml distilled water and
mix it with 2.2 ml 0.8 *M* NaBH, using the same round flask in which
the hydrolysate was dried. Put a small magnetic bar in the flask
and stir it for 2 h at room temperature. At the end of 2 h, stop
the reduction by the addition of 0.5 ml concentrated acetic acid.
Check the pH, which at that point should be approximately 3.5.

5. Remove the stirring bar from the mixture and evaporate the
solution using a Büchi evaporator. Use a water aspirator to
produce the vacuum for this evaporation, since motor driven vacuum
pumps will be badly corroded by the acetic acid vapors. After
the reaction mixture is dry, add 20 ml methanol-acetic acid =
200:1 mixture and dry it again. Repeat the addition of 20 ml
methanol acetic acid and vacuum distillation to dryness three
more times. Place the flask in a vacuum desiccator and keep it
over P_2O_5 overnight.

6. The total acetylation of the reduced carbohydrates is
done by adding 10 ml acetic acid anyhdride (freshly distilled
before use) to the completely dried sample. Attach the flask to
a reflux condenser provided with 24/40 ℔ joint and keep the
reaction mixture at 100°C (oil bath) for 2 h. After this, cool
the reaction mixture and remove the excess acetic anhydride under
vacuum (water aspirator), again using the Büchi vacuum distillation apparatus. To facilitate the removal of acetic acid anhydride,
3 × 5 ml toluene (freshly distilled before use) should be added
to the mixture of alditol acetates.

7. Add 5 ml chloroform to the round flask and scrape off the
crusty deposit on the inside of the flask with a bent spatula.
The chloroform will extract the alditol acetates from this mixture. To facilitate this, immerse the flask in warm water and
shake it gently for a few minutes. Transfer the chloroform with
a Pasteur pipette into a 50 ml centrifuge tube. Repeat the
chloroform extraction twice more. Combine the chloroform layers

and mix with an equal volume of distilled water. Stir vigorously
with a glass rod or, better, shake it if the centrifuge tube is
provided with a glass stopper. Centrifuge at 3000 g for 15 min.
Lift the water phase and discard it. Transfer the chloroform
phase into a 125 ml round flask and evaporate it, using the
Büchi evaporator, to complete dryness.

8. Add 0.5 ml pure acetone to the dried alditol acetates.
Facilitate the solution of the dried residue by warming it in a
water bath for a few seconds. The sample is now ready for gas-
liquid chromatographic analysis. It may be transferred to a 1 ml
test tube for storage, tightly closed with a glass stopper. To
prevent possible evaporation, samples should be kept in a -20°C
freezer.

9. Carry out the chromatographic analysis by injecting 10 µl
of the acetone-dissolved alditol acetate sample into the gas
chromatograph at 100°C. Program the instrument at a rate of 4°
or 5°C per minute to a maximal temperature of 250°C and keep it
constant at that point for approximately 30 min longer. Adjust
the carrier gas flow rate to the same value before every experiment at the 100°C starting temperature. This can be achieved
relatively easily. The carrier gas flow rate will slow down at
elevated temperatures. This is beyond the control of most GLC
instruments, therefore it is not possible to compensate for it.
This slowdown, however, will be quite reproducible provided the
starting flow rate was the same and the rate of programming of
the temperature is the same between each experiment.

Evaluation

For the identification of unknown carbohydrates in a hydrolysate
as described here, the relative retention time (RRT) values of
standard alditol acetate samples can be used. These values can
be obtained by preparing such preparations as described here
using pure monosaccharides. To determine their RRT values,
inositol hexaacetate is used as an internal standard. This can
be prepared using inositol and acetylating it as described in
this Exercise. The RRT for inositol is taken arbitrarily as 1.00.
The 0 time is established by the appearance of the solvent peak
(in our case, acetone) which appears a few seconds after injection. The carbohydrates which have an RRT shorter than inositol
will have a value smaller than 1.00, those which leave the column
after inositol will be greater than 1.00. For reproducible results
each laboratory should establish its RRT values using its instrumentation. Only experiments carried out under entirely identical conditions can be compared, and RRT values can be used to
identify compounds of an unknown mixture only if the unknown
sample has been treated and chromatographed under conditions identical to those used for the standard samples. Here again, as in
every chromatographic procedure, co-chromatography of known alditol
acetates with the unknown mixture provides the best way to identify individual components in the GLC pattern. The RRT values can
indicate only what the individual peaks may be. A few micrograms
of authentic alditol acetate of the suspected carbohydrate should
be added to the unknown mixture and the chromatography repeated.
If this addition shows an increase of the height of the questioned
peak, the assumption was correct. If the addition appears somewhere

else or results in "shoulder" formation of the peak in question,
the assumption was incorrect.

Use and Limitations

While the development of the GLC procedure for alditol acetates
of hexoses and pentoses was relatively easy, it was more difficult
to establish the method for aminosugar separations. The RRT for
aminosugars is very long at 200°C temperature. On the other hand,
higher temperatures led to constant "bleeding" of the columns
and to their rapid deterioration. The new column which was developed for this purpose is the OV-225 used in this experiment.
At 250°C final temperature limit, the aminosugars will leave the
column shortly after the inositol internal standard.
    The RRT values of standard compounds must be determined from
time to time, since the characteristics of any GLC column will
change with use. This is due to constant slow deterioration of
the stationary phase of the column at elevated temperatures. It
is recommended that the standard mixture of the alditol acetates
be injected first on the day when the analysis is carried out,
the RRT values of the known components tabulated and this table
used to attempt to identify components is unknown hydrolysates.

References

Bishop, C.T., Cooper, F.P.: Can. J. Chem. *38*, 793 (1960)
Kannan, R., Seng, P.N., Debuch, H.: J. Chrom. *92*, 95 (1974)
Kircher, H.W.: Anal. Chem. *32*, 1103 (1960)
Lonngren, J., Pilotti, A.: Acta. Chem. Scand. *25*, 1144 (1971)
McInnes, A.G., Ball, D.H., Cooper, F.P., Bishop, C.T.: J. Chrom *1*, 556 (1958)
Sawardeker, J.S., Sloneker, J.H., Jeanes, A.: Anal. Chem. *37*, 1602 (1965)
Sweeley, C.C., Bentley, R., Makita, M., Wells, W.W.: J. Am. Chem. Soc. *85*,
    2497 (1963)

Exercise No. 41

Gas-Liquid Chromatographic Analysis of Amino Acids

Automatic amino acid analyzers separate the components on columns
filled with ion exchange resins. The equipment is quite expensive,
the operation is cumbersome, and requires a well trained operator.
A somewhat less quantitative and less accurate but far simpler
and much less expensive procedure for qualitative and quantitative
determination of amino acids has been provided by the method developed first by Graff et al. (1963) and further elaborated by
Coulter and Hann (1968). The procedure converts amino acids into
their n-propyl-N-acetylated esters. These derivatives are sufficiently volatile for separation in GLC. The method described in
this Exercise is based on the work of Coulter and Hann.

Materials and Equipment

Hydrolysate of 1 mg Concanavalin A. The hydrolysis is carried
 out in 5 N HCl as described in Exercise No. 43
L-amino acid standards (chromatographically homogeneous purity
 is required)
Propanol-HCl (See Procedure)
Pyridine
Acetic anhydride
Ethyl acetate
N_2 gas
16 × 160 test tubes with 12/40 ℲSℲ joint
Reflux condenser with 12/40 ℲSℲ joint
Oil bath, adjusted to 100°C
GLC equipped with Chromosorb column for amino acid derivatives
 (available from Accurate Chemical & Sci Corp., Hicksville,
 NY 11801)

Procedure

1. Propanol-HCl can be prepared in the laboratory. First,
reflux 500 ml reagent grade propanol with 2 to 3 g calcium hydride
for 1 h on a 100°C oil bath. Distil it into a dry container.
The major fraction distilling between 96 and 99°C should be
collected and used for this Exercise, the other fractions should
be discarded. Dry HCl gas can be generated in the laboratory by
adding concentrated sulfuric acid dropwise to NaCl. The generated
gas has to be bubbled through two gas washers filled with concentrated H_2SO_4 to remove traces of water, and led into the redistilled propanol. Saturation of propanol with HCl gas should
be continued until the HCl concentration in the propanol is approximately 8 N. This can be determined by taking aliquots from
the propanol, diluting them with water and titrating with normal
NaOH.
   Instructors: It is recommended that inexperienced students
should not carry out this step. Propanol-HCl should be prepared
and provided for this exercise by the teaching staff.
2. Prepare the completely dry and acid-free hydrolysate of 1 mg
Concanavalin A according to Exercise No. 43. This dry and acid-
free residue must now be dissolved in 1 ml glass distilled water
and transferred from the hydrolysis tube into a 16 × 160 mm test
tube provided with a 12/40 ℲSℲ joint. From a nitrogen tank blow
gas in a gentle stream over the surface of the hydrolysate through
a pipette, while heating the outside of the test tube with a hair
dryer from a distance of approximately 20 cm. While the drying
goes on, rinse the hydrolysis tube with another 1 ml distilled
water and add this also to the test tube, after the first milliliter has dried. Dry it again and continue the drying and heating
for 15 more min. Repeat this step with another 1 ml water. If you
cannot continue the exercise now, immediately put the tube with
the dried samples in a vacuum desiccator and keep it over P_2O_5
under a good vacuum overnight.
3. In this step the esterification with n-propanol will be
carried out. Add 0.5 ml propanol-HCl to the dry residue, attach
the tube to a dry reflux condenser and immerse it in an oil bath
adjusted to 100°C (±2°C) for 10 min. Disconnect the tube and dry

it down with nitrogen gas. Do not heat it this time. Again add
0.5 ml propanol-HCl, reflux it for 10 min and dry it down with
nitrogen gas.

   4. The N-acetylation also requires completely dry substances.
Add 0.5 ml freshly prepared pyridine:acetic anhydride = 4:1 mixture to the residue. Attach the tube to the reflux condenser
again and heat it up in the oil bath for approximately 2 min.
Let it cool for 10 min. Now chase away the pyridine and the
acetic anhydride excess with nitrogen gas. Use the hair dryer
to warm the tube as in step 2. This time the residue will be
oily, since the n-propyl-N-acetyl esters of most amino acids are
liquids. It is sufficiently dry if no more odor of pyridine and
acetic acid can be detected.

   5. Dissolve the residue in 0.5 ml pure ethyl acetate. First
inject 5 μl of this into the GLC. If this quantity is not sufficient, concentrate 0.2 ml of the ester solution in N_2 gas stream,
redissolve it in 0.02 ml and inject 5 μl of this. If the concentration in the first sample was too high, lower the sensitivity
of the GLC.

   The chromatographic conditions will require programmed temperature starting from 100°C at the time of injection and level-
ing off at 240°C. It is recommended that the temperature rise be
6°/min. The injection port should be heated at 240°C.

   6. 100-1000 μg samples of amino acids can be treated in an
identical fashion to obtain their n-propyl-N-acetyl ester standards. If a mixture of various amino acids is required, 100 μg
of each pure amino acid can be mixed in the test tube described
in step 2 and converted together into the volatile esters.

Evaluation

As an internal standard, several laboratories include methyl
stearate in the standard mixture of amino acid esters as well
as in the unknown hydrolyzed sample. The position of amino acid
esters relative to the internal standard greatly facilitates the
identification of amino acid esters in an unknown mixture.

   Table 1 in this Exercise, which gives retention times for the
amino acid esters, listing them in order of their appearance in
the detector system, has been taken from the publication of
Coulter and Hann (1968). The same table also indicates that the
response of the individual amino acids to this method of detection varies greatly. Quantitative evaluation with this procedure
or with any other modifications of similar systems for amino acid
GLC analysis will give quantitative values only if certain correction factors are applied.

Use and Limitations

The major limitations of this very attractive method lie in the
low recovery of some amino acids. The reasons for this are twofold. The first, as was already discussed in Exercise No. 33, is
the degradation of amino acids during their liberation from their
peptide linkage by strong acidic hydrolysis. This can be reduced
by applying various hydrolytic procedures to liberate them, but
this source of error can never be totally eliminated. The same
applies to samples analyzed by ion exchange column chromato-

Table 1. Retention times of n-propyl N-acetyl amino acids

Amino acid	Retention time (min:s)
Alanine	4:48
Valine	5:06
Isoleucine	6:12
Leucine	6:24
Glycine	6:51
Proline	8:48
Threonine	9:36
Serine	11:00
Aspartic acid	12:06
Histidine (as aspartic acid)	12:06
Methionine	13:00
Cysteine	13:48
Phenylalanine	14:12
Hydroxyproline	14:24
Glutamic acid	14:36
Tyrosine	20:12
Ornithine	21:06
Lysine	22:00
Methionine sulphone	23:00
Tryptophan	28:45

graphy in automatic amino acid analyzers. The second reason for
poor recovery of some amino acids is the incomplete conversion
of the liberated amino acids into their volatile esters. As a
consequence of these two factors, the quantitative measurement
of the amino acids in certain proteins will give erratic results,
since the analytic data of some acid-sensitive amino acids and
of those which cannot be quantitatively converted to esters will
show a much lower percentage than their actual percentage in the
intact protein. Some amino acids will not show up at all on the
GLC chart, although they were part of the protein structure.

Several procedures were used to correct these, a full description of them is beyond the scope of this Manual, but re-
aders are advised to read the publications of Hediger et al.
(1973); and Gehrke and Stalling (1967) for the procedures as well
as for correction factors which reduce the errors arising from
the shortcomings of this method discussed above.

Another limitation originates from the fact that most proteins
consist not only of polypeptide chains, but may have a considerable percentage of carbohydrates or lipids bound to them. Treatment of such complex molecules with the chemical procedures described here or other methods to convert amino acids to their
volatile derivatives may in some cases result in side products
which will appear as unidentified peaks on the chart of the recorder. This applies particularly to glyco- or lipoproteins where
the percentage of nonamino acid components of the macromolecule
structure may be quite high. The best use of the procedure described here is for pure proteins and particularly for their
isolated enzymatic breakdown products.

We should also emphasize that several other amino acid derivatives are known which were used for gas-liquid chromatographic
separations. Most of these procedures have their advantages and
limitations. The N-trifluoroacetyl amino acid n-butyl esters
were first used by Zomzely et al. (1962). N-acetyl n-butyl esters
of some amino acids were separated first by Youngs in 1959.
There are several other methods described in the literature,
well reviewed in *New Techniques in Amino Acid, Peptide and Protein Analysis*
(1971) edited by Niederweiser and Pataki.

<u>References</u>

Coulter, J.R., Hann, C.S.: J. Chrom. *36*, 42 (1968)
Gehrke, C.W., Stalling, D.L.: Separ. Sci. *2*, 101 (1967)
Graff, J., Wein, J.P., Winitz, M.: Fed. Proc. *22*, 244 (1963)
Hediger, H., Stevens. R.L., Brandenberger, H., Schmid, K.: Biochem. J. *133*,
    551 (1973)
Gehrke, C.W., Stalling, D.L.: Separ. Sci. *2*, 101 (1967)
Niederweiser, A., Pataki, G. (eds.): New techniques in amino acids, peptide
    and protein analysis (Ann Arbor: Science Publishers, 1971)
Youngs, C.G.: Anal. Chem. *31*, 1019 (1959)
Zomzely, C., Marco, G., Emery, E.: Anal. Chem. *34*, 1414 (1962)

Exercise No. 42

Demonstration of Thin-Layer Chromatography

Thin-layer chromatography (TLC) is also called "open-column
chromatography", indicating the similarities between TLC and
regular column chromatography. In the latter the adsorbent forms
a column in a glass tube, but in the TLC system it is evenly
spread on a glass plate, forming a thin layer. The method was
developed and reviewed by Stahl (1965).
All the principles of the chromatographic system can be applied
to TLC. In the ascending TLC system, the stationary phase is
usually silicic acid, but aluminum trioxide or other adsorbents
may also be used. The mobile phase may be any of a wide variety
of solvent mixtures.
In this exercise the lipids and pigments of *Serratia marcescens*
bacteria will be separated together with a model mixture of synthetic dyes. The lipid extract, which also contains the pigments
(called Prodigiosins) may be obtained as described in this exercise.

<u>Materials and Equipment</u>

    Silicic acid G, specially prepared for thin-layer chromato-
       graphy
    Mixture of synthetic dyes containing Nigrosine, Gentian violet,
       Rhodamine 6G, and Saffranin, equal amounts dissolved in
       methanol, 1% solution of each
    Extract of *Serratia marcescens* obtained by chloroform:methanol =
       1:1 mixture

Chloroform
Methanol
Graduated cylinder, 250 ml
TLC chromatographic jar
20 cm × 20 cm glass plates
Equipment for spreading adsorbent onto the plates (see Fig. 29)
Disposable capillary pipette (see Fig. 30)

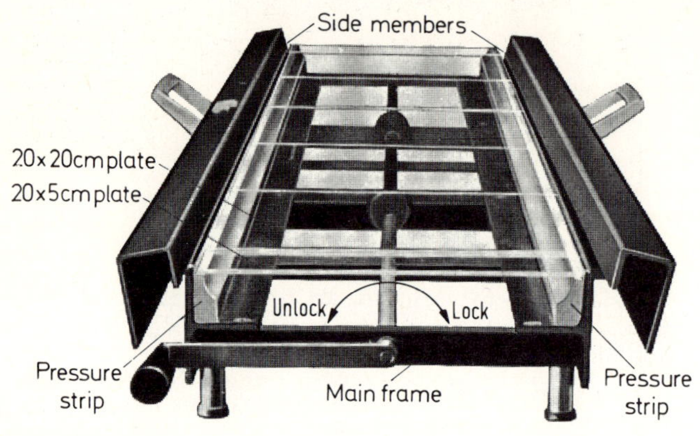

Fig. 29. Thin-layer chromatographic equipment manufactured by Quickfit and Quartz, Ltd., Stone, Staffordshire, England

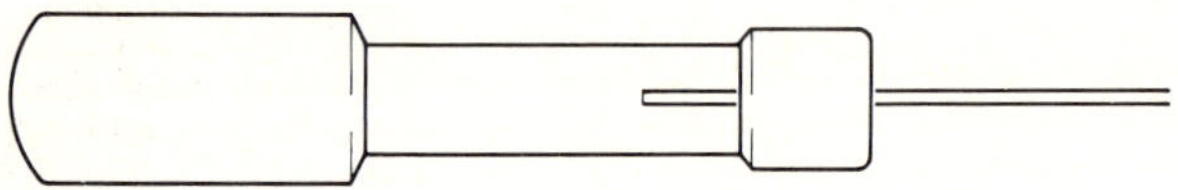

Fig. 30. Disposable micropipette, Drummond type, available from Kensington Scientific Corp., Oakland, CA

Ultraviolet light cabinet (see Fig. 31)
Spray bottle (Fig. 32)
40% Sulfuric acid containing 5% potassium dichromate
Acid resistant drying oven

Procedure

1. To obtain a lipid extract from bacterial cells, or from any other tissues, the best solvents to use are the methanol-chloroform mixtures. Most of the bacterial or tissue lipids are soluble to a certain degree in a 1:1 mixture of methanol:chloroform, but some require other ratios. Firmly bound lipids, such as bacterial lipopolysaccharides, or some lipid complexes, such as lipoproteins, are either not soluble or barely soluble in any ratios, and for their preparation, more complicated procedures are necessary.

Take 5 g freeze-dried *Serratia marcescens* cells (or any other tissues), place them in the extraction thimble of a Soxhlet type extractor. Add 50 ml methanol:chloroform = 1:1 mixture into the boiling flask of the Soxhlet and, by heating the solvent, start

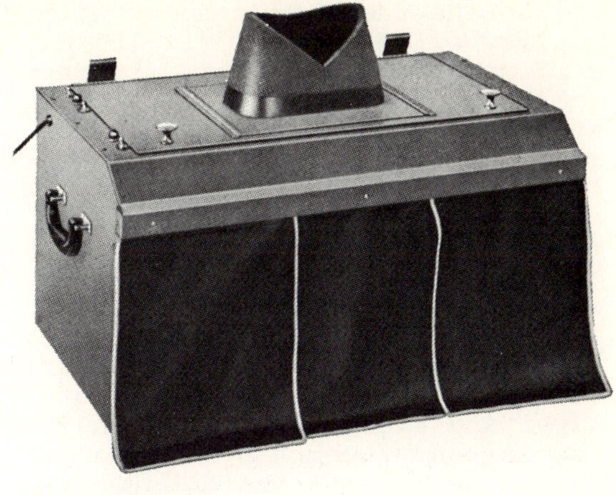

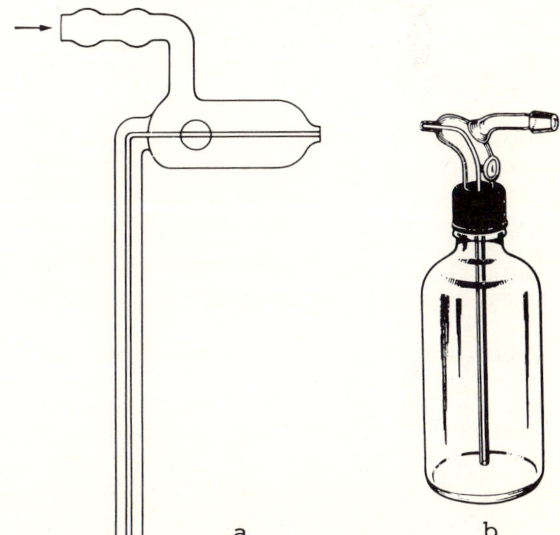

Fig. 32.
(a) Chromatographic sprayer
available from Metaloglass,
Inc., Boston, MA
(b) Chromatographic spray
bottle (courtesy of Arthur
H. Thomas, Philadelphia, PA)

the extraction of lipids and pigments from the cells. In approximately 2 h the solvent will be dark purple. This 50 ml solution
may be concentrated in a vacuum distillation unit to approximately
5 ml. The extract is ready for thin-layer chromatography.

2. Add 30 ml water to 50 g silicic acid in the mortar with constant mixing. Pour the thick slurry into the spreading container
and apply the layer by moving the container over the surface of
the glass plates in the spreading tray. To obtain an even thin
layer of the wet adsorbent on the glass plates, the spreader must
be pulled at a uniform rate without interruption. The glass plates
must be carefully cleaned and kept dust-free. Invisible organic
materials will show up if sulfuric acid is used for detection.
After the plates have dried at room temperature, transfer them
to the drying oven for final drying at 120°C for 60 min.

3. Line the TLC chambers with heavy but clean, analytic
quality filter paper. Mix 160 ml chloroform, 40 ml methanol and
2 ml water in a graduated cylinder and pour the solvent into a
rectangular TLC jar. Close it tightly. It should remain there
at least 8 h before use to allow the atmosphere of the chamber
to become saturated with the solvent.
4. During this time, a capillary tube will be used to place
a small drop, approximately 2 to 3 μl, of the dye mixture on
the adsorbent-coated side of one plate approximately 3 cm from
the lower edge. Apply the *Serratia marcescens* extract in the same
way.
5. Put the plate into the TLC chamber so that the lower edge
is immersed in the chloroform-methanol-water solvent mixture.
The solvent will immediately start to rise in the dry adsorbent
layer and the development should be allowed to continue until
the solvent front has moved about 12-14 cm beyond the points at
which the pigment and dyes were applied. This should take approximately 1 h, and during this time the separation of the visible
components can be observed through the glass walls of the TLC
jar.

Evaluation

Remove the plate and let it dry at room temperature. The synthetic
dye mixture will show a number of different-colored spots. Put
the plate into a UV cabinet (Fig. 31) and observe the pattern.
Take a pin or a sharp pencil and scratch marks under UV light
into the silica layer around the spots which became visible.
Finally, place the plate on a large sheet of plastic under a
fume hood. Fill a chromatographic sprayer (Fig. 32) with H_2SO_4 +
$K_2Cr_2O_7$ and spray the plate evenly. Transfer it to the drying
oven (or simple househould electric frying pan) and heat it at
150-180°C for approximately 10 or 15 min. Open the oven and observe the appearance of brown-to-black colored charred spots on
the plate. You will see that some of the previously visible spots
are faded and new components have appeared which were visible
neither under UV light nor in ordinary light.

Use and Limitations

The great advantages of TLC as opposed to column or paper chromatographic systems are: (1) The development time is much shorter
and the experiment can be carried out in 1 or 2 h. (2) This system allows the use of sulfuric acid and heat for the detection
of all kinds of organic compounds which could escape detection
unless their specific reactions were used. For example, fatty
acids, carbohydrates, glycolipids, or other components can all
be detected by one sulfuric acid spray. The detection of the same
components on filter paper would require specific procedures.
(3) Small amounts of separated pure components can be obtained
by preparative TLC, as described in Exercise No. 58.
There are nondestructive detection methods which can be used
in both paper and thin-layer chromatography. Detection of compounds by these procedures can be followed by isolation of the
separated bands, frequently in several hundred microgram quanti-

ties from a single run. One of these methods, spraying lipids
separated by TLC with water, is described in Exercise No. 58.

Lipids from erythrocytes, cardiolipin from heart muscles,
lipids from other tissues such as the brain, etc., can be separated under the conditions described here. Materials especially
difficult to separate because of their complexity, such as
"firmly bound" bacterial glycolipids, could be separated by
TLC (Kasai, 1964 and 1966). Needless to say, the composition and
components of the solvents may be varied, as well as the developers to detect components.

Not only silicic acid, but other adsorbents such as apatite
or Al_2O_3 may be used for TLC. Silicic acid is unquestionably the
most versatile with regard to its use for the separation of different natural products. Fine cellulose slurry may also be used
to prepare a thin adsorbent layer on glass plates. Good separation of carbohydrates could be obtained in such systems (Jeanloz
and Hakomori, 1965). Textbooks of TLC have been written by Stahl
(1965) and by Randerath (1966). Touchstone edited a book on
quantitative thin layer chromatography (1973).

In TLC an especially important factor is the proper saturation
of the chamber with solvent vapors. It is recommended to line
the inside walls of the jar with filter paper impregnated with
the chromatographic solvent. This will ensure good and fast
saturation within a few hours.

A slight disadvantage of the TLC system is that certain silicic
acid preparations require a binder such as $CaSO_4$; otherwise the
adsorbent layer may not adhere to the plates. Recently, very
fine silicic acid powders (2-10 micron particle size) have been
manufactured which can be used for coating glass plates without
a binder. It must be stressed that the quality of the silicic
acid is of utmost importance in successful TLC. Often a new
package of the same silicic acid purchased from the same manufacturer will not give the same results as the previous batch. If a
hitherto undescribed substance is to be the subject of TLC
separation, all available silicic acids have to be tried out
first using only two or three solvents. After a promising pattern
has been achieved with a certain silicic acid preparation, the
best solvent combinations may be further explored.

References

Jeanloz, R.W., Hakomori, S.I.: Personal communication (1965)
Kasai, N.: Ann. N. Y. Acad. Sci. *133*, 486 (1966)
Kasai, N., Yamano, A.: Jap. J. exp. Med. *34*, 329 (1964)
Randerath, K.: Thin layer chromatography. New York: Academic Press 1966
Stahl, E.: Thin layer chromatography. New York: Academic Press 1965
Touchstone, J.C.: Quantitative thin layer chromatography. New York: John
    Wiley & Sons, 1973

2.1.3   Some Other Qualitative Methods for Structural Studies

Exercise No. 43

Qualitative Analysis of the Constituents of an Unknown Natural Product

In chemical studies of natural products the first question the in
vestigator wants to know is which commonly occurring constituents
of natural products are present in the preparation. Qualitative
analysis has to precede quantitative studies because it demonstrates not only the absence or presence of certain detectable
components, but also gives information about the relative amounts
of constituents present in the preparation. The presence of
unusual components in hydrolysates investigated by paper electrophoresis or paper chromatography has led in the past to the
discovery of many new compounds.

In this exercise, a procedure is described which can be used
without alteration for many organic natural products. Two hydrolysates will be prepared from the preparation and each hydrolysate
will be investigated in two different systems for separation; in
paper electrophoresis and in paper chromatography. The chromatogram obtained will be cut into several narrow strips and the
strips will be sprayed with specific reagents to detect the constituents of the preparation.

Materials and Equipment

    M/20 Phosphate buffer, pH 7.0
    Alkaline silver nitrate reagent (see procedure)
    Phosphorus reagent (see procedure)
    Ninhydrin reagent (see procedure)
    Amino sugar reagent (see procedure)
    2-Keto-3-deoxycarbohydrate reagent (see procedure)
    Reagent for acidic compounds (see procedure)
    Concentrated, 1 N and 0.05 N sulfuric acid
    Concentrated, 5 N and 1 N hydrochlorid acid
    Potassium hydroxide, 50% solution
    Pyridine
    Boron trifluoride (14% in methanol)
    Acetic acid
    n-Butanol
    Ethanol, anhydrous
    Barium carbonate
    Hexane
    Heptane
    Lutidine
    Acetone
    Acetyl acetone
    Ethylene glycol
    0.02 M Sodium periodate
    4% Ammonium molybdate
    Perchloric acid (70%)
    Copper (I) nitrate
    Nitric acid, conc

Ammonium hydroxide, conc
Sodium thiosulfate
Bromphenol blue
Methyl red
6% Na-2-thiobarbiturate
Potassium hydroxide pellets
Drying oven
High voltage paper electrophoretic chamber (Fig. 27)
Paper chromatographic jar (Fig. 23)
Hydrolysis tubes (Fig. 33)
Centrifuge
5 × 100 mm centrifuge tubes
Boiling water bath
Alundum
Whatman No. 1 or similar quality filter paper
Marbles or "cold fingers"
20 and 100 lambda micropipettes
Hot air fan
UV lamp
Reagent sprayer (Fig. 32)

Procedure

1. Weigh approximately 5 mg of your unknown material into a
hydrolysis tube and add 0.5 ml 5 N HCl. Seal the tube in an
oxygen flame. Hydrolyze for 5 h in a boiling water bath as described in Exercise No. 33. After the hydrolysis is over, scratch

Fig. 33. Vacuum hydrolysis tube manufactured by
Kontes Glass Co., Vineland, NJ

the top of the sealed tube with a glass cutter and press the
red-hot end of a glass rod against this scratch. This will crack
the test tube open. Place the hydrolysate in a vacuum desiccator
containing KOH pellets and evacuate the tubes. The O.5 ml HCl
will evaporate overnight.

If hydrolysis tubes are available such as shown in Fig. 33,
carry out the hydrolysis under vacuum. The same amounts of
material and acid are added to the hydrolysis tube as above, and
the tube is then immersed in a mixture of salt and ice. Connect
the hydrolysis tube to a vacuum pump and evacuate the tube. If a
vacuum pump is used for the evacuations during these experiments,
protect it with an acid-trap, gas-drying tube inserted between
the pump and the acid-containing sample. Turn off the stopcock,
and the material is ready for hydrolysis. Place the tube in a
boiling water bath and carry out the hydrolysis for 5 h at that
temperature. Thereafter, connect the tube through the proper
adapter to a flash evaporator. Remove the acid by vacuum distillation. To prevent loss of material through sudden boiling, place
a few alundum crystals in the hydrolysis tube. Cool the tube in
ice water before connecting it to the flash evaporator, then turn
on the vacuum. Start rotating the sample slowly, do not apply
heat during the first 5 min. Then carefully immerse the end of
the tube into an approximately 30° to 40°C water bath. The acid
will evaporate in a few minutes, leaving a colored dry residue.

2. Another hydrolysate must be prepared which will be more
likely to leave the sensitive constituents present in the unknown material intact after hydrolysis. Again weigh approximately
5 mg of material, add 2 ml 1 N H_2SO_4. Cover the tube with a
marble and wrap the top of the tube with the marble in aluminum
foil. The hydrolysis may be carried out overnight in a boiling
water bath. If hydrolysis tubes are available, add to approximately 5 mg material 2 ml 1 N HCl. Cool the contents of the tube
and evacuate it with an efficient vacuum pump as above. Turn off
the stopcock and carry out the hydrolysis for 8 h in a boiling
water bath or oil bath at 100°C.

After the hydrolysis is complete, carefully add small amounts
of solid barium carbonate with a spatula into the tube which contained H_2SO_4. If no CO_2 formation is seen after the addition of
further barium carbonate, transfer the entire contents of the
test tube into a centrifuge tube. Rinse the contents of the tube
with 1 or 2 ml of water. Centrifuge the suspension and take off
the supernate. Add 2 ml distilled water to the sediment, stir
with a glass rod, and centrifuge again for 10 min at 3000 g.
Combine the washings with the supernate. Ion exchangers may also
be used for neutralization (see Exercise No. 34).

If a hydrolysis tube and HCl was used for the hydrolysis,
connect the tube with the proper adapter of a flash evaporator
and concentrate it to dryness under vacuum distillation. As above,
take care not to lose material through sudden boiling.

3. Some information can be obtained about the fatty acids
present in the analyzed sample if the acid hydrolysate is extracted with hexane. After the tubes have been opened, add 2 ml
reagent grade hexane to the hydrolysate. Stir the heterogeneous
mixture with a glass rod intensively for at least 5 min. Then
let the two phases separate. Take a capillary pipette and lift
as much of the hexane phase as possible. Repeat the extraction

with hexane twice more. Transfer the hexane phase to a 16 × 150 mm
test tube equipped with a 12/30 $\mathbb{T}$ joint. Dry the sample by blowing a gentle stream of nitrogen gas over the surface of the hexane
in the tube at room temperature. After the hexane has been removed, add 0.5 ml BF_3-methanol reagent to the residue, a few granules
of alundum, and reflux it on a condenser (with a similar joint)
for 15 min. Cool the tube before disconnecting it from the condenser, then add to it 0.5 ml reagent grade hexane and 2 ml water.
Shake the contents well, then let separate. The upper phase will
contain the methyl esters of the long-chain carboxylic acids
present in the analyzed sample. This material is ready for GLC
analysis (see Exercise No. 39) or for quantitative ester determination (see Exercise No. 63).

The acidic phase of the hydrolysate, which has been extracted
with hexane, must be dried in a vacuum desiccator over KOH pellets.

4. To each dried hydrolysate add 0.5 ml distilled water. Try
to dissolve the contents of the tube; use a glass rod to remove
all dried drops from the inside walls of the tube. Approximately
30 min after the addition of water to the dried residue, transfer
the contents of the tube with a capillary pipette into a 5 × 100 mm
centrifuge tube and spin it down at 3000 g for 15 min. The clear
supernate of the hydrolysis is ready for analysis.

5. Prepare a strip of Whatman No. 1 paper approximately 15 cm
wide. This will be used for high-voltage paper electrophoresis.
Draw a line across the center of the paper strip parallel to its
shorter edge. This will be the starting line. Mark the positive
end of the paper with a pencil. Fill a 100 μl micropipette (0.1 ml)
with the hydrolysate and apply it on the starting line by gently
touching the paper with the tip of the pipette and drawing it
rapidly along the starting line. Using a hot air fan, dry the
wet strip immediately. Repeat the application in the same manner
until the entire 100 μl hydrolysate has been applied on the starting line. Stretch the paper on the rack of the high-voltage electrophoresis apparatus, moisten the paper with pyridine-acetic acid
buffer, pH 5.4 (9% pyridine and 1% acetic acid and 90% water), and
proceed as described in Exercise No. 37. Immerse the paper in the
heptane of the high-voltage electrophoresis chamber, apply 2000 V
and carry on the electrophoresis for 60 min. After electrophoresis,
dry the paper in a 90°C oven for 10 min. Cut each 15 cm wide
strip into five narrow strips. Number the strips from 1 to 5.
Be sure that the positive end mark is visible on each narrow
strip, and also be certain that all the necessary information
is written on each narrow strip. Use easily visible marks made
with a soft pencil.

6. Spray and develop each strip with the specific reagents,
using the procedure described below. After all strips have been
developed, place them side by side in their original position
before they were cut.

7. Descending paper ιchromatography may be carried out with
both hydrolysates for the same purposes. Cut a piece of Whatman
No. 1 paper strip approximately 15 cm wide as before and draw a
pencil line across the paper approximately 12 cm from and parallel
to the narrow edge. Apply only 20 μl of the hydrolysate on the
line of application, using the same technique as before. Be sure
that the applied hydrolysate never runs out forming a band of application wider than 5-6 mm. Dry the applied spots and carry out

descending chromatography in the jar shown in Fig. 23. First
fold the paper as shown in Fig. 25, then place one end of it in
the empty trough of the chromatography jar.

Different paper chromatography solvent systems may be used
in this assay, but the most recommended is n-butanol:pyridine:
water = 6:4:3. This solvent system gives a very nice separation
among carbohydrates. Another very generally used solvent system
is the lower phase of phenol saturated with water. This solvent
is especially good in separating amino acids. Similarly popular
is n-butanol:acetic acid:water = 2:3:5.

After paper chromatography is completed, the paper strips
must be dried very thoroughly at 90°C. If phenol solvent has been
used, it is advisable to let the paper hang, after drying in the
oven, under a well-ventilated fume hood, for 24 h.

Cut the paper chromatogram into five narrow strips as before,
and by using the different staining procedures, develop the strips.
After all have been stained, place them side by side in their
original position.

Staining Procedures

The procedures described here can be used for any paper strips.
They are characteristic for certain groups of materials. The
preparation and use of these developers is described below.

Ninhydrin reagent. Dissolve 0.6 g ninhydrin in 90 ml n-butanol
and add 10 ml lutidine. Keep the reagent in a dark bottle. Spray
the dried paper strips lightly but evenly. Never use so much of
any spray reagent that it makes the paper dripping wet. Put the
paper strip in an oven and develop it at 90°C for 5 min. Different
colors will appear. Store the paper strip in a dark cabinet overnight. The intensity of the spots will increase. The pattern obtained will fade in a few days. To obtain a permanent paper chromatogram, prepare the following reagent: Dissolve 0.5 ml saturated
$CuNO_3$ solution in 100 ml acetone. Add one drop of concentrated
HNO_3. Dip the developed paper chromatogram into this solution
once, and dry it by hanging in the air. All spots will turn red.
These colors are stable for years. It is worthy to mention that
these red complexes may be eluted in 70% methanol and measured
spectrophotometrically. If proper controls made from standard
amino acids are available, a quite reliable quantitative measurement can be achieved, according to the procedure of Bode et al.
(1952). Primary amines react with ninhydrin.

Alkaline silver nitrate. Take 20 ml 5% $AgNO_3$ dissolved in water,
put it on a magnetic stirrer and add concentrated NH_4OH dropwise
under constant mixing. A brown precipitate will form which will
be later dissolved in the presence of a slight excess of NH_4OH.
The reagent cannot be kept longer than one day. Spray the paper
with the reagent as above, heat at 100°C for 5-8 min. To "fix"
the developed spots, and to achieve a permanent chromatogram,
dip the paper into a 5% $Na_2S_2O_3$ solution for 2 min. Rinse it with
tap water for 30 min; dry at room temperature. The pattern is
stable for years. Another alkaline $AgNO_3$ procedure was described
by Trevelyan et al. (1950). Reducing carbohydrates will react
with these reagents.

Phosphorus reagent. Phosphorylated derivatives and inorganic phosphorus compounds can be detected by the method of Hanes and

Isherwood (1949). Spray the paper lightly but thoroughly with
the following reagent: Mix 60 ml water, 10 ml 1 N HCl, 25 ml 4%
$(NH_4)_6Mo_7O_{24} \cdot 4H_2O$ (ammonium molybdate) and 5 ml 70% $HClO_4$. The
reagent is stable for approximately one week. Heat the paper at
85°C for 7 min. A higher temperature or longer treatment will
destroy the paper. Hang the paper at room temperature for 30 min,
then irradiate it with a UV lamp for 30 min. Blue spots will
appear on a gray-brown background. Exposure to sunshine also in-
tensifies the colors. The test is not very sensitive, and the
spots on the fragile paper will fade in a week. An improved version has been published by Charalampous and Mueller (1953).

Amino sugars can be detected by the Elson-Morgan reagent modified by Partridge (1948). Prepare the following reagents: Reagent No. 1: Dissolve 0.5 ml acetyl acetone in 50 ml n-butanol.
Reagent No. 2: Add 5 ml 50% aqueous KOH to 20 ml anhydrous
ethanol. Reagent No. 3: Dissolve 1 g p-dimethylaminobenzaldehyde
in a mixture of 30 ml ethanol, 30 ml concentrated HCl and 180 ml
n-butanol. Mix 0.5 ml of reagent No. 1 with 10 ml of reagent
No. 2. Spray the paper and heat it for 5 min at 150°C. Spray the
paper with reagent No. 3 and heat it for 5 min at 90°C. Pink
spots show the presence of amino sugars. Those derivatives which
are substituted in C-4 position will give a very weak yellowish
color.

2-Keto-3-deoxysugars and sialic acid derivatives can be detected by
thiobarbituric acid after periodate treatment. Use the following
reagents. Reagent No. 1: 0.02 M $NaIO_4$. Reagent No. 2: Ethylene
glycol 50 ml, acetone 50 ml, and concentrated H_2SO_4 0.3 ml.
Reagent No. 3: 6% Na-2-thiobarbiturate. Spray the paper first
with reagent No. 1, leave at room temperature for 15 min. Spray
with reagent No. 2, let it stand at room temperature for 10 min,
then spray it with reagent No. 3. Heat for 5 min at 100°C. Red
fluorescence can be seen under UV light. In visible light, red
spots will show the positions of ketodeoxyheptonic acid, ketodeoxy-
octonic acid, or 2-deoxyribose. For the detection of sialic acid,
the 0.02 M $NaIO_4$ reagent must be dissolved in 0.05 N H_2SO_4. Reagents No. 2 and No. 3 are the same. The method has been elaborated by Warren (1960).

Acidic compounds, organic or inorganic, can be detected by spraying them with bromphenol blue indicator. Dissolve 100 mg bromphenol blue in 100 ml ethanol. Mix with 30 ml 0.3% methyl red
dissolved in ethanol, add 70 ml pH 7.0 M/20 phosphate buffer. It
is very important to apply a very light but even spray in this
method. Too much indicator may cover weakly acidic components.
Acidic components will show as pink spots on a pale gray background. The color is unstable. This method has been described
by Fewster and Hall (1951.).

Evaluation

If the two hydrolysates have been analyzed both in paper electrophoresis and in paper chromatography in the manner described
above, a great deal of information may be obtained by applying
different developing reagents. Table 1 lists only a few compounds
which may occur in natural products. The R_F values and the electrophoretic mobilities are dependent upon the experimental conditions.
The presence of large amounts of certain components may influence

Table 2. R_F values of amino acids

Substances	Paper chromatographic R_F values				Relative electrophoretic mobilities[a]	
	Solvents					
	1	2	3	4	5	6
Alanine	0.58	0.30	–	–	+1.58	0.00
Arginine	0.56	0.16	–	–	+1.91	-0.78
Aspartic acid	0.20	0.22	–	–	+1.00	+1.00
Asparagine	0.38	0.12	–	–	–	–
Cysteic acid	0.08	0.08	–	–	–	+1.15
Glutamic acid	0.34	0.26	–	–	+1.10	+0.75
Glycine	0.39	0.23	–	–	+1.80	0.00
Histidine	0.63	0.12	–	–	+1.94	-0.82
Hydroxyproline	0.68	0.22	–	–	–	+0.02
Isoleucine	0.85	0.65	–	–	+1.22	–
Leucine	0.85	0.70	–	–	+1.22	-0.07
Lysine	0.46	0.12	–	–	+2.20	-0.85
Methionine	0.83	0.52	–	–	+1.16	0.00
Ornithine	0.17	0.13	–	–	–	-0.94
Phenylalanine	0.88	0.61	–	–	–	0.00
Proline	0.90	0.34	–	–	–	-0.05
Serine	0.35	0.22	–	–	+1.33	-0.03
Threonine	0.48	0.25	–	–	–	0.00
Tryptophan	0.77	0.52	–	–	–	-0.20
Tyrosine	0.62	0.44	–	–	+0.87	-0.10
Valine	0.77	0.50	–	–	–	0.00
Arabinose	0.54	0.21	0.43	0.28	–	–
Desoxyribose	0.74	–	–	–	–	–
Fructose	0.50	0.23	0.49	0.29	–	–
Fucose	0.54	0.27	0.54	–	–	–
Galactosamine	0.65	0.12	–	–	–	-0.72
Galactose	0.44	0.16	0.43	0.19	–	–
Glucosamine	0.62	0.13	0.35	–	–	-0.72
Glucose	0.38	0.18	0.48	0.23	–	–
Glucuronic acid	0.12	0.14	–	0.15	–	–
KDO	–	–	–	0.10	–	+0.48
Lyxose	0.45	0.26	–	0.30	–	–
Mannose	0.45	0.20	0.58	0.29	–	–
Rhamnose	0.58	0.37	0.70	0.50	–	–
Ribose	0.57	0.31	0.59	0.33	–	–
Xylose	0.44	0.28	0.52	0.35	–	–
Phosphate ion	–	–	–	–	–	+1.30
Chloride ion	–	–	–	–	–	+3.30
Ethanolamine	–	0.33	–	–	–	-1.80
Choline	–	0.29	–	–	–	-1.70
Histamine	–	0.19	–	–	–	-2.10

Solvents: No. 1 - Phenol saturated with water. For paper chromatography of carbohydrates, add 0.5 ml conc. ammonia to 100 ml solvent No. 1. (Basic compounds have higher R_F values in phenol + ammonia than in No. 1); No. 2 - n-BuOH: Acetic acid:Water = 12:3:5; No. 3 - Pyridine: n-BuOH:Water = 4:6:3; No. 4 - Ethyl acetate: Pyridine:Water = 2:1:2; No. 5 - pH 2.0 (1 *M* Formic acid:Acetic acid = 2:1); No. 6 - pH 5.4 (Pyridine:Acetic acid:Water = 9:1:90).

the electrophoretic or chromatographic mobility of other constituents. The figures given, therefore, may not be considered
as absolute values, but merely as guides to finding the possible
location of certain components on the paper strips.

Use and Limitations

The limitations of various hydrolytic procedures apply here as
well. Several components will not be liberated during this
hydrolysis, while others which are more sensitive to acidic
cleavage will be destroyed. For the detection of the components
in the hydrolysate, obviously other staining procedures than
those few listed here may be applied, including all methods used
in paper chromatography or in paper electrophoresis. Frequently
the colors developed by specific spray reagents help in identification of the detected spots. Such a method has been described
for carbohydrates by Bailey and Bourne, using diphenylalanine-
aniline spray (1960).
  Some unknown components may be isolated after paper electrophoresis or paper chromatography in sizable amounts for classical
chemical analysis. Such a method is described in Exercise No. 36
and Exercise No. 37.

References

Bailey, R.W., Bourne, E.J.: J. Chromatog. *4*, 206 (1960)
Bode, F., Hubener, H.J., Bruckner, H., Hoeres, K.: Naturwissenschaften *39*, 524 (1952)
Charalampous, F.C., Mueller, G.C.: J. Biol. Chem. *201*, 161 (1953)
Fewster, M.E., Hall, D.A.: Nature London *168*, 78 (1951)
Hanes, C.S., Isherwood, F.A.: Nature London *164*, 1107 (1949)
Partridge, S.M.: Biochem. J. *42*, 238 (1948)
Trevelyan, W.E., Procter, D.P., Harrison, J.S.: Nature London *166*, 444 (1950)
Warren, L.: Nature London *186*, 237 (1960)

Exercise No. 44

Identification of N-Terminal and C-Terminal Amino Acids

In the structural investigation of proteins important information
can be obtained by hydrolysis and analysis of the liberated amino
acids. With this procedure it is possible to discover which amino
acids are present in a particular protein, but the biologic activities of proteins are determined by the sequence of the amino
acids in the polypeptide chain and by the folding of these chains.

---

[a] The electrophoretic mobility values are relative to aspartic acid,
which has been taken as 1.00. The 0.00 line is the line of application at
pH 2.0 runs, but at pH 5.4 the position of glycine has been taken as the
0.00 value. Negative values mean migration towards the negative end of the
paper strip.

The physico-chemical properties of a protein molecule are determined, among others, by the number of free amino or carboxyl groups. Those amino acids which have free carboxyls because they are at the end of the polypeptide chain are called C-terminal amino acids, while those which are at the beginning of the chain will have their NH_2 group free, and are called N-terminal amino acids. Of course, diamino acids or monoaminodicarboxylic acids may have one amino or one carboxyl group free even if they are not in terminal position.

For the analysis of the N-terminal amino acids, the dinitrophenyl (DNP) method is widely used. The reaction takes place as follows:

$$NO_2\text{-C}_6H_3(F)(NO_2) + NH_2\text{-CH(R)-CO-protein} \longrightarrow NO_2\text{-C}_6H_3(NO_2)\text{-NH-CH(R)-CO-protein} + HF \xrightarrow{HCl}$$

$$NO_2\text{-C}_6H_3(NO_2)\text{-NH-CH(R)-COOH} + \text{Amino acids.}$$

The dinitrofluorobenzene (DNFB) reacts with the free amino groups of a polypeptide under slightly alkaline conditions where the peptide bond is quite stable. On hydrolysis of the proteins, the N-terminal residues are liberated in the form of DNP-amino acids. These are bright yellow compounds which can be extracted with organic solvents, estimated colorimetrically and identified by chromatography. Akabori and associates (1953) described hydrazinolysis for the determination of C-terminal amino acids. The reaction takes place as follows:

$$NH_2\text{-CH(R}_1)\text{-CO-NH-CH(R}_2)\text{-COOH} + NH_2NH_2 \longrightarrow$$

$$NH_2\text{-CH(R}_1)\text{-CO-NH-NH}_2 + NH_2\text{-CH(R}_2)\text{-COOH}$$

The reaction products are reacted in turn with dinitrofluorobenzene and the DNP derivatives of the amino acids and amino acid hydrazides will be extracted with ethyl acetate. The C-terminal amino acids as well as the dicarboxylic amino acid DNP derivatives will be extractable from the ethyl acetate with dilute $NaHCO_3$ solution. They can be purified and identified by paper chromatography. The other hydrazides will remain in ethyl acetate. A simplified version of the procedure is described here.

In the first part of this Exercise, the DNP derivatives of bovine gamma globulin will be prepared and both the native gamma globulin and its DNP derivative will be hydrolyzed. The analysis of the hydrolysate obtained from the parent bovine gamma globulin will give us qualitative information regarding the amino acid composition of this protein; hydrolysis of the DNP gamma globulin

will liberate the DNP derivative of those amino acids which are
in N-terminal position. Identification of the DNP amino acids
will be carried out by paper chromatography. In the second part
of this exercise, the C-terminal amino acids will be liberated.
In the third part of this exercise, the paper chromatography of
the amino acids and their DNP derivatives will be described.

Part A. N-Terminus

Materials and Equipment

    Bovine gamma globulin
    2,4-Dinitrofluorobenzene
    Ethyl alcohol
    Ethyl acetate
    Diethyl ether
    5 N and 1 N Hydrochloric acid
    pH 9 carbonate buffer. Add 0.1 M Na_2CO_3 to 0.1 M $NaHCO_3$ until
      the pH reaches 9
    Test tube with Teflon lined screw cap
    Glass filter, medium
    Filter flask
    Gas-oxygen torch
    Boiling water bath with water level regulator
    Aluminum foil
    Vacuum desiccator with potassium hydroxide pellets
    18 × 150 mm Pyrex test tubes (or vacuum hydrolysis tubes, as
      shown in Fig. 33)
    Centrifuge, explosion proof, refrigerated

Procedure

    1. Dissolve 100 mg lyophilized bovine gamma globulin in 2 ml
pH 9 carbonate buffer. Add 2 ml 10% 2,4-dinitrofluorobenzene in
ethanol. Wrap the tube in aluminum foil and shake the mixture at
room temperature for 2 h.
    2. As the next stage of the preparation involves the use of
ether, certain precautions must be taken. Burners and cigarettes
must be extinguished and any pieces of electric eqipment which
could cause a spark must be turned off. It should be remembered
that ether vapor is heavier than air and can form a layer on the
bench or floor so that a flame or spark several yards away from
an open bottle of ether can result in an explosion.
    The DNP derivative of the bovine gamma globulin forms a pre-
cipitate. Transfer the contents of the tube to a medium fine glass
filter and wash the precipitate on this filter five times with
10 ml cold water, three times with 10 ml cold ethanol + ether =
1:1 mixture and finally allow approximately 30 ml cold ethyl
ether to flow through while agitating the precipitate with a
glass rod. Sedimentation of the precipitate and its washing with
the above solvents may be carried out in a refrigerated and explosion-proof centrifuge. The yellow, dried preparation should
be kept in a dark vial or dark vacuum desiccator for further investigation.

3. Weigh 20 mg bovine gamma globulin into a Pyrex test tube.
Add 2 ml 5 N HCl. Seal the tube in a gas-oxygen flame. Check the
tube for leakage. Mark the tube properly. Do the same with 20 mg
DNP bovine gamma globulin.

4. After 6 h of hydrolysis at 100°C cut the tubes open and
place them in a vacuum desiccator over KOH pellets and evacuate
the desiccator. The hydrolyzed samples will dry completely in
the desiccator in approximately 24 h, and may be stored in this
form in the dark.

5. The hydrolysate of the DNP bovine gamma globulin must be
extracted three to five times with ethyl ether. Add 2 ml ethyl
ether to the dried residue, close the tubes and let them stand
at room temperature for 30 min. Shake occasionally during this
time. Remove the yellow ether extract by puring it out. Replace
the ether, and repeat the same procedure twice more. Discard the
ether which contains side products of the dinitrophenylation.

Evaluation see *Part C*.

Use and Limitations see *Part C*.

Part B. C-Terminus

Materials and Equipment

Hydrazine, anhydrous
2% Sodium bicarbonate solution
Vacuum desiccator with concentrated sulfuric acid

Procedure

1. To approximately 50 mg of protein add 1 g of anhydrous
hydrazine (99% to 100%). This has to be prepared according to
the procedure of Kusama (1957). If the water content of the hydrazine is higher than 1%, the reaction will not take place. The
hydrazinolysis can be carried out in a sealed tube at 100°C for
5 h. Open the tube and transfer it to a vacuum desiccator and
dry it completely over concentrated H_2SO_4.

2. Dissolve the dry residue in 2 ml water and shake it for
1 h very vigorously with 1 ml freshly distilled benzaldehyde.
The acid hydrazides will form an oily phase with the benzaldehyde.
This oily phase has to be separated from the supernate by sharp
centrifugation. Extraction of the oily phase with one more ml
water will enhance the yield. The supernate contains the free
amino acid from the C-terminus. This has to be lyophilized without
delay. The C-terminal amino acids may be analyzed as done by paper
chromatography or by automatic amino acid analyzers. A more reliable procedure, employing dinitrophenylation of the liberated
amino acids and identification of these by chromatography, is
described below.

3. Dissolve the residue from Step No. 1 in 15 ml of 2% sodium
bicarbonate and add to it 30 ml of a 0.01 mg/ml DNFB solution in
ethanol. Shake the mixture in the dark at room temperature for
3 h, then add to it 50 ml water and slowly add 10 ml normal HCl.
Mix and extract the acidified mixture three times with 10 ml
ethyl acetate, using a separatory funnel. Combine the ethyl
acetate phases and extract with 3 portions of 10 ml 2% $NaHCO_3$.

This extract will contain the DNP derivate of the C-terminal
amino acid, together with the DNP derivatives of the dicarboxylic
acid hydrazides. Ethyl acetate extraction of the bicarbonate solution will yield these DNP derivatives, which can either be identified by paper chromatography or separated by column chromatography on silicic acid. These two groups can be differentiated
on the basis of their resistance to 5 N HCl. The DNP derivatives
of hydrazides can be hydrolyzed thus liberating free amino acids.
The hydrolysis may take place at 100°C for 6 h. The DNP derivative
of the C-terminal amino acid will hardly be affected by this
treatment.
   Evaluation see *Part C.*
   Use and Limitations see *Part C.*

Part C. Paper Chromatography of the Amino Acids and Their Derivatives

Materials and Equipment

   Reference mixture of known amino acids and DNP amino acids
   Paper chromatography solvents (a) phenol saturated with water
    and (b) normal butanol saturated with water
   Ninhydrin spray reagent (see Exercise No. 43)
   Two 40 × 40 cm sheets cut from Schleicher and Schuell No. 2041
    paper, or any other make of comparable quality
   Chromatographic spray bottle (Fig. 32)
   Micropipettes
   Two paper chromatographic chambers (Fig. 24)
   Chromatographic drying oven
   Hair dryer
   Rubber gloves
   Face shield
   Wooden clamps

Procedure

   1. Pour 100 ml of the prepared paper chromatographic solvent
into the chamber, cover it with a glass plate in order to obtain
saturation of the water phase. Handle the phenol carefully and
wear a face shield. If it is splashed on your skin, wash it with
much water and alcohol. Prepare another chamber with butanol
water.
   2. Add 0.5 ml distilled water to the dried residue of the
parent gamma globulin hydrolysate and add 0.5 ml ethyl acetate
to the DNP amino acid preparation. Agitate with a glass rod at
room temperature for a few minutes, then allow time to stand.
   3. Mark the starting line on the paper with a pencil approximately 6 cm from the bottom edge. Mark the points of application
using one sheet for DNP and one sheet for parent gamma globulin
paper chromatography.
   4. With a micropipette, apply 10 microliter parent gamma
globulin hydrolysate to the starting line. Apply 10 µl of the
DNP hydrolysate on the other sheet. Use the hair dryer from a
distance of approximately 10-20 cm to facilitate the drying of
the spots.

    5. Place the sheets in the spiral glass rod holders. It is
advisable for one person to hold the spiral while another inserts
the paper sheet.
    6. Carefully immerse the spiral with the prepared sheet into
the chamber. The solvent in the chamber must reach the bottom
edge of the sheet at all points, but may not directly reach the
spot of the applied sample. The chamber will be covered, and from
time to time the ascending solvent front must be observed through
the walls of the chamber. Note irregularities, if any, which
occur. The chromatogram in phenol must run overnight; the butanol
solvent ascends faster.
    7. The chromatography sheet must be removed from the chamber
if the solvent front is close to the upper edge of the paper.
Wearing clean rubber gloves, remove it carefully from the spiral.
Hang the paper with the amino acids of the parent gamma globulin
in a preheated chromatographic drying oven with clean wooden
clamps and dry at a maximum of 100°C for 10 min. Do not heat it
for an unnecessarily long time. Dry the DNP amino acid paper
chromatogram at room temperature under a fume hood.

Evaluation

The DNP spots are visible directly without further treatment.
To detect the free amino acids, dissolve 0.3 g of ninhydrin
in 90 ml n-butanol and 10 ml lutidine. Spray the amino acid
chromatogram evenly with this reagent. Spray thoroughly, but do
not soak the paper. Hang it in the chromatographic drying oven
with wooden clamps; develop the color at 90°C for 5-10 min. Mark
the solvent front with a pencil.
    Locate the spots, encircle them with pencil, and mark their
center. Measure their distance from the starting line in millimeters, also measure the distance of the solvent front from the
starting point in millimeters. Calculate the R_F value, and using
a table of R_F values of amino acids (see Exercise No. 43), try
to identify them. For the identification of DNP-amino acids, the
use of Table 3 may be helpful.
    Basic diamino acids within a peptide chain may have a free
amino group which will react with DNP. The differentiation of
these DNP derivatives from the di-DNP derivatives of basic amino
acids takes place on the basis of their R_F values. Some of the
mono-DNP derivatives of basic amino acids may also react with
ninhydrin.
    Unsubstituted amino sugars will also give yellow DNP products.
These compounds can be easily differentiated from amino acids on
paper chromatogram, because these will react with alkaline $AgNO_3$
reagent used for the detection of reducing carbohydrates.

Use and Limitations

Attention must be called to the fact that the procedure described
here for the determination of N-terminal and C-terminal amino
acids is the simplified approach. For quantitative studies, a
more thorough procedure must be applied. The references given
below accurately describe the necessary methodology.

Table 3. R_f values of DNP-amino acids

DNP-amino acid	n-Butanol saturated with water
DNP-alanine	0.50
α-DNP-arginine	0.36
DNP-aspartic acid	0.12
DNP-glutamic acid	0.14
DNP-glycine	0.36
Di-DNP-histidine	0.34
DNP-isoleucine	0.72
DNP-leucine	0.73
α-DNP-lysine	0.33
ε-DNP-lysine	0.32
Di-DNP-lysine	0.70
DNP-methionine	0.63
DNP-phenylalanine	0.69
DNP-proline	0.48
DNP-serine	0.32
DNP-threonine	0.43
DNP-tryptophan	0.68
DNP-tyrosine	0.78
DNP-valine	0.66

References

Akabori, S. Ohno, K., Ikenaka, T. Nagata, A., Haruna, I.: Proc. Jap. Acad. *29*, 561 (1953)
Block, R.J., Durrum, E.L., Zweig, G.: A manual of paper chromatography and paper electrophoresis. New York: Academic Press 1958, pp. 37-43
Brenner, M., Niederwieser, A., Pataki, G.: In: New biochemical separations. James and Morris (eds.): pp. 123-156 Princeton, N.J.: Van Nostrand Co., Publ., 1964
Koch, G., Weidel, W.: Hoppe Seyler's Z. Physiol. Chem. *303*, 213 (1956)
Kusama, K.: J. Biochem. Tokyo *44*, 375 (1957)
Lederer, E., Lederer, M.: Chromatography. New York, N. Y.: Elsevier 1957
Porter, R.R., Sanger, F.: Biochem. J. *52*, 287 (1948)

Exercise No. 45

Staining of Gels for Carbohydrates

The periodic acid-Schiff (PAS) staining procedure for carbohydrate-containing macromolecules, such as glycoproteins, is given here, based on the publication of Zacharius et al. (1969), which is a modification of the method published earlier by Clarke (1964). At some points we applied modifications used by Kapitany and Zebrowski (1973).

Materials and Equipment

    15% trichloroacetic acid (TCA) dissolved in water
    1.2% periodic acid in 5% acetic acid, prepared freshly before
    use
    1% silver nitrate in water
    0.5% metabisulfite ($K_2S_2O_5$) in water, freshly prepared before
    use
    15% acetic acid in water
    The preparation of the Schiff reagent is described here ac-
    cording to the procedure of McGukin and McKenzie (1958).
    Dissolve 3 g of basic fuchsin in 1000 ml water. After complete
    dissolution, add 6 g K metabisulfite and 30 ml 6 N H_2SO_4.
    Mix overnight in a dark cold room. The next day add 6 g Darco
    charcoal. Continue mixing for 30 min longer, then filter
    through two layers of Whatman No. 1 or similar quality paper.
    The filtrate must be almost colorless. Keep it in a well-
    closed dark bottle in a cold room.

Procedure

1. Fill a suitable glass tray (butter dish or Petri dish)
with 15% TCA. Immerse the microslide with the gel on its surface
into the dish. If the gel comes from disc electrophoresis, as described in Exercise No. 10, simply fill a test tube with TCA and
drop the extruded gel into it. Keep it there for approximately
30 min. After this time, proteins may appear as white precipitated
bands. The use of elevated temperatures such as 50°-70°C will
facilitate the precipitation of proteins.

If the gel contains SDS (see Exercise No. 10) a very thorough
washing of the fixed gel must follow. Soaking for one or two days
in large volume distilled water may remove SDS.

2. Pour off the TCA and rinse the gel once with cold distilled
water. Do not shake or stir, pour off the water after a few
seconds.

3. Fill the same container you used in step 1 with freshly
prepared periodate solution in acetic acid. Keep the gel there
for 120 min, then pour off the periodate.

4. Pour off the reagent, rinse the gel, discard the washing
and fill the container with 15% acetic acid. Wait 15 min and
replace the acid again. Repeat the 15 min washing with fresh 15%
acetic acid four to five times. Now transfer the gel to a larger
container (at least 200 ml capacity), fill it with 15% acetic
acid and let it soak for 1 h. The purpose of the extensive washing is to remove all traces of excess periodate from the gel.
Repeat washing until the washing fluid does not react any more
with the silver nitrate reagent.

5. Place the gel in the continer used in step 1, fill it
with Schiff's reagent and let it stand in the dark for 120 min
in a refrigerator.

6. Discard the reagent and wash the stained gel at least
three times with freshly prepared 0.5% K metabisulfite. The carbohydrate-containing components in the gel will appear as pink
bands in a colorless background. If the gel is still colored,
continued washing is recommended. Kapitany and Zebrowski (1973)
recommended extensive washing in 15% acetic acid, instead of K
metabisulfite.

Evaluation

By placing the stained gels on an opaque glass plate illuminated
from below, the carbohydrate-containing bands will be visible.
Measure the migration of the bands in millimeters and record the
results. All the necessary information of the experimental conditions of the electrophoresis have to be recorded also. These
should include (a) voltage/cm applied (b) current in mA per gel
(c) time of electrophoresis (d) temperature, (e) pH and concentration of the buffer, and (f) sample size (mg dry weight applied).

Use and Limitations

The periodate-Schiff staining procedure was developed originally
by Köiw and Grönwall in 1952 for the staining of glycoproteins
separated by paper electrophoresis (see Exercise No. 9). The
method described here can also be applied to paper strips. The
extensive washing process may be shortened in the case of paper
strips.

It is essential that the periodate be completely removed from
the supporting medium, whether it be paper or gel. If this has
not been achieved, the gel will have a pink or purple background,
which will make the detection of pink bands difficult or impossible. The washing procedure must be kept as described here and
no shortcuts can be taken. If the sample or any one of the reagents used for the electrophoretic separation contained SDS or
carbohydrates, such as sucrose or Ficoll, one overnight washing
may not be sufficient to wash these out of the gel, 24 h or
longer washes after step 4 may be required.

If the stained bands are too faint, this may be due to either
low carbohydrate content in the components detected or insufficient
oxidation with periodate. In the first case, larger amounts must
be separated by electrophoresis. In the second case, either longer
oxidation time will be needed or new periodate reagent should be
prepared. If the Schiff reagent becomes slightly colored, fresh
reagent should be prepared. *Caution:* If the reagent gets on your
skin, wash it off immediately. It may not only stain you purple,
but also has a mild toxic effect.

If both PAS staining for carbohydrates and Coommassie blue
staining for proteins are to be carried out on the same gel, the
PAS should be finished first, and the gel photographed, since the
much more intensive protein stain with Coommassie blue may mask
the pink color of PAS. It is highly advisable to place the stained
gel next to a millimeter scale with the origin of the migration
exactly at 0 mm. The Coommassie stain can be carried out after
this, in the same fashion as described in Exercise No. 10. Repeat
the photography as above.

For permanent record of the gels on microscope slides, one may
dry them out as follows: Immerse the gel in a 3%-4% glycerol
solution in water for 15 min. Place the gel, after removing it
from the glass, between two sheets of Cellophane, put them on a
glass plate. Dry the plate in a ventilated oven at 37°C overnight.
A thin film, showing the stained bands, will be obtained from the
gel. Otherwise the stained gels can be stored for a few weeks in
10% acetic acid, without much change in color intensity.

A more recent version of this PAS method has been published
by Kapitany and Zebrowski in 1973. It has been reported that this
method has a sensitivity superior to the one described by Zac-
charius and co-workers.

References

Clarke, J.T.: Ann. N. Y. Acad. Sci. *121*, 428 (1964)
Kapitany, R.A., Zebrowski, E.J.: Anal. Biochem. *56*, 361 (1973)
Köiw, E., Grönwall, A.: Scand. J. Clin. Lab. Inc. *4*, 244 (1952)
McGukin, W., McKenzie, B.F.: Clin. Chem. *4*, 476 (1958)
Zacharius, R.M., Zell, T.E., Morrison, J.H., Woodlock, J.J.: Anal. Biochem.
     30, 148 (1969)

Exercise No. 46

Methylation of Carbohydrates

Several methods have been described for the permethylation
(methylation of all available OH groups) of different carbohydrate
derivatives. Most of them apply dimethyl sulfate in an alkaline
medium, which may lead to complete methylation if repeated
several times. The disadvantage of this procedure, in addition to
the incompleteness of the methylation, is that the alkaline
medium may set free hydroxyl groups which were substituted in
the native product. Other procedures suffer from similar disadvantages and/or may require the use of liquid ammonia, which
makes these procedures unfit for use as exercises for students.
The procedure described here is quite efficient for the permethyla
tion of carbohydrates with fewer undesirable side effects on the
investigated material. The method was developed by Hakomori (1964)
Raffinose will be used as a model substance in this method.
    The reaction takes place in two steps:

$$R - OH + CH_3 - SO - CH_2^- + Na^+ \rightarrow R - ONa + CH_3SOCH_3$$

$$R - ONa + CH_3I \rightarrow R - O - CH_3 + NaI$$

Materials and Equipment

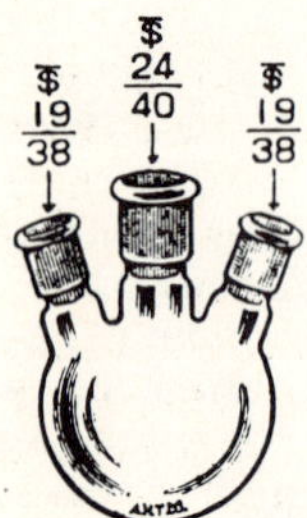

Fig. 34. Three neck reaction
flask (courtesy of Arthur H.
Thomas, Philadelphia, PA.)

Raffinose
Dimethyl sulfoxide (b.p. 189°C at
   760 mm Hg)
N_2 gas
Sodium hydride (analytic grade)
Methyl iodide (analytic grade)
Chloroform (analytic grade)
5% hydrochloric acid containing
   methanol
Ether:petroleum ether = 1:1 mixture
1 *N* hydrochloric acid
Reaction flask, 50 ml capacity, shown
   in Fig. 34

    Separatory funnel, 200 ml capacity
    Vacuum distillation equipment (Fig. 12)
    Glass filter, coarse
    Anion exchange resin, HCO_3^- form
    Water bath
    Oxygen torch
    Indicator paper

Procedure

1. Weigh 100 mg raffinose into the reaction flask and dissolve it in 10 ml dimethyl sulfoxide. A slow stream of nitrogen
gas is bubbled through the solution. Add 50 mg moisture-free,
undecomposed NaH. This is in slight excess of the number of
hydroxyl groups in the raffinose. Due to the effect of NaH on
the dimethyl sulfoxide, methylsulfinyl carbanion is formed. Continue the stirring of the reaction mixture by the nitrogen stream
for 10 min. Add 300 mg methyl iodide to the solution and continue
the stirring for 30 min.

2. Transfer the reaction mixture to a 200 ml separatory funnel,
rinsing the contents of the flask three times with approximately
10 ml water. Add 20 ml chloroform and extract the methylated
carbohydrate by vigorous shaking. Separate the lower chloroform
phase; repeat the extraction of the upper water phase twice more
with 20 ml chloroform each time. Discard the water phase and
pool the chloroform extracts. Wash the chloroform phase three
times with 50 ml water and discard the washings. Evaporate the
chloroform to dryness in vacuum. Dissolve the residue in 50 ml
ether:petroleum ether mixture and wash the extract with water
using three 20 ml washings. Discard the washings. Evaporate the
ether:petroleum ether extract. The residue contains the methylated
raffinose.

3. For further analysis of this product (or of other methylated
oligo- or polysaccharides), the O-methyl hexose components have
to be liberated by methanolysis. Place approximately 10 mg in a
regular test tube and add 1 ml 5% HCl-containing methanol. This
reagent can either be obtained commercially or prepared in the
laboratory by leading dry HCl gas from a HCl generator (NaCl +
concentrated H_2SO_4) into anhydrous methanol. The HCl content
can be determined by titration and adjusted by further HCl gas
addition or by dilution with anhydrous methanol. Quickly seal the
tubes in an oxygen flame as described in Exercise No. 33. Wrap
the tubes in several layers of strong fabric and place them in a
boiling water bath for 6 h.

4. This methanolysis yields methyl glycosides (in addition to
methyl esters from O-acyl groups, if present in the preparation).
This can be cleaved by regular acid hydrolysis. Therefore, the
top of the tube has to be carefully opened and the HCl-methanol
evaporated to dryness in a 40-45°C water bath. Blowing a slow
stream of nitrogen into the tubes through a gas capillary hastens
the evaporation. Care must be taken to prevent loss of material
by sudden or intensive boiling of the methanolysate. After the
HCl-methanol has been removed, add 4 ml 1 N HCl and seal the tube
again. For the hydrolysis of the methyl glycoside, immerse the
tube in a boiling water bath for 6 h.

5. Open the cooled tubes and trasfer their contents quantitatively into a 200 ml beaker. Any anion exchange resin in hydrocarbonate (HCO$_3^-$) form may be added to neutralize the solution.
Check the pH with indicator paper after each addition of ion
exchange resin. Filter the solution through a coarse glass filter
and wash the resin with approximately 20 ml water three times.
Evaporate the filtrate in vacuum to a syrup. This contains the
methylated carbohydrates ready for paper or thin-layer chromatography.

Evaluation

For the separation, any of the described paper chromatographic
equipment and procedures can be used. The recommended solvent is
the upper phase of n-butanol:ethanol:water:concentrated NH$_3$ =
40:10:49:1. As spray reagent, ammoniacal AgNO$_3$ (Exercise No. 43)
is suitable. Using 3% p-anisidine-HCl in butanol gives different
colors with methylated aldohexoses, methylated aldopentoses, free
aldohexoses, ketohexoses, and uronic acids. This has been describe
by Hough and co-workers (1950). Gas-liquid chromatography may also
be used for the identification of methylated monosaccharides.
For the separation of methylated carbohydrates Hakomori and associates used gas-liquid chromotography on a 3% ECNSS-M column
(Applied Sci. Labs., State College, PA. 16802). More recently the
separation is carried out by 2% OV 17 or OV 226 columns (Applied
Sci Labs) which gives a significantly better recovery of ammino
sugars (Hakomori, personal communication). GLC combined with mass
spectrometry for the analysis of methylated olygosaccharides is
certainly the most powerful procedure in structural studies.

Use and Limitations

Permethylation is one of the most useful methods in the structural
analysis of carbohydrates. Those hydroxyl groups in natural products which could be methylated were neither bound to substituents
nor involved in the formation of glycosidic linkages with another
carbohydrate unit. The identification of the methylated products
indicates the structure of the analyzed trisaccharide.
    It is obvious that if O-substituents are present, the evalua-
tion of the results obtained is more complicated. Positions of
substituents may be mistaken for places of branching points, but
the most important source of misinterpretation may lie in the fact
that some labile substituents can be cleaved from the carbohydrates
during the methylation process itself, yielding a free hydroxyl
group for methylation. Incomplete methylation is obviously another
frequent source of error. A methylation procedure which is free
of these shortcomings has not yet been developed.
    This procedure has been successfully applied to cerebrosides
by Hakomori (1964), to sulfatides by Malone and Stoffyn (1965),
to gangliosides by Handa (1965), and to polysaccharides of *Aero-
bacter* by Sandford and Conrad (1966), using conditions practically
identical to those described here. Polysaccharides such as glycogen
require a considerably longer reaction time. The yield of methylated glycogen was significantly lower than theoretical. Ovomucoid
and other investigated mucoproteins presented even more difficulties, probably due to the precipitation of denatured proteins.

These facts show that the application of the method to unknown
biopolymers is limited and obviously requires preliminary studies
to elaborate the optimal conditions.

More recently, Hakomori and associates studied the amino sugar
linkages in ceramide pentasaccharide of erythrocyte antigens
(Stellner et al., 1973). The structure was analyzed by methylation, enzymatic degradation, and mass spectrometry. Using the
same procedure, glycolipids in transformed cells were studied
in the laboratories of Hakomori and co-workers (Watanabe et al.,
1975, 1976; Hakomori, 1975).

References

Hakomori, S.-I.: J. Biochem. (Tokyo) *55,* 205 (1964)
Hakomori, S.-I.: B.B.A. *417,* 55 (1975)
Handa, N., Handa, S.: Japan. J. Exp. Med. *35,* 331 (1965)
Hough, L., Jones, J.K.N., Wadman, W.H.J.: J. Chem. Soc. 1702 (1950)
Malone, M., Stoffyn, P.: Biochim. Biophys. Acta *98,* 218 (1965)
Sandford, P.A., Conrad, H.E.: Biochem. *5,* 1508 (1966)
Stellner, K., Saito, H., Hakomori, S.-I.: Arch. Biochem. Biophys. *155,* 464
    (1973)
Watanabe, K., Laine, R.A., Hakomori, S.-I.: Biochem. *14,* 2725 (1975)
Watanabe, K., Matsubara, T., Hakomori, S.-I.: J. Biol. Chem. *251,* 2385 (1976)

2.2   Quantitative Analytical Determinations

Exercise No. 47

Dry Weight and Inorganic Ash Determinations

One of the most frequently needed pieces of information in working with various extracts or fluids is the total dry weight content which includes both organic, inorganic, small and macro-
molecular substances. For any quantitative studies, this information is the first one we need. Working with biologically active
extracts, one should be aware of the fact that inorganic salts
may be present which will contribute to the dry weight of the
sample. In such cases, either dialysis against water should precede the dry weight determination or they should be determined
as inorganic ash. *Part A* of this exercise describes a microprocedure for total dry weight determination followed by inorganic ash
measurement which is described in *Part B.* The samples used can be
any one of the preparations described in this Manual.

Part A. Total Dry Weight

Materials and Equipment

Extract from Exercise No. 21, containing endotoxic glycolipid
    dissolved in chloroform:methanol and an unknown amount of
    small SiO_2 particles

    Cahn electrobalance
    Aluminum foil
    Infrared (IR) drying oven
    Scissors
    Forceps
    50 µl pipettes
    Agglutination plates, glass. (Available from Scientific Pro-
       ducts, McGaw Park, IL 60086, Cat. No. M6200)

Procedure

1. Either wear clean disposable gloves or do not touch anything
with your bare hands. Your fingerprints may contribute to the dry
weight your are going to determine.

2. Cut small squares approximately 15×15 mm from aluminum
foil. Place the foil between two layers of filter paper, put it
into your palm and push it down gently with your finger to form a
shallow cup. Transfer these small pieces of foil with forceps to
individual compartments of the glass agglutination plate. Number
the positions on the glass plate with crayon. Dry the aluminum
cups under the IR lamp for 10 min, then determine their weight
one by one on a Cahn electrobalance with the greatest possible
accuracy. Record the weights. Put the cups back into their numbered positions.

3. Pipette 50 µl solution into the aluminum cups, place them
under an IR drying lamp for 30 min. The distance of the lamp from
the samples should be between 15 and 20 cm. If they are too close,
organic macromolecules may be charred.

4. Turn off the lamp and wait 5 min to let everything cool
down. Determine the weight of the aluminum cups with the dried
samples on them by reweighing them on the Cahn balance. Calculate
the dry weight per milliliter of the analyzed samples.

Part B. Inorganic Ash

Materials and Equipment

    Vacuum desiccator
    P_2O_5
    Chrome-sulfuric acid
    2 ml pipette (high precision)
    Porcelain crucibles, 2-3 cm diam
    Furnace, adjustable up to 800°C
    Semi-micro or micro analytical balance. Required sensitivity:
      10 µg or less with a loading capacity of 20 g
    3 to 4 cm diameter glass Petri dishes

Procedure

1. Clean the porcelain crucibles by soaking them in chrome-
sulfuric acid overnight. The next day they should be rinsed very
thoroughly with glass distilled water. Do not touch them with
your fingers, always use forceps. Dry the crucibles at 100-120°C
in an oven or under the IR lamp. Place the crucibles individually

into 3-4 cm diameter glass Petri dishes and number the Petri
dishes. The role of the dishes is just to carry the number,
since one may not number the crucibles. Using forceps, transfer
the crucibles to the IR drying oven for 30 min, then into a vacuum desciccator for approximately 1 h. The desiccator should
contain 5-10 g P_2O_5 in a small beaker. After drying, determine
the weight of the crucibles on a semimicro- or microbalance.

2. Pipette exactly 1.00 ml solution into the crucible and dry
it under the IR lamp. Inspect the crucible after approximately
60 min. In case of any doubt about complete drying, continue
drying for another 30 min. Most solutions, unless with extremely
high dry weight content, will be dry in 1 h. Now transfer the
crucibles with their Petri dishes into a vacuum desiccator and
let them dry overnight under a vacuum less than 0.1 mm Hg.
Determine their weights, and the total dry weight in the crucibles.

3. To determine the inorganic salt content, one must burn
away the organic matter. Preheat the furnace to $700^\circ C$. Transfer
the crucibles into the furnace but without Petri dish saucers.
Glass will melt completely at this temperature. Place the crucibles into the furnace in a sequence so that you will be able to
identify them by their position. Leave them in the furnace for
2 h. Turn off the furnace and wait 1 h. Transfer the crucibles
into the vacuum desiccator and place them back into their numbered
Petri dishes. Keep them in the desiccator overnight. The following
day determine the weights again.

Evaluation

Most materials dry rapidly within an hour, and complete information about dry weight content can be obtained in 2 h using the
microprocedures.

The total dry weight, the organic matter dry weight, and the
inorganic dry weight can be calculated at this point. It is important that all dry weight determinations be carried out at
least in triplicate. The difference between the measurements
cannot be more than 5% in the case of the micromethod, and should
be less than that working with crucibles and 1 ml samples.

Use and Limitations

Some samples with high dry weight content will be more difficult
to dry than others. In case of any doubt, it is highly recommended
to return the small cups or porcelain crucibles to the IR oven
for another 60 min and reweigh them. If there is a greater than
1% or 2% difference between the measurements, a third drying
period is needed. This should be continued and repeated until a
constant dry weight is reached.

Some substances, particularly inorganic salts, are highly
hygroscopic. This can be noticed if the samples gain weight while
they are on the scale. In such case, one has to put them back
into the desiccator, evacuate it thoroughly, and repeat the measurement the next day. Sometimes it is sufficient to transfer
them back to the IR lamp oven for another 60 min drying. In any
case, the samples have to be transferred from the desiccator
directly to the analytical balance and the weighing must be carried out as fast as possible.

One should keep in mind that some inorganic salts are volatile. Ammonium compounds may disappear completely if exposed to elevated temperatures, other salts will form anhydrous oxides. In case such inorganic salts are present, handbooks of inorganic analytical chemistry should be studied and the necessary corrections made.

Some organic samples are sensitive to heat, they may lose water and form internal anhydrides. This is sometimes noticed if the dried sample turns a brown or darker color. Such substances should not be dried completely under IR lamp, but transferred while still slightly moist into a good vacuum desiccator for 24 h or longer for complete drying and constant weight.

If the sample is too dilute, it will be difficult to obtain accurate information about the dry weight content as described above. In such a case, the drying down of several 50 µl samples will be needed in the case of the microprocedure. Working with crucibles, not 1 but 5 or 10 ml should be dried down. Another approach, if feasible, is to concentrate the dilute sample before the dry weight determination is carried out. Such concentration can be achieved by vacuum distillation (see Exercise No. 32) or by the use of the procedure described in Exercise No. 8, provided that the drying agent (Dextran or Carbowax 6000 or Lyphogel) will not release soluble components or particles into the solution to be concentrated. Lyophilization of large volumes into almost invisible quantities of dry substances is another approach (see Exercise No. 32). Dissolving the total residue in accurately measured 5 or 10 ml glass distilled water or other clean solvent may give a solution ideally suitable for the dry weight determination procedures described here.

Exercise No. 48

Microdetermination of Nitrogen

For the determination of total nitrogen in biologic samples, the material must be digested with sulfuric acid, during which all the nitrogen will be transformed into ammonium sulfate. The ammonia from this solution may be liberated with an excess of highly concentrated sodium hydroxide.

Different catalysts are necessary to facilitate the digestion of the nitrogen-containing samples. Selenium dioxide is frequently used; mercury salts are also described. In the present Exercise, copper sulfate and potassium sulfate are added to facilitate and enhance the rate of digestion.

Materials and Equipment

    Concentrated sulfuric acid, reagent grade, saturated with anhydrous copper sulfate
    Potassium sulfate
    Sodium hydroxide, 50%
    Hydrogen peroxide, 30%
    Hydrochloric acid, 1/70 N
    Indicator mixture (80 mg methyl red + 20 mg methylene blue in
      100 ml ethanol)

    Boric acid solution, saturated
    Kjeldahl flasks, 100 ml capacity
    Alundum or glass beads
    Pipettes
    Digestion rack (Fig. 35)
    Steam distillation unit (Fig. 36)
    Erlenmeyer flask, 125 ml capacity
    L-leucine standard. Dissolve 47.0 mg leucine in 50 ml water.
     This will contain 100 ug N/ml

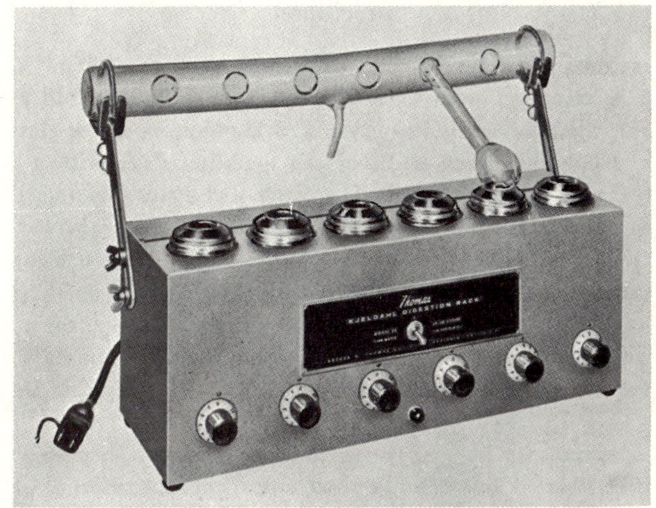

Fig. 35. Kjeldahl
digestion rack
(courtesy of Arthur
H. Thomas, Philadelphia,
PA.)

Procedure

1. With maximum accuracy, pipette the sample into the 100 ml
Kjeldahl flask. Take 0.1 ml aliquots from the L-leucine standard
also into separate flasks. These will serve as controls of the
efficiency of the digestion applied. Add 0.2 ml H_2SO_4 saturated
with $CuSO_4$. With a small spoon, add approximately 0.2-0.4 g
K_2SO_4 crystals. Add a few crystals of alundum or a few glass
beads to the digestion flask. Rinse the inside walls of the
digestion flask with a few drops of distilled water, then start
the digestion. The water will evaporate first, therefore only
moderate heating should be applied. After the water has been
removed, white sulfur dioxide fumes will appear. At the same time,
the sample will start to turn dark. With continued digestion, this
charring should clear up. If this does not happen, cool the solution and carefully add a few drops of 30% H_2O_2. Heat the digestion
flask again. Continue the boiling of the sulfuric acid-containing
sample approximately 20 min, until the dark color of the digested
sample disappears. Discontinue the heating.

2. Two procedures may be used for the determination of nitrogen
in the digested sample. A well elaborated and reliable procedure
is the liberation of ammonia from the acidic solution with concentrated NaOH and steam distillation of the liberated ammonia
into saturated boric acid. The absorbed ammonia in this boric

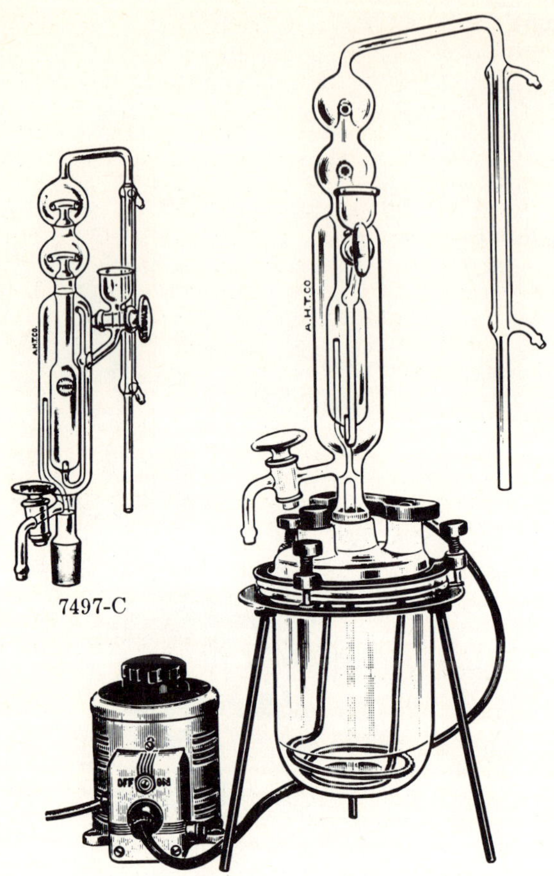

Fig. 36. Kjeldahl distilling apparatus (courtesy of Arthur H. Thomas, Philadelphia, PA.)

acid sample may be titrated with HCl. This procedure is accurate
and reliable but somewhat more time-consuming than the colorimetri
determination of nitrogen with the help of the Nessler reagent,
which gives a dark yellow color with ammonia.

Fill the steam distillation apparatus with distilled water
and let it boil for 10 min. Pipette approximately 10 ml nitrogen-
free distilled water and 10 ml 50% NaOH through the funnel of the
steam distillation equipment into the compartment which will later
receive the nitrogen-containing sample. Place a 125 ml Erlenmeyer
flask at the effluent end of the condenser and adjust the height
of the Erlenmeyer flask so that the tip of the condenser is submerged in the 15 ml boric acid-indicator mixture (95 ml boric
acid + 5 ml indicator). Start the distillation and continue it
until approximately 5 min have elapsed. After 5 min, lower the
position of the Erlenmeyer flask without interrupting the steam
distillation so that the effluent end of the condenser is approximately 1 cm above the surface of the receiving boric acid mixture.
Continue the distillation for 3 more min. Discontinue the heating
of the water container of the steam distillation equipment. This
is your nitrogen blank. Repeat the same procedure once more,
pipetting distilled water and NaOH into the sample-receiving
compartment.

3. Through the funnel, which served for the introduction of
sample into the steam distillation chamber, transfer approximately 20-40 ml nitrogen-free distilled water into this compartment. This rinse water can be drained from the equipment. Turn
off the stopcock of the drain, and now transfer the contents of
the digestion tube into the sample compartment. Use approximately
10 ml water for the quantitative transfer of the digested sample
and add 10 ml 50% NaOH. Place another Erlenmeyer flask with fresh
15 ml boric-acid-indicator mixture under the effluent end of the
condenser as described above. Start the distillation and continue
it for approximately 5-6 min. Without interrupting the distillation, lower the Erlenmeyer flask and continue the process for
3 more min. Rinse off the tip of the condenser with a few drops
of distilled water. The distillation of the sample is now complete.
    Rinse the equipment as described above and proceed with dis-
tilling the next digested sample.
    4. The titration is carried out by using $N/70$ HCl from a micro-
buret. The end point of the titration is reached when the color
of the sample turns light greyish-blue.

Evaluation

The milligrams of nitrogen content in the sample equals the milliliters of $N/70$ HCl used for the titration.

Use and Limitations

Two sources of error may be introduced by this method, first by
imperfect digestion of the nitrogen-containing sample, and second
through loss of ammonia during the distillation procedure. Both
of these sources of error may be eliminated with some practice.
It is recommended to take a reagent grade nitrogen-containing
sample such as highly purified amino acids and determine the
nitrogen content of a standard solution by the above described
procedure. The value found must be within ±0.5% of the theoreti-
cally calculated value.
    In serologic experiments, usually large numbers of samples
have to be digested in order to determine their nitrogen content.
A modification of the Kjeldahl reaction is described by Campbell
and coworkers (1963). In their modification, the digestion of the
samples is carried out in Folin-Wu tubes. The digested samples
are filled to 50 ml in the same tube and aliquots are taken from
the hydrolysate for Nesslerization. The sensitivity of this reaction is approximately 1 µg and it can be used within a range
of 2-50 µg N. Another Nesslerization method was published by
Lang (1958), which has a range of 2-1000 µg N. If larger amounts
of nitrogen are present in the hydrolysate, the Nessler reagent
will form a slight turbidity in the tubes. This turbidity cannot
always be observed unless the investigator looks down through the
tube longitudinally. This light turbidity may result in high
extinction values. To avoid this disturbance, a lower dilution
of the digestion sample must be used. The reproducibility of the
method is within ±2%. For details of the use of Nessler's reagent,
see the references given below.

References

Campell, D.H., Garvey, J.S., Cremer, N.E., Sussdorf, D.H.: Methods in immunology, p. 53. New York: W. A. Benjamin, Inc. 1963
Lang, C.A.: Anal. Chem. *30*, 1692 (1958)

Exercise No. 49

Microdetermination of Phosphorus

Several different procedures are recommended for the determination of total phosphorus in biologic samples. The method described
here has the greatest sensitivity, being able to demonstrate the
presence of 0.10 µg phosphorus per ml. It is based on the reduction of phosphomolybdate complex with ascorbic acid. The procedure
described here is essentially identical to the method published
by Chen, Toribara and Warner (1956).

Materials and Equipment

    2.5% ammonium molybdate dissolved in distilled water
    10% ascorbic acid dissolved in distilled water
    6 *N* sulfuric acid
    Concentrated sulfuric acid
    Perchloric acid
    O-phospho-L-serine, cryst.
    KH_2PO_4, reagent grade
    Concentrated nitric acid
    Digestion rack (Fig. 35)
    Pipettes
    Spectrophotometer
    $37^{\circ}C$ water bath

Procedure

   1. In order to determine the total phosphorus content of a
biologic sample, the material must be digested in concentrated
H_2SO_4. Prepare an accurate stock solution of the unknown material
and pipette an amount of this, which contains approximately 1 mg
dry weight, into a Kjeldahl digestion flask. The amount of the
material pipetted must be known with maximum accuracy. If dry
material is used, use a microbalance to weigh the sample. Add
0.2 ml concentrated H_2SO_4 and 0.2 ml concentrated HNO_3. Heat the
samples on the digestion rack until the brown vapors of nitrous
oxide disappear. This may be enhanced by repeated addition of
0.5 ml water to the cooled digestion tube. Heat the tube further
until white fumes of sulfur trioxide fill the digestion tube.
Cool the tube, then add 2 drops 72% $HClO_4$ to the tube and heat
again until the liquid becomes clear. Pipette exactly 9.8 ml
distilled water into the digestion tube.

2. Take different aliquots of the digested sample in order
to find those which are within the usable range of this determination. Take 0.5, 1.0, 2.0, and 4.0 ml of the digested sample. Add
water to make up the volume of all samples to 4 ml.

3. Prepare an inorganic phosphorus standard containing 10 µg
P/ml prepared from analytic grade KH_2PO_4 by dissolving 4.394 mg
KH_2PO_4 in 100 ml distilled water. Prepare the standard curve.
Take five different amounts from 0.2 to 1.0 ml of the 10 µg/ml
phosphorus-containing standard solution. Add water to each tube
to make up its volume to 4 ml.

4. Prepare the reagent. Mix 1 vol 6 N H_2SO_4 with 2 vol distilled water. Add 1 vol 2.5% $(NH_4)_3MoO_4$ and finally add 1 vol 10%
ascorbic acid. Mix well. The reagent must be prepared fresh
daily.

5. The development of the blue color is done as follows: Pipette
4 ml of reagent into each 4 ml P-containing sample. Do the same
with the tubes containing different amounts of phosphorus standard.
Mix the contents of the tubes and place the tubes in a 37°C water
bath or incubator for 90-120 min.

Evaluation

Read the optical densities of the reagents samples in a spectrophotometer at 820 nm against a blank which contains all the reagents except the phosphorus-containing sample. Plot the calibration curve. Read the amount of phosphorus in the unknown samples
at the same wavelength. Readings below 0.05 and above 0.6 should
be disregarded. Calculate the amount of total phosphorus in your
entire digested sample.

Use and Limitations

The critical point of the determination is the proper digestion
of the sample. Several laboratories use a standard organic
phosphorus compound which will be digested in the same experiment.
For such purpose it is recommended to use O-phospho-L-serine or
other compounds which are commercially available in reagent grade
quality.

The same reagents as described here may be used for the determination of inorganic phosphorus in any biologic materials. In
such case, the material will not be digested. The incubation at
37°C with the reagent should be restricted to 60 min because
longer incubation with the highly acidic reagent may liberate
organically bound phosphorus.

Reference

Chen, P.S., Jr., Toribara, T.Y., Warner, H.: Anal. Chem. *28*, 1756 (1956)

Exercise No. 50

Protein Determination by the Biuret Method

The biuret reaction, used for the quantitative determination of
proteins, is based upon the formation of a copper chelate with
the peptide bonds of the protein at alkaline pH. In this complicated reaction one copper atom forms a complex with four
peptide nitrogens.

This reaction, being simpler and faster than other procedures,
is widely used, especially in immunochemistry where quite often
a large number of specific precipitates must be analyzed. The
sensitivity of the biuret reaction described here is 25 µg to
350 µg protein. This Exercise demonstrates the preparation of a
calibration curve between the above limits.

Materials and Equipment

    Normal human serum with known nitrogen content
    Physiologic saline
    0.2 N sodium hydroxide (CO_2-free, standardized)
    Sodium potassium tartrate
    Copper sulfate · 5 H_2O
    Potassium iodide
    Volumetric flask, 100 ml capacity
    Bausch & Lomb Spectronic 20 photometer

Procedure

1. Carefully pipette 0.5 ml serum and 9.5 ml saline for a
dilution of 1 to 20. Determine the protein N content of this
dilution by the Kjeldahl procedure (Exercise No. 48).

2. Prepare the biuret reagent as follows: Dissolve 4.5 g
sodium potassium tartrate in 200 ml standardized 0.2 N NaOH.
Pour the pulverized 1.5 g $CuSo_4$ · 5 H_2O into it and wait until it
dissolves, then add 2.5 g KI. After it is dissolved, make up the
solution to 500 ml with 0.2 N NaOH.

3. Pipette 0.1, 0.2, 0.3, ... 1.0 ml amounts of diluted serum
into properly marked test tubes, make them up to 1 ml with water,
add 1.5 ml biuret reagent to each tube. Take two additional tubes
add 1.0 ml water and 1.5 ml reagent to each. These will be the
blanks.

4. Mix and incubate the tubes at 37°C for 30 min and read the
developed color against the blanks at 550 nm in the spectrophotometer. The two blanks read against each other should not show a
higher difference than 0.02 extinction.

Evaluation

Plot the protein concentration of the samples against optical
densities read with the spectrophotometer. This is your calibration curve.

Use and Limitations

The method is fast and simple, but in the analysis of certain
antigen-antibody precipitates gives erroneous results. Proteins
absorb light in UV at 280 nm. This also serves as a sensitive
and simple measurement for protein-containing solutions, particularly in monitoring chromatographic effluents (see Exercises
Nos. 2, 3, 6, 7, and 8). The Kjeldahl method is more reliable for
accurate measurement of protein content. The biuret method or the
micromodification of the Folin reaction by Herriott (1941) are
also useful in column chromatographic experiments where a large
number of collected samples must be analyzed and the measurement
of exact quantities is not the primary aim. A popular procedure
for similar purposes is the method of Lowry and co-workers (1951),
which can be applied for the detection of as little as 0.2 µg
protein. A comparative study of methods used for the analysis of
specific precipitates in immunochemistry has been made by McDuffie
and Kabat (1956).

References

Herriott, R.M.: Proc. Soc. Exp. Biol. Med. *46*, 642 (1941)
Lowry, O.H., Rosebrough, N.J., Farr, A.L., Randall, R.J.: J. Biol. Chem.
   193, 265 (1951)
McDuffie, F.C., Kabat, E.A.: J. Immunol. *77*, 193 (1956)

Exercise No. 51

Determination of Primary Amino Compounds With Ninhydrin

Ninhydrin reacts with substances containing primary amino groups,
such as amino acids, amino sugars, aliphatic amines, etc. Moore
and Stein (1954) developed a ninhydrin reagent which can be used
for quantitative determination of these compounds in neutralized
hydrolysates and other solutions. The Exercise described here is
based on their procedure.

Materials and Equipment

    Glycine (reagent grade kept in desiccator under vacuum)
    Sodium acetate · 3 H_2O
    Acetic acid, glacial
    Ninhydrin
    Hydrindantin
    Ethanol
    Cellosolve (Ethylene glycol monomethyl ether)
    Nitrogen gas
    Boiling water bath with a dark cover
    Electric pH meter
    Spectrophotometer
    Magnetic stirrer
    Test tubes with Teflon lined screw caps

Procedure

1. Prepare a 4 mol sodium acetate buffer, pH 5.5. Dissolve 272 g sodium acetate · 3 H_2O in 400 ml distilled water. Use hot water bath to enhance the dissolution. Cool to room temperature. Add 50 ml glacial acetic acid. The pH of this solution should be 5.5 ± 0.1. The pH may be adjusted with careful addition of concentrated acetic acid or by adding dropwise 50% sodium hydroxide solution. The pH adjustment must be carried out under vigorous stirring of the solution and the measurement of the pH must be done with an electric pH meter. After pH 5.5 has been reached, dilute the buffer to 500 ml.

2. The ninhydrin reagent is prepared by dissolving 2 g ninhydrin and 0.3 g hydrindantin in 75 ml Cellosolve, in a 200 ml dark screw-cap bottle. Then add 25 ml 4 M sodium acetate buffer. Bubble nitrogen gas through the solution, then close the container and shake gently to enhance the dissolution of the reagents. The reagent must be kept in a dark place in the cold room. If the atmosphere above the reagent in the flask consists of nitrogen, the reagent is stable for two weeks.

3. Make a 1 N sodium acetate buffer by diluting one volume of the concentrated buffer with 3 vol distilled water. Prepare 120 ml. Using a micro- or semimicro analytical balance, weigh 1.501 mg glycine into a 100 ml volumetric flask. Make up to 100 ml with the 1 N sodium acetate, pH 5.5, buffer. This standard amino acid solution (0.2 mM) serves for the preparation of the calibration curve.

4. Make all determinations in triplicate. Take the screw-capped test tubes and pipette 1 ml of your unknown into each of the first three tubes. Pipette 1 ml of 1 N buffer into two additional tubes. These will serve as blanks. Pipette the following amounts of the glycine standard solution into five additional tubes: 0.2, 0.4, 0.6, 0.8, and 1.0 ml. Add the necessary amounts of 1 N buffer in the last five tubes to make up the volumes to exactly 1 ml. To each tube add 1 ml ninhydrin reagent, dissolved in Cellosolve and sodium acetate buffer. Close the tubes tightly with Teflon-lined caps and shake them for approximately 10 sec. Put the tubes into a boiling water bath and cover them to exclude light. Keep them in the water bath for exactly 15 min. Remove the tubes and cool them in cold water. Unscrew the caps and add to each tube 15 ml 50% ethanol in water. Close the tubes tightly again with the Teflon caps and shake them thoroughly for a few seconds. Read the developed color at 570 nm, within 30 min.

Evaluation

Plot the optical densities of the standard glycine samples against the millimoles amino NH_2-content. Note that 1 ml of above 0.2 mM glycine solution will have 0.2 microequivalent (µEq) NH_2- (or 0.2 µM glycine) per milliliter. Read the µEq NH_2-content of the unknown samples from this standard curve. Calculate the number of NH_2-µEq per mg dry substance.

Use and Limitations

The amino sugars and other aliphatic primary amino compounds
will react with ninhydrin. Ammonia or ammonium salts will also
develop color. Therefore, the samples to be analyzed should either
be free of these constituents or their content has to be determined in separate analytic procedures. For quantitative determination of amino sugars, a procedure is given in Exercise No. 55.
The amount of amino sugar NH_2-μEq has to be calculated and deducted from the total NH_2-μEq values obtained by the ninhydrin
procedure.

Reference

Moore, S., Stein, W.H.: J. Biol. Chem. *211*, 907 (1954)

Exercise No. 52

Carbohydrate Determination by Phenol-Sulfuric Acid

The carbohydrate content of unknown biologic products can be
determined by a very simple and fairly reliable procedure. This
method does not include previous hydrolysis of the sample and
liberation of the constituting monosaccharides, but uses concentrated sulfuric acid in the presence of phenol. Carbohydrates
form either furfuraldehyde or its homologs with strong acid.
These derivatives of the carbohydrates produce colored compounds
by polymerization or condensation with aromatic phenols. This is
the basis of the Molisch reaction and also of Bial's orcinol
assay (1902) which was originally designed for pentoses and
hexuronic acids. The anthrone reaction is perhaps the best known
version. Certain SH-containing molecules also react with the
furfural derivatives. The cysteine-sulfuric acid reaction developed and elaborated by Dische (1949 and 1955) was most studied
in this group of reactions. A version of the cysteine-sulfuric
acid reaction is described in this Manual for quantitative determination of heptoses (Exercise No. 56) in the presence of other
carbohydrates.

Dubois and co-workers (1956) elaborated the phenol-sulfuric
acid procedure which is probably the most sensitive among these
types of reaction; therefore this is included in this Manual in
the form of a brief exercise.

Materials and Equipment

    5% phenol dissolved in distilled water
    Dextrane
    Concentrated sulfuric acid
    Spectrophotometer
    Pipettes
    Test tube mixer (see Fig. 37)
    Rubber gloves

For large numbers of carbohydrate determinations, e. g.,
samples from column chromatographic fractionation, the use of all
glass automatic reagent distributors is highly recommended. Such
an automatic pipetter is shown in Fig. 38.

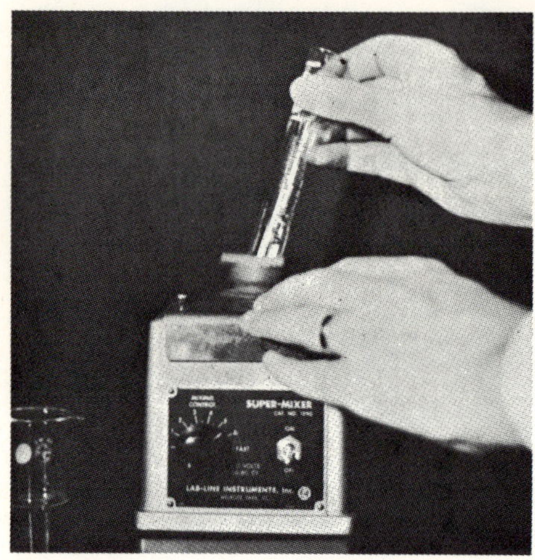

Fig. 37

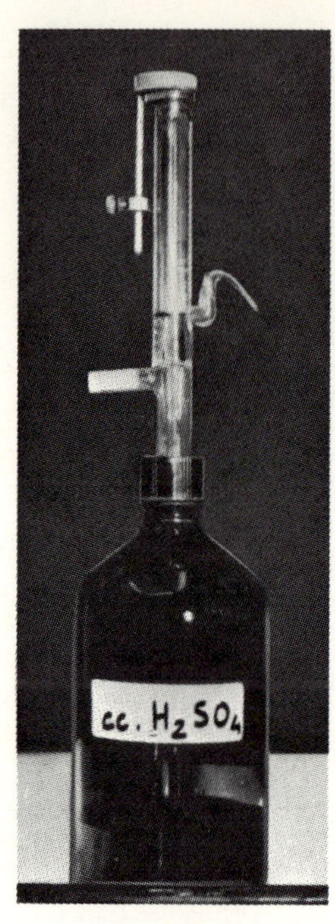

Fig. 37. "Vortex"-type test tube mixer (courtesy
of Scientific Industries, Inc., Springfield, Ma.)

Fig. 38. Glass automatic pipetter, available from
Labindustries, Berkeley, CA.   Fig. 38

Procedure

1. Pipette 0.5 ml carbohydrate-containing solution into a
test tube and add 0.5 ml 5% phenol. Add to it quickly 2.5 ml
concentrated H_2SO_4. Protect your hands with rubber gloves from
possible spilling of hot acid. Work carefully. Mix the contents
of the tube immediately; the use of a Vortex-type shaker shown
in Fig. 37 is recommended. Transfer the tubes into boiling water
bath for 15 min, then chill them in ice water. Read the developed
color at 490 nm, against a blank which contains 0.5 ml distilled
water and 0.5 ml phenol + 2.5 ml H_2SO_4.
2. Prepare a calibration curve from a standard polysaccharide
solution. Use dextrane which has been dried in vacuum desiccator
and prepare a stock solution containing 100 µg polysaccharide per
ml. Pipette different amounts varying from 5 µg up to 50 µg
into different test tubes and make up the volume of each tube to
0.5 ml. Treat these standards in the same manner as the unknown
samples containing carbohydrate.

Evaluation

Plot the OD values against µg dextrane. Use this calibration
curve to calculate the amount of carbohydrate in your unknown
samples.

Use and Limitations

The phenol-sulfuric acid reagent does not react with sugar alcohols or with amino sugars. Pentoses show a lower molecular extinction in this reaction than hexoses.

The procedure is especially useful for rapid analysis of large
numbers of samples. For example, in the analysis of chromatographic effluents, the investigator is not concerned with the
absolute value of the carbohydrate content as much as with the
distribution of different carbohydrate-containing zones among the
collected test tubes. This procedure is the most suitable for such
purposes.

If the phenol-sulfuric acid procedure is used for quantitative
studies, it is recommended to add the sulfuric acid to a prechilled
sample and keep the reaction mixture in an ice bucket. In that
case, the development of the color takes place when the tubes are
immersed in a boiling water bath for 15 min. After exactly 15 min,
the tubes must be cooled down, again using an ice-water bath.

The developed color is stable at cold room temperature for
24 hours.

References

Bial, M.: Dtsch. med. Wschr. *28*, 253 (1902)
Dische, Z.: J. Biol. Chem. *181*, 379 (1949); see also Dische, Z.: Meth.
   Biochem. Anal. *2*, 313 (1955)
Dubois, M., Gilles, K.A., Hamilton, J.K., Rebers, P.A., Smith, F.: Anal.
   Chem. *28*, 350 (1956)

Exercise No. 53

Ultramicrodetermination of Reducing Carbohydrates

One of the most sensitive methods for the determination of reducing carbohydrates was elaborated by Park and Johnson (1949).
Reducing mono- and oligosaccharides will form potassium ferro-
cyanide with potassium ferricyanide in alkaline solution. The
addition of ferric ammonium sulfate in a detergent such as sodium
lauryl sulfate (or Duponol) will form a colloidal dispersion of
ferric-ferrocyanide. The intense blue color of this compound can
be measured quantitatively on a spectrophotometer.

Materials and Equipment

   Potassium ferricyanide (reagent grade)
   D-glucose (reagent grade)

Sodium carbonate (reagent grade, anhydrous)
Potassium cyanide (reagent grade)
Ferric ammonium sulfate (reagent grade)
Sodium lauryl sulfate (Duponol)
Sulfuric acid (reagent grade)
Acid-washed regular test tubes
Boiling water bath
Vacuum desiccator containing phosphorus pentoxide
Spectrophotometer

Procedure

1. Dissolve 0.5 g potassium ferricyanide in 1000 ml water.
Store in a brown bottle. This is solution A.
2. Dissolve 5.3 g sodium carbonate and 0.65 g potassium cyanide
(KCN) in 1000 ml water. This is solution B.
3. Dissolve 1.5 g ferric ammonium sulfate in 1000 ml 0.05 N
H_2SO_4 by heating gently and with constant stirring. Cool and add
1 g sodium lauryl sulfate. This is solution C. It must be free
of any turbidity. Filter it, if needed.
4. Pipette the test solution containing not more than 10 µg
of reducing monosaccharide or the corresponding amount of reducing
oligosaccharide into a test tube, and make up the volume to 3 ml
with water. Add 1 ml solution B and 1 ml solution A. Mix together
gently, and heat for 15 min on a boiling water bath, covered with
"cold fingers". (see Fig. 16). Cool in ice water and add 5 ml
solution C. Mix well. Let stand at room temperature for 15 min,
then read at 690 nm against a blank which does not contain glucose.
5. Prepare a standard glucose solution by dissolving 10 mg
glucose in 1000 ml water. The glucose must be reagent grade and
must be dried in a vacuum desiccator over P_2O_5 at room temperature
for at least three days. Use greatest accuracy in preparing this
standard solution. Pipette 1, 2, 3, etc. µg glucose-containing
aliquots up to 10 µg glucose into acid-washed test tubes, make
up the volume of these samples to 3 ml with water. Proceed as
above and develop the colors. The preparation of the standard
curve must be done simultaneously with the determination. At each
new determination, recheck the calibration curve you obtained at
a minimum of four different points.

Evaluation

Plot the optical density readings shown by the standard glucose
samples against the amount of glucose present in these determinations. This is your calibration curve. To determine the amount
of reducing carbohydrates present in your samples, read the amount
of reducing carbohydrates from the calibration curve.

Use and Limitations

This method is often used to determine the carbohydrate content
of natural products. Less than 0.1 mg sample is sufficient for
quantitative carbohydrate analysis. If this amount is hydrolyzed
in 1 ml 1 N H_2SO_4 for the necessary length of time, (see Exercise
No. 33) the sample must be neutralized by adding 1 ml 1 N NaOH.

Add 1 ml distilled water and proceed by adding the further reagents as described in the Procedure.
   This method gives similar but not identical values with all
reducing carbohydrates. In contrast to the phenol-sulfuric acid
determination of carbohydrate content (see Exercise No. 52), this
method is suitable for quantitative analysis of common amino
sugars and aldopentoses as well as aldohexoses.
   The sensitivity of this determination is between 1 and 10 µg.
A higher amount of reducing carbohydrates will produce turbidity
and precipitation of the blue ferric-ferrocyanide complex. If
this is observed, use higher dilutions of your carbohydrate
samples. Since the blue complex is a stabilized colloid system,
high salt concentrations can cause precipitation. Further dilution
of the carefully neutralized hydrolyzate may be necessary to avoid
this.
   Because of the extreme sensitivity of the determination, it is
imperative to use acid-washed test tubes and pipettes in the
entire operation.

Reference

Park, J.T., Johnson, M.J.: J. Biol. Chem. *181*, 149 (1949)

Exercise No. 54

Enzymatic Determination of Carbohydrates

A large number of different carbohydrate determination methods
described in the literature are useful for the determination of
reducing carbohydrates in biologic samples, but only a few methods
are described to differentiate between individual carbohydrates
in a more or less specific manner, eliminating the disturbance
caused by the presence of other constituents in natural products.
   Keston (1956) developed a specific colorimetric reagent for
enzymatic glucose determination. A simplified scheme of the reaction can be given as follows:

$$\text{D-glucose} + H_2O + O_2 \xrightarrow{\text{glucose oxidase}} H_2O_2 + \text{gluconic acid}$$

$$H_2O_2 + \text{a reduced chromogen} \xrightarrow{\text{peroxidase}} \text{oxidized chromogen}$$

   The chromogen used in this reaction as well as the enzyme
preparation are available from Worthington Biochemical Corp.,
Freehold, NJ 07728, or from Boehringer, Mannheim, West Germany.
   The most useful procedure for applying the enzymatic approach
to quantitative glucose determination is given here in *Part A*.
   Galactose oxidase is used for the quantitative determination
of galactose in the presence of other hexoses. Galactose + galactose
oxidase produces H_2O_2 which will oxidize a reduced chromogen and
produce a color which can be measured spectrophotometrically. The
reaction is a follows:

$$\text{D-galactose} + H_2O + O_2 \xrightarrow{\text{galactose oxidase}} \text{D-galacto-hexadialdose} + H_2O_2$$

$$H_2O_2 + \text{reduced chromogen} \xrightarrow{\text{peroxidase}} \text{oxidized chromogen}$$

Commercially available preparations produced by Worthington Biochemical Corp. will be used in this determination, under *Part B*.

Part A. Glucose

Materials and Equipment

Glucostat (available from Worthington Biochemical Corp. Freehold, NJ 07728)
Chromogen for use with Glucostat (available from the same company)
D-glucose, standard solution, containing 500 µg glucose/ml
4 *N* hydrochloric acid
37°C water bath
Electrometric pH meter
Spectrophotometer
100 ml graduate cylinder

Procedure

1. Dissolve the commercial reagent as follows: Open the Glucostat chromogen vial, dissolve its contents in distilled water and add it to a graduate cylinder containing approximately 50 ml distilled water. Dissolve the contents of the Glucostat vial in distilled water and mix it with the chromogen. Make the solution up to 100 ml with distilled water. The commercial enzyme preparation contains phosphate buffer to maintain its pH at 7.0.

2. Prepare glucose standards from your stock solution. Take six tubes and pipette different amounts of your glucose standard into them, starting with 0.2 ml and increasing the amount by 0.2 m. in each tube. Tube No. 5 will contain 1.0 ml standard glucose solution; tube No. 6 will contain 1.0 ml distilled water. Make up the volume of the glucose solution in each tube to 1.0 ml with distilled water.

3. Take 1.0 ml, 0.5 ml, and 0.2 ml of your unknown sample. Make them up to 1 ml with distilled water. If your material is a product of acidic hydrolysis, previously neutralize the hydrolysat to pH 7.0 using a pH meter. The presence of salts does not disturb the enzymatic reaction provided the ionic strength is not too high. The amount of glucose in the unkown samples should be within the range of 0.05 mg to 0.3 mg glucose/ml.

4. Pipette 9.0 ml reagent into each tube. After exactly 10 min at room temperature, add 1 drop of 4 *N* HCl to stop the reaction and stabilize the developed color. The reaction time must be kept standard. The developed color may be read in a spectrophotometer at 400 nm against the reagent blank, which did not contain carbohydrates.

Evaluation

Read the glucose content of your unknown sample from the calibration curve. To obtain the calibration curve, plot the amount of
glucose versus optical density.
Use and Limitations see *Part B.*

Part B. Galactose

Materials and Equipment

Galactostat reagent (available from Worthington Biochemical
  Corp.)
Chromogen preparation for use with Galactostat (available from
  the same company)
D-galactose, standard solution, containing 50 µg galactose/ml
Methanol
0.25 *M* glycine buffer, pH 9.7
Triton X-100, 2% in water (commercially available from Rohm &
  Haas Co., Philadelphia PA 19105)
Other equipment is the same as in *Part A*

Procedure

1. Add 0.5 ml methanol to the Galactostat chromogen vial. Dilute
it with water to 30 ml. Take the vial containing Galactostat, add
to it approximately 5 ml water, and add it to the chromogen solution. Add 5.0 ml 2% Triton X-100 to the mixture. Rinse the vials
and bring the total volume of the chromogen + Galactostat up to
50 ml. The solution is light sensitive, therefore it has to be
kept in a dark bottle. The Galactostat vial contains galactose
oxidase prepared by the Worthington Corp., together with small
amounts of phosphate buffer. This phosphate buffer keeps the pH
of the dissolved reagent at 7.0. The chromogen used in the enzymatic galactose determination is o-toluidine.
2. Take seven test tubes and pipette 0.4 ml D-galactose standard into the first one, 0.8 ml into the second, 1.2 ml into the
third, 1.6 ml into the fourth, and 2.0 ml into the fifth. The
sixth and seventh tubes will receive 2 ml of distilled water.
Add distilled water to each tube to make up the volume to 2 ml.
3. Take 2, 1, and 0.5 ml of the unknown sample in other test
tubes for galactose determination. Make them up to 2 ml. Place
the tubes with the standard D-galactose samples and with the unknown material in a 37°C water bath, and wait approximately 15 min
until all the samples reach the temperature of the water bath. Add
to them at 0 time 2 ml Galactose reagent. Mix and incubate the
tubes for 1 h. After the incubation time expires, pipette into
each tube 6 ml of the pH 9.7 glycine buffer. Addition of this
alkaline buffer will stop the reaction. The reaction may also be
stopped by the addition of EDTA (0.2 ml of 0.5 *M* EDTA), according
to the method of Sempere and co-workers (1965). The color developed may be read at 425 nm against the blank.

Evaluation

Read the amount of D-galactose in your unknown samples from the
calibration curve. Include a standard calibration curve in each
determination in order to check whether all your reagents are
functioning properly.

Use and Limitations

The same reaction may be used for determination of serum glucose
or galactose. In such cases, the sample must be deproteinized wit
zinc hydroxide as described by Somogyi (1945). The method can als
be used for quantitative determination of urine glucose or galactose. In that case, the constituents of the urine must be removed
by ion exchangers. The reaction is disturbed by reducing material
such as ascorbic acid, glutathione, uric acid, and bilirubin. Roth
and associates (1965) incubated the unknown sample with galactose
oxidase in the presence of peroxidase and benzidine. This method
has been applied to numerous biologic fluids.

References

Keston, A.S.: A.C.S. Abstracts, 129th meeting, p. 31 c (1956)
Roth, H., Segal, S., Bertoli, D.: Anal. Biochem. *10*, 32 (1965)
Sempere, J.M., Gancedo, C., Asensio, C.: Anal. Biochem. *12*, 509 (1965)
Somogyi, M.: J. Biol. Chem. *160*, 61 (1945)
Worthington Biochemical Corp.: Enzymatic diagnostic reagents (1964)

Exercise No. 55

Determination of Amino Sugars

Free amino sugars react with acetylacetone in alkaline solution
and form a pink-colored complex with p-dimenthylaminobenzaldehyde
This reaction was used for the quantitative determination of amin
sugars by Morgan and co-workers. Bound amino sugars, as present i
some natural products, must be liberated first by hydrolysis, as
described in Exercises 33 or 34. The procedure given here describ
one of the numerous modifications based on the publication of
Rondle and Morgan (1955).

Materials and Equipment

    Acetylacetone, analytic grade
    0.5 *N* sodium carbonate (53.0 g into 1000 ml)
    p-Dimethylaminobenzaldehyde, analytic grade
    D-glucosamine (kept in vacuum desiccator)
    Ethanol
    Boiling water bath
    65°C water bath
    Spectrophotometer
    Marbles or "cold fingers" (Fig. 16)
    Hydrochloric acid, concentrated, analytic grade

Procedure

1. Dissolve 1 ml acetylacetone in 50 ml 0.5 N Na$_2$CO$_3$. This is reagent A, which must be freshly prepared daily.

2. Prepare the so-called Ehrlich reagent by dissolving 0.8 g p-dimethylaminobenzaldehyde in 30 ml ethanol and add 30 ml conc. HCl. Store in the cold room. This is reagent B.

3. Prepare a standard solution using D-glucosamine hydrochloride. Weigh accurately 215.64 mg D-glucosamine hydrochloride into a 1000 ml volumetric flask and add a few hundred milliliters water to dissolve the amino sugars. Make up to 1000 ml. Of this 1 ml contains 179 µg = 1 µM D-glucosamine (or 1 amino sugar NH$_2$ microequivalent).

4. Take 0.5 ml of the unknown solution which must have a pH close to neutral. In another tube pipette 0.1 ml and in a third tube add 0.02 ml of the same unknown solution. Measure the volumes with maximum accuracy. Fill tubes Nos. 2 and 3 to 0.5 ml with distilled water.

5. Take five tubes and pipette 0.1, 0.2, 0.3, 0.4, and 0.5 ml standard glucosamine solution into the tubes. Make them up to 0.5 ml with distilled water. A sixth tube containing 0.5 ml distilled water will serve as control blank.

6. Add 0.5 ml of reagent A to all tubes. Mix gently, then add 0.5 ml distilled water. Place a marble or "cold finger" on the top of the tubes and place them in a boiling water bath for 20 min. Cool the tubes in cold water. Add 3.5 ml methanol and 0.5 ml of reagent B. Mix gently.

7. Place the tubes in a 65-67°C water bath for 10 min to expel the CO$_2$. Cool to room temperature and read the optical density at 530 nm against the blank.

Evaluation

Three different amounts of the unknown sample have to be used to ensure that at least one will fall in the right range of determination. If more accurate results are required, the determination must be repeated within closer dose intervals in the optimal range, which is now known from the previous run. Carry out the determinations in duplicate.

The optimal range of this determination is from 10 to 60 µg hexosamine per sample.

Plot the optical densities of the standard glucosamine samples against their amino sugar NH$_2$-microequivalent content. This latter is necessary if the amino acid content of the unknown material is to be determined. In Exercise No. 51 the amino acid NH$_2$-microequivalents have been determined by ninhydrin. Hexosamines give a similar reaction with ninhydrin, therefore this must be deducted from the total if the amino acid content is to be determined. If both the amino acid NH$_2$- and the amino sugar NH$_2$-content is expressed in microequivalents, the difference between the two values will give the amino acid NH$_2$-microequivalents (if no other ninhydrin reactive substances are present in the sample). For practical purposes, it is customary to express the amino acid content as glycine, or, more often, leucine, in which case the NH$_2$-microequivalent must simply be multiplied by the molecular weight of the corresponding amino acid. The amino sugar content is usually

expressed as D-glucosamine. Here again the amino sugar NH_2-micro-
equivalent value found, multiplied by 179, will give the microgram
amino sugar content of the analyzed sample expressed as D-glucosamine.

Use and Limitations

Only free hexosamines give a molecular extinction close to the
glucosamine standard in the above assay. Amino sugars with any
substituents on C-4 position are negative in this determination.
Other amino sugars will have slightly different molecular extinc-
tions, which must be determined using pure substances if available.
The method described here, as well as most similar group analyses,
will give only an average value of all amino sugars present in the
hydrolysate.

Reference

Rondle, C.G.M., Morgan, W.T.J.: Biochem. J. *61*, 586 (1955)

Exercise No. 56

Determination of Heptoses and Pentoses

Bacterial O-antigens often contain heptoses in addition to pentose
and hexoses. *Part A* described here is based on the work of Dische
(1953) for the quantitative determination of heptose. *Part B* is
for pentose determination according to Dische (1949).

Part A. Heptose

Materials and Equipment

    85% sulfuric acid, reagent grade
    3% cysteine hydrochloride dissolved in water
    Heptose standard, stock solution containing 200 µg/ml L-glycero
     D-mannoheptose
    Boiling water bath
    Pipettes
    Test tubes
    Spectrophotometer

Procedure

    1. With maximum accuracy, pipette between 4 and 8 mg aliquots
of the unknown material dissolved (or dispersed) in 0.5 ml water
into regular test tubes. Prepare three tubes in the identical
manner. Add to the solutions (or suspensions) 2.25 ml 85% H_2SO_4
while keeping the tubes in an ice water bath. Add the acid very
slowly while continuously shaking the tubes. Leave the tubes in
the bath for 5 min, then remove them and transfer them to a boil-

ing water bath and keep them there for exactly 10 min. After that immerse the tubes in ice cold water again. A few minutes later add 0.05 ml 3% cysteine hydrochloride to the samples.

2. Prepare a blank by measuring into a tube the same amount of unknown sample as used in the assay. The final volume of this sample must also be 0.5 ml. Add the H_2SO_4 to this sample in exactly the same manner as was done for the other samples. Do not add cysteine to the blank, but pipette 0.05 ml water into it. Prepare both the blank and the standard heptose solutions simultaneously with the unknown sample.

3. Take five test tubes and pipette different amounts of standard heptose into them, measuring increasing amounts from 10 µg to 100 µg. The final volume of the standard must be 0.5 ml in each tube. Pipette identical amounts into another five tubes; also adjust the volume of each to 0.5 ml with water. Pipette 2.25 ml H_2SO_4 into each; treat as above. This second set will give the blanks for the calibration curve. Add 0.05 ml 3% cysteine hydrochloride to the first five tubes, but add 0.05 ml distilled water to the second set. After the incubation time is over, read the different concentrations against their corresponding blanks.

4. Measure the optical density of the samples in a spectrophotometer exactly 2 h after the addition of cysteine against the blank to which no cysteine was added. Two readings must be carried out, one at 505 nm and another at 545 nm. OD_{505}-OD_{545} values will give the amount of heptose present in the samples.

Evaluation see *Part B*.

Use and Limitations see *Part B*.

Part B. Pentose

Materials and Equipment

D-arabinose, stock solution containing 200 µg/ml
Conc. H_2SO_4
Equipment as in *Part A*

Procedure

1. For the determination of pentoses a very simple procedure may be applied. To the same amount of sample (4-8 mg dissolved or suspended in 0.6 ml water) add very slowly, under ice-water cooling, 2.4 ml concentrated H_2SO_4. Stir the mixture constantly while adding the acid. Keep in ice water for approximately 15 min, then let it stand at room temperature for 1 h. Add to it 0.1 ml cysteine hydrochloride solution, mix, and measure the OD of the solution after exactly 20 min at 390 nm and 425 nm. The difference between OD_{390} and OD_{425} will give the pentose content. As a blank, take the same amount of sample, add sulfuric acid but do not add cysteine. Read the developed color of the unknown samples against this blank.

2. For a calibration curve, prepare a 200 µg/ml D-arabinose solution in water. Take 0.1, 0.2, ... ml amounts up to 0.6 ml and carry out the determination as above. For the standard calibration curve prepare the blanks by taking 0.1, 0.2, etc. amounts of arabinose, make up to 0.6 ml with water, add sulfuric acid, but

do not add cysteine. Read the different concentrations against
the corresponding blanks.

Evaluation

Plot the optical densities of your standard samples against the
heptose or pentose content. Read the heptose or pentose content
of your unknown samples from these calibration curves.

Use and Limitations

Although the absorption maxima of the heptoses and pentoses are
relatively far from each other, the presence of one in a sample
may influence the determination of the other. This is usually
low, however. Another method has been developed for pentose
determination by Dische and Schwartz (1937) using the Bial reaction. While this is less disturbed by the presence of heptoses,
other carbohydrates show a stronger influence on this reaction
than on the sulfuric acid-cysteine procedure. The possible influence of other different carbohydrates present in a polysaccharide has to be carefully studied and considered, if accurate
quantitative determinations are required.

References

Dische, Z.: J. Biol. Chem. *181,* 379 (1949)
Dische, Z.: J. Biol. Chem. *204,* 983 (1953)
Dische, Z., Schwartz, K.: Microchim. Acta *2,* 13 (1937)

Exercise No. 57

Determination of Sialic Acid and 2-Keto-3-Deoxyoctonic Acid Derivatives

Deoxyribose reacts with thiobarbituric acid after periodate oxydation, as discovered by Warawdekar and Saslaw (1957). Other
deoxycarbohydrates, such as 2-keto-3-deoxy (KDO) derivatives as
well as sialic acid will also give a pink colored reaction product after above treatment. Exercise No. 43 in this Manual gives
a paper- and thin-layer chromatographic spray reagent for the
detection of such carbohydrates. This Exercise is a colorimetric
method for the quantitative determination of sialic acid *(Part A)*
and of 2-keto-3-deoxyoctonic acid *(Part B)* contents in hydrolysates. The procedure is based on the method of Warren (1963)
for sialic acid, and of Weissbach and Hurwitz (1959) for 2-keto-
3-deoxyoctonic acid determination.

Part A. Sialic Acid

Materials and Equipment

    2 ml citrated blood
    Sodium metaperiodate reagent (Dissolve 4.3 g $NaIO_4$ in 100 ml
      50% H_3PO_4)
    Sodium arsenite reagent (Dissolve 10 g $NaAsO_2$ and 7 g Na_2SO_4
      anhydrous in 100 ml water)
    2-Thiobarbituric acid (TBA) solution (Dissolve 6 g TBA and 7 g
      anhydrous Na_2SO_4 in 100 ml water)
    N-acetylneuraminic acid (Sigma Chemical Co., St. Louis, MO.
      63178)
    Cyclohexanone
    Water bath, 100°C
    Vortex
    16 × 100 mm test tubes with $ stopper
    10 ml centrifuging tube (Corex)

Procedure

1. Sialic acid has to be liberated by careful acidic hydrolysis
for the TBA determination procedure. Take 2.0 ml citrated human
blood and prepare from it erythrocyte membranes (stroma) as described in Exercise No. 27. Take the wet stroma sediment and make
it up with distilled water to 5 ml. Add to the suspension 5 ml
0.2 N H_2SO_4, Vortex. Hydrolyze the sample on boiling water bath
for 30 min. Centrifuge at 3000 g for 15 min.

2. Pipette 0.25 ml of the supernate into a test tube (with
glass stopper) and add to it 0.1 ml periodate solution. Vortex
and let it stand at room temperature for 20 min. Pipette 1 ml
arsenite solution to the mixture to stop the oxydation. Vortex
for 2 min. Add to the mixture 3 ml of TBA reagent, Vortex again
and place the tube into a boiling water bath for exactly 15 min.
Cool the tubes in ice water.

3. Extract the reaction mixture with 4 ml cyclohexanone. Close
the test tube tightly with the glass stopper and shake it very
vigorously for 60 s. Transfer the tube's content into a 10 ml
centrifuge tube and separate the two phases at 1000 g for 15 min.

4. Lift the upper phase (cyclohexanone) and measure the OD at
548 nm.

Evaluation

Prepare a calibration standard curve from a pure, commercially
available sialic acid (N-acetyl-neuraminic acid) sample. Make
an 0.50 mg/100 ml standard solution. Take 0.05, 0.1, 0.15, 0.20,
and 0.25 ml aliquots, adjust the volume of all samples to 0.25
with water. Carry out the reaction as described above. Plot the
OD readings against the sialic acid content to make your calibration curve. For more accurate measurements it may be necessary
to take concentrations in addition to those listed. According
to Warren (1963) the molecular extinction coefficient for sialic
acid is 57,000.

Part B. 2-Keto-3-Deoxyoctonic Acid

Materials and Equipment

Lipopolysaccharide (LPS) or endotoxic glycolipid (EGL) prepared by procedures given in Exercise No. 21
Periodic acid (Dissolve 800 mg HIO_4 in 100 ml 0.1 N H_2SO_4)
Sodium arsenite solution (Dissolve 3 g $NaAsO_2$ in 100 ml 0.5 N
  HCl)
2-thiobarbituric acid reagent (Dissolve 0.3 g TBA in 100 ml
  water)
2-keto-3-deoxyoctonate (Sigma Chemical Co., St. Louis, MO
  63178)
0.025 N H_2SO_4
Water bath, 100°C
10 ml centrifuge tube (Corex)

Procedure

1. Weigh 2 mg liophylized LPS (or EGL) into a 10 ml centrifuge
tube and disperse it in 5 ml 0.025 N H_2SO_4. Put the tube into a
boiling water bath for 20 min. Cool in ice water and centrifuge
at 5000 g for 30 min in a refrigerated centrifuge. Use the supernate for KDO determination.
2. Take a small test tube and mix 0.25 ml supernate with 0.25 m
periodic acid reagent. Let it stand at room temperature for 30 mi
Stop the oxydation by mixing 0.5 ml arsenite solution to the samp
Vortex for 2 min. Pipette into the mixture 2 ml TBA reagent,
briefly Vortex again and immerse the test tube into a boiling
water bath for exactly 20 min. Cool the sample and read the OD
at 548 nm.

Evaluation

KDO is now available commercially, use this for the preparation
of calibration curves, as described in *Part A*.

Use and Limitation

Several carbohydrates with similar structure of the first four
carbon atoms will yield periodate oxydation products which will
give the characteristic pink color with TBA. Besides 2-deoxyribose
sialic acid, KDO and other carbohydrates, such as 2-keto-3-deoxy-
arabonic, galactonic, and heptonic acids, will all react under
the here described reaction conditions. Kasai and Nowotny (1967)
found three TBA-positive components in an endotoxic glycolipid
from *S. minnesota* R595 bacteria. The identification of these compounds cannot be done by the TBA method above. Gas-liquid chromatography has been preferred by Suttajit and Winzler (1971) for
the determination of sialic acid. Paper chromatography, paper
electrophoresis, or a combination of both can be used for identification of the TBA-positive compounds in extracts and hydrolysate
    One should also keep in mind that not all derivatives of neu-
raminic acid will react with TBA. Blix (1962) described the isolation of five different kinds of sialic acids from submaxillary

glands. Bovine diacetyl and triacetyl-neuraminic acids do not
react with TBA, but they can be detected by other colorimetric
procedures, such as the resorcinol method of Svennerholm (1957).
   KDO and related compounds are very sensitive to acidic
hydrolyses. One should not attempt to find them in those hydrolysates which were obtained by using 1 N or stronger acids. The
hydrolytic conditions as described here are about optimal for
their liberation.

References

Blix, G.: Methods in carbohydrate chemistry. Vol I. p. 246, Whistler, Wolfrom
   (eds.) New York Academic Press 1962
Kasai, N., Nowotny, A.: J. Bacteriol. *94,* 1824 (1967)
Suttajit, M., Winzler, R.J.: J. Biol. Chem. *246,* 3398 (1971)
Svennerholm, L.: Biochim. Biophy. Acta *24,* 604 (1957)
Warawdekar, V.S., Saslaw, L.D.: Biochim. Biophy. Acta *24,* 439 (1957)
Warren, L.: In: Methods of enzymology Vol VI, p. 463, Colowick, Kaplan (eds.)
   Berlin-Heidelberg-New York: Springer 1963
Weissbach, A., Hurwitz, J.: J. Biol. Chem. *234,* 705 (1959)

Exercise No. 58

Chemical and Biologic Analysis of Compounds Separated by Preparative Thin-Layer Chromatography

TLC is particularly suitable to separate a few hundred micrograms
of material containing aliquots if it is applied in the form of
a 160-180 mm long streak to the bottom of a TLC plate. If the
thickness of the SiO_2 layer increased to 0.5 mm or 0.75 mm, even
more material can be separated and recovered from the plates.
Since the chrome-sulfuric acid spray destroys the separated
compounds, it is replaced in this Exercise by spraying the TLC
plates with bis-distilled water. Hydrophobic bands will remain
white, while background or hydrophilic components will pick up
water and will become translucent gray. The visible bands may be
marked and scraped off together with the silicic acid carrier.
Part A of this Exercise describes procedures, developed in our
laboratories, to determine molar ratios of nitrogen, phosphorus,
hexosamines, long-chain carboxylic acids, 2-keto-3-deoxy octonate
(KDO), and heptose in such scrapings without attempting to elute
the components from the SiO_2 carrier (Chen et al., 1974). In
Part B methods are described to determine some of the biologic
activities present in the separated components. The procedure has
been applied by us to studies of endotoxic glycolipids (Chen et
al., 1975). These methodologies are described here.

Part A. Chemical Analyses

Materials and Equipment

Endotoxic glycolipid, extracted from lyophilized *S. minnesota*
R595 Re rough mutant. The extraction procedure is described
in *Part C* of Exercise 21. For further details see Chen et al.,
(1973)

TLC system, consisting of 20 × 20 cm plates, developing tanks
and other accessories as described in Exercise No. 42

Chromatographic solvent. Mix chloroform:methanol:concentrated
NH_4OH:bis-distilled water = 100:100:8:4 ratios. The organic
components in this solvent mixture should be redistilled in
the laboratory before use. Only the highest purity solvents
can be applied

Silicic acid, microgranular. (BioRad 2μ-10μ without binder or
other similar quality may be used. The purity of SiO_2 is very
critical. It has to be carefully analyzed and, if needed,
further purified, as described below)

Reagents and equipment for chemical analysis: For nitrogen
determination see Exercise No. 48, for phosphorus determination the same as in Exercise No. 49, for hexosamine determination see Exercise No. 55, for long-chain carboxylic acid
content measurement see Exercise No. 64, for heptose determination the same as in Exercise No. 56, and for 2-keto-3-
deoxyoctonic acid (KDO) determination see Exercise No. 57

Sample applicator for preparative TLC is available from Applied Science Laboratories, State College, PA 16802. Large
desiccator, ID 30 cm or larger, with Drierite or $CaCl_2$ desic-
cant

Thermometer, rabbits and restraining cages for pyrogenicity
measurements, as in Exercise No. 95

Rabbits (New Zealand Albino), needles and syringe for Shwartzman assay, as listed in Exercise No. 91

Limulus Lysate (available from Associates of Cape Cod, P. O.
Box 249, Woods Hole, MA 02543), water bath and test tubes

Micro N distillation apparatus (Labconoco Corp., Kansas City,
MO 64132)

Procedures

1. Prepare 1 ml (10 mg/ml) endotoxic glycolipid solution by
dissolving the preparation in chloroform-methanol = 4:1.

2. Analyze the purity of the silicic acid before use. Take
1 g aliquots and subject them to the chemical analytic procedure
described below. The silicic acid should not contain detectable
quantities of the components you wish to determine. If it does,
the following washing procedure must be used. First wash with
0.1 *N* NaOH at room temperature for 60 min, followed by distilled
water, then by 1 *N* HCl at 80°C, and finally again with distilled
water to bring the pH to neutral. Constant stirring must be applied during each step. Dry the washed silicic acid at 80°C under
vacuum.

3. Prepare the plates by mixing 30 g silicic acid with 56 ml
distilled water +4 ml concentrated NH_4OH. If nitrogen content
will be measured, adjust the pH of the slurry carefully with

10 *N* NaOH to 7.4, using a pH meter. This silicic acid slurry will
be spread on the plates with the piece of equipment identical or
similar to the one shown in Fig. 29 in Exercise No. 42. The plates
should be air dried, first at room temperature, then activated
at 120°C for 1 h in a vacuum oven, then cooled in a large vacuum
desiccator.

4. Apply 0.1 ml (1 mg) of the endotoxic glycolipid to each
plate, using the sample applicator or disposable calibrated micropipettes (Fig. 30), forming a continuous and approximately
160 mm–180 mm long streak. Dry the plates at room temperature,
then transfer them into the above vacuum desiccator and keep it
under good vacuum for at least 4 h. This treatment will give even
chromatographic runs and reproducible results.

5. Now transfer the plates into the TLC tank and develop them
in chloroform:methanol:concentrated NH_4OH:water = 100:100:8:4
solvent mixture. For nitrogen determination omit NH_4OH. Remove
the plates from the tank when the solvent front reaches the upper
one-tenth of the plate.

6. To make the separated hydrophobic bands visible, spray the
plates carefully but thoroughly with bis-distilled water, using
an all-glass spray, as shown in Figure 32. Do not spray so heavily
that the SiO_2 layer will be washed away or the separated bands
start running together again. If the spray is too light, no sharp
difference will be visible between background and glycolipid
components. Now take a needle and mark the contours of the separated bands by scraping into the SiO_2 layer. Dry the plates at
room temperature.

7. With the aid of a thin spatula or a razor blade scrape off
the bands to be analyzed, one by one. Identical bands from several TLC plates can be pooled, mixed well and dried at 70°C in
vacuo over P_2O_5. The dried silicic acid scrapings can be stored
in vials in a vacuum desicator over P_2O_5 at room temperature.

Scrape off an approximately 20 mm wide 160 mm–180 mm long
strip from below the starting line, where no sample was applied.
This scraping should be handled in the same fashion as other
scrapings and will serve as a blank control.

Whenever a portion of the samples will be taken for biologic
or chemical assays, the samples should always be mixed very thoroughly before weighing, and the weighing done as quickly as possible to avoid change in moisture content.

8. Determination of nitrogen. A 500 mg amount (the weighing
should be carried out as rapidly as possible) of the scrapings
is transferred into a 100 ml Kjeldahl digestion flask. Concentrated sulfuric acid (2 ml) saturated with anhydrous copper (II)
sulphate is added and the digestion is carried out in the presence of alundum particles, as described in Exercise No. 48. The
distillation can be performed in a Labconoco microdistillation
apparatus. Add 5 ml 10 *N* NaOH to the digested SiO_2 scraping,
after it has been quantitatively transferred to the distillation
equipment. Steam distil the liberated ammonia into a 25 or 50 ml
graduate cylinder which contains 10 ml 0.02 *N* H_2SO_4. After
distillation, fill the cylinder accurately to 25 ml and determine the NH_3 content by Nesslerization. Measure the color at
440 nm. Prepare a calibration curve as described in Exercise
No. 48.

9. Phosphorus determination in the scrapings can be carried out using a modified method of Chen et al. (1956). Weigh 50 mg SiO_2 scraping accurately and add 0.8 ml 6 *N* H_2SO_4 in a stoppered glass tube. The mixture is kept at 100°C (oil bath) for 16 h, then mixed with 4 ml water. Add 0.8 ml 2.5% ammonium molybdate, 1.6 ml water, and 0.8 ml 10% ascorbic acid to the sample. Mix the contents of the tubes thoroughly and incubate them at 37°C for 2 h. Centrifuge the mixture at 10,000 g for 15 min to sediment the silicic acid, and determine the color intensity of the supernatant at 820 nm. The calibration curve can be prepared by applying standard KH_2PO_4 solution to blank TLC plates. Applying five different amounts of KH_2PO_4, the plates can be scraped and the same procedure as described above carried out to determine the phosphorus content. With the aid of this procedure, a standard calibration curve can be established. As we found in our laboratories, this procedure can be replaced by simply preparing the calibration curve with standard KH_2PO_4 solutions as described in Exercise No. 49 in this Manual.

10. For hexosamine determination 300 mg scraping has to be taken and hydrolyzed with 2 ml 3 *N* HCl in a ♀ stoppered test-tube at 100°C for 16 h. The contents of the tube can be filtered into a round flask through a glass filter, washed several times with 2 ml aliquots of bis-distilled water, and the entire filtrate dried quickly and thoroughly on a rotating Büchi evaporator (see Fig. 12). The flask with the dry hydrolysate is placed in a vacuum desiccator and dried over KOH pellets overnight. The residue is dissolved in exactly 2 ml water and aliquots can be taken for hexosamine determination. The procedure for the color development is essentially the same as that developed by Rondle and Morgan (1955) and described in this Manual in Exercise No. 55. The calibration curve can be prepared as described above for phosphorus or as described in Exercise No. 49.

11. 2-Keto-3-dexyoctonic acid (KDO) is quite sensitive to acidic treatment but, as we found when the sample coats silicic acid particles, the KDO seems to be protected against acidic destruction. Hydrolyze 100 µg scraping with 1 ml 0.025 *N* H_2SO_4 in a boiling water bath for 20 min. Centrifuge the cooled hydrolysate at 1000 g for 15 min. Use 0.4 ml amount of the supernatant for KDO determination. The determination applies the procedure of Weissbach and Hurwitz (1959) and is described in Exercise No. 57 of this Manual.

12. Mix 100 mg silicic acid with 0.5 ml distilled water and determine the heptose content by the procedure of Dische (1953), as described in Exercise No. 56 of this Manual. Centrifuge the samples at 1000 g for 15 min after color development and read the supernatant at 505 and 545 nm using a microcuvette. D-gluco-heptose can be used to prepare the standard curves.

13. For the determination of long-chain carbocylic acids, the transesterification with BF_3 in methanol can be done directly on 500 mg silicic acid scraping in an 80°C oil bath. The procedure is the same as developed by Duron and Nowotny (1963) and described in Exercise No. 64 of this Manual. Extract the methylated fatty acid with hexane four times as described earlier, except that after each extraction allow enough time for the silicic acid to settle out. Evaporate the pooled hexane phase to dryness under a stream of nitrogen at room temperature. Redissolve the

fatty acid methyl esters in exactly 2 ml hexane and determine
them quantitatively by the hydroxylamine method (see Exercise
No. 63). The calibration curve can be made using pure tristearin.
  It should be emphasized here that the methyl esters present
in this hexane extract are also ready for gas chromatographic
analyses.

Evaluation

The molar ratios of the above analyzed components can be determined without making any effort to elute the separated components
or to determine the total organic matter content of the scrapings.
The results from the above determinations will give the percentage
of the analyzed components present in the silicic acid scraping.
The amounts of nitrogen, phosphorus, or other components are to
be converted to micromols (31 µg P = 1 µM P; 179 µg D-glucosamine = 1 µM, etc.). The molar ratio can now be calculated by
taking the relative amount of one of the components (for example
phosphorus) as 1 µM, and calculating from the available analytic
data how many micromols of hexosamine, fatty acid, KDO etc. we
have per 1 µM of phosphorus or nitrogen in the chromatographically separated and homogeneous band.

Part B. Measurements of Biologic Activities

Materials and Equipment

  Pyrogen-free water and glassware, pyrogen-free sonicator tip
  must be used throughout the entire procedure
  Equipment and animals for the determination of biologic activities include the same as described in Exercise No. 91 for
  local Shwartzman skin reactivity and in Exercise No. 95 for
  pyrogenicity

Procedure

  1. Since the scraped bands will be used for sensitive biologic
assays, the procedure for TLC separation must be kept as pyrogen-
free as possible. Silicic acid, pipettes and centrifuge tubes
should be heated at 180°C for 3 h. The glass plates should be
soaked in 1 N NaOH overnight and then washed with pyrogen-free
water to eliminate pyrogenic contaminants.
  2. Take 500 µg scraping and suspend in 5 ml 0.5 M triethylamine in water. Sonicate the suspension for 1 min at 1.7 A under
ice water cooling. Centrifuge at 10,000 g for 30 min. Take the
supernatant from which the triethylamine can be removed as follows: accurately pipette 2 ml supernatant into a calibrated test
tube (or centrifuge tube) and blow nitrogen gas through a sterile
filter at the surface of the sample at approximately 30°-35°C,
until the sample dries. The dry residue can be redispersed in
2 ml pyrogen-free saline, again using sonication under cooling
and by using the microtip attachment of the sonicator. Aliqots
of this extract can be used for biologic activity measurements
as described below.

3. For pyrogenicity, 1 ml of such supernatant should be sterilized using a syringe-attached filter. Make 10-, 100-, and 1000-fold dilutions from this filtrate, using pyrogen-free saline and pyrogen-free glassware. Determine the pyrogenic response of rabbits to undiluted as well as diluted extracts, giving 0.1 ml aliquots intravenously as described in Exercise No. 95.

4. For the *Limulus* lysate clotting activity measurements, take 0.1 ml supernatant and add this to 9.9 ml pyrogen-free saline. Prepare five 10-fold dilutions from this. The final dilution will be 10^7. Use strictly pyrogen-free saline as well as glassware throughout. Disregarding this precaution will inevitably give useless results in the extremely sensitive *Limulus* lysate assay.

Pipette 0.1 ml of the various dilutions into small test tubes (8 × 80 mm or similar size). Add to each tube 0.1 ml *Limulus* lysate which should be prepared according to the instructions of the manufacturer. The samples and the lysate should be mixed thoroughly with the use of a Vortex and incubated at 37°C in a water bath. Remove the tubes every 5 min and check for increased viscosity and gelation by slowly tilting them. Do not shake them! Record the time when viscosity and later complete gelation occur for every sample. The activity of the various extracts can be compared by establishing the highest dilution which still gives solid gel formation within 2 h.

5. The local Shwartzman skin assay requires 2 µg-20 µg endotoxin injected intradermally. Therefore, for this assay 0.1 ml undiluted supernatant should be injected intradermally, which treatment is followed 24 h later by injecting the rabbits intravenously with 20 µg of any standard Gram-negative endotoxin preparation. Details of this procedure are described in Exercise No. 91. Further dilutions of the undiluted supernatant may be prepared. It is recommended to use two-fold dilutions here, to compare the activity of the various extracts.

Evaluation

The procedure as described here is eminently suitable for the localization of endotoxic activity among the separated bands in a qualitative fashion. To convert this procedure to a quantitative comparison of activities of the separated components, one must carry out quantitative determinations of total organic matter per milliliter supernatant. Simple dry weight determination should be carried out first, using a microprocedure as described in Exercise No. 47. This dry weight value will also contain microscopic SiO_2 particles which cannot be removed by the above procedure. Therefore, the dried sample has to be incinerated in a platinum crucible at 700°C for 2 h. (See also Exercise No. 47). Cool the crucible with its inorganic ash content in a vacuum desiccator over P_2O_5 and weigh it. From the weight of the crucible containing the dried sample one has to deduct the incinerated crucible weight. The difference will give the organic material content of the sample. From this, the amount of material injected in the different dilutions can be calculated.

According to our experience the presence of small amounts of silicic acid in the filtered supernatants does not interfere with the determination of endotoxic activity in the above described biologic parameters.

Use and Limitations

While this procedure has been elaborated by us for endotoxic
glycolipids, it is applicable to any other natural or synthetic
product which can be separated by TLC and which can be identified
by chemical as well as biologic assays. For example, it could be
used in hormone research, in antibiotic studies, in toxicology,
or in drug separation, just to mention a few. Obviously, the TLC
developing system must be worked out first, second a nondestructive method is needed to locate the separated bands on TLC.

Other nondestructive methods of detection include UV analysis,
which has been successfully applied to compounds which showed UV
absorption or reemission of incident light at a different wavelength, thus making them visible under a UV lamp (see Fig. 31).
Iodine vapor can be used to detect unsaturated compounds as well
as a number of amino acids, indoles steroid derivatives and many
others. The process of iodine staining seems to be reversible in
some instances. Biologic methods, other than those described here,
were widely used in microbiology, using the "bioassays" or "bio-
autographic" procedures for bactericidal compounds. A detailed
review of the nondestructive detection methods was published by
G.C. Barrett in 1974.

An even more sensitive method for the detection of endotoxin
or endotoxin-containing bacteria has been elaborated by Watson
and co-workers (1977). These authors developed a spectrophotometric method to read the reaction between the *Limulus* lysate
and endotoxin-containing samples at 360 nm.

References

Barett, G.C.: Adv. Chrom. *11,* 145 (1974)

Chen, C.H., Chang, C.M., Nowotny, A.M., Nowotny, A.: Anal. Biochem. *63,* 183
    (1975)

Chen, C.H., Johnson, A.J., Kasai, N., Key, B.A., Levin, J., Nowotny, A.: Inf.
    Dis. *128*S, 43 (1973)

Chen, C.H., Nowotny, A.: J. Chrom. *97,* 39 (1974)

Chen, P.S. Jr., Toribara, T.Y., Warner, H.: Anal. Chem. *28,* 1756 (1956)

Dische, Z.: J. Biol. Chem. *204,* 983 (1953)

Duron, O.S., Nowotny, A.: Anal. Chem. *35,* 370 (1963)

Rondle, C.J.M., Morgan, W.T.J.: Biochem. J. *61,* 586, 1955

Watson, S.W., Novitsky, T.J., Quinby, H.L., Valois, F.W.: Appl. Envir.
    Microbiol. *33,* 940 (1977)

Weissbach, A., Hurwitz, J.: J. Biol. Chem. *234,* 705 (1959)

Exercise No. 59

Oxydation of Carbohydrates With Periodate. Quantitative Spectrophotometric Determination

Periodate oxidation is one of the most valuable procedures for
structural analysis in carbohydrate chemistry. It was observed
by Malaprade (1928) that periodate ions have a selective oxidizing
power on adjacent hydroxyl functional groups (vicinal diols such

as -CHOH-CHOH-) in acidic, neutral, or weakly alkaline solutions
at room temperature. Periodate does not oxidize monohydroxyl compounds or those hydroxylated derivatives in which the hydroxyl-
carrying carbon atoms are separated from each other by other
atoms. It has been observed that quantitative consumption of 1 mol
of periodate by one pair of adjacent hydroxyl groups takes place
in this reaction. The stoichiometric relationship of the oxidation of D-glucose by IO_4^- is as follows:

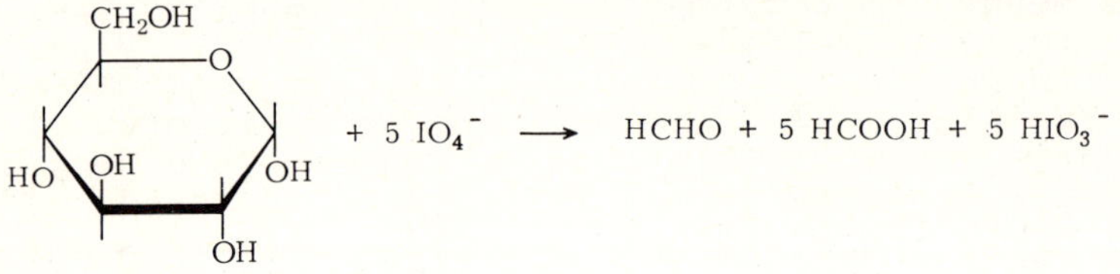

$$\text{(glucopyranose)} + 5\ IO_4^- \longrightarrow HCHO + 5\ HCOOH + 5\ HIO_3^-$$

As the equation shows, 1 mol of formaldehyde is formed from
primary alcohol groups. 1 mol of formic acid is the product of
the reaction from secondary alcohol and from aldehyde groups. The
reaction itself is rather complicated and not fully understood.
Earlier and more recent results show that a cyclic ester is formed between the vicinal hydroxyl groups and the periodate ion as
an intermediary reaction product. Because a detailed discussion
is beyond the scope of this manual, the review articles by Dyer
(1956) or Pigman (1957) or chapters 75,76,77,78,79,80 in *Methods
of Carbohydrate Chemistry* edited by R.L. Whistler, Acad. Press 1965,
Vol. V, should be studied for further information.
In addition to qualitative and quantitative determination of
carbohydrates, the periodate oxidation method can be used not
only for the measurement of the number of vicinal OH groups in
an unknown compound, but also for the study of positions of substituents and/or linkages in derivatives of known materials.
Three general periodate oxidation procedures are used in carbohydrate chemistry to elucidate structures: (a) Measurement of
consumed periodate per mol or per weight unit, expressed as number
of vicinal diol equivalents per mol or per weight unit. (b) Measurement of the rate of periodate consumption. (c) Isolation,
identification, and measurement of oxidation products such as
glycerolaldehyde, acetaldehyde, formaldehyde, formic acid, etc.
Exercises are described here using the most practical procedures.
The simplest way to measure periodate consumption by carbohydrates is spectrophotometric measurement of the amount of excess
periodate at 223 nm. At that wave-length, the periodate ions have
a very intensive light absorption between pH 4 and pH 6 (Crouthamel et al., 1949). The Exercise given here is based on the
publication of Dixon and Lipkin (1954).

Materials and Equipment
____

     12.5 m*M* sodium periodate solution
     10 m*M* D-glucose
     10 m*M* α-methylglucoside (Methyl-α-D-glucopyranoside)
     UV spectrophotometer
     5 ml pipettes
     1 ml pipettes

        20 ml pipettes
        50 ml graduate cylinders
        100 ml volumetric flasks

Procedure

1. Dissolve 2.6736 g periodate in 1000 ml distilled water.
Use only analytic grade preparation which was kept in a vacuum
desiccator in a dark container.
2. Take 0.5 ml carbohydrate sample and add 2.0 ml of the above
periodate solution. This mixture will contain 10 μM sodium periodate per ml. Do all pipetting with analytic accuracy. Swirl,
place the tubes in a dark container immediately. Prepare a control by adding 2.0 ml $NaIO_4$ to 0.5 ml distilled water. Handle
this in the same way as the test samples.
3. To follow the oxidation process, remove samples at different time intervals. It is recommended to take samples of the above
carbohydrates used as model substances in this Exercise at $\frac{1}{2}$, 1,
2, 4, and 6 h. If oligosaccharides or more complex polysaccharides
are subjected to oxidation, 12, 24, 48, 72, and 96 h samples should
be measured in addition to the above time intervals.
Very accurately pipette 0.2 ml of the reaction mixture into
40.0 ml distilled water. If micropipettes are available, add
0.02 ml (20 lambda) to 4.0 ml distilled water and read the optical
density immediately at 223 nm. The blank readings will be between
0.46 and 0.48 OD.
4. Prepare a calibration curve using the 12.5 mM sodium periodate solution. Take 20.0 ml of this and add 5.0 ml water. This
is a 10 mM solution. Take five 100 ml volumetric flasks; pipette
into them 0.2, 0.4, 0.6, 0.8, and 1.0 ml with maximum accuracy. Fill
the flasks to 100 ml. Mix and read the optical densities as above.
The optical density readings will be close to 0.2, 0.4, 0.6, 0.8,
and 1.0.

Evaluation

Use the calibration curve to determine the molarity of the analyzed samples. Note that molarity divided by 1000 gives the
number of mols per milliliter sample.
Compute the difference in mols per ml values between the samples
and the controls. Plot the values obtained against time in hours.
Analyze the rate of periodate consumption, determine the time
required to consume the first, second, etc. mols of periodate.
Also calculate the number of mols of periodate consumed by 1 mol
of carbohydrate. If an unknown carbohydrate is being analyzed,
express the periodate consumption as μmol periodate used by mg
dry substance. This value, in the case of free aldohexoses, for
example, is 27.7.

Use and Limitations

The range of periodate determination by spectrophotometry is
from 0.01 mM to 0.1 mM. This means that 0.01 μM periodate in
1 ml can be accurately measured.

An obvious limitation of this procedure is that it can be used only for substances which do not absorb light at this range.

Periodate will also oxidize quantitatively carbonyl adjacent to hydroxyl and amino adjacent to hydroxyl groups, 1,2-diketo groups as well as 1,2-diamines, disulfides, α-keto acids will also be oxidized, but under nonstoichiometric conditions. If these groups are present, they may give misleading values of consumed periodate.

References

Crouthamel, C.E., Meek, H.V., Martin, D.S., Banks, C.V.: J. Am. Chem. Soc. *71*, 3031 (1949)

Dixon, J.S., Lipkin, D.: Anal. Chem. *26*, 1092 (1954)

Dyer, J.R.: In: Methods of Biochem. Anal. *3*, 111 (1956), New York: Glick, Interscience Publ.

Malaprade, L.: C. R. Acad. Sci. Paris *186*, 382 (1928); and Bull. Soc. Chim. *43*, 683 (1928)

Pigman, W.W. (ed.): The carbohydrates. New York: Academic Press 1957

Exercise No. 60

Measurements of the Products of Periodate Oxydation

The products of the periodate oxidation reaction are important in the structural analysis of natural products. There are quantitative analytic procedures by which the amount of acetaldehyde, formaldehyde, or formic acid formed can be quantitatively measured.

Part A. Formaldehyde Determination

The mechanism of the chemical reaction of chromotropic acid with formaldehyde in the presence of concentrated sulfuric acid is not known, but the development of the characteristic purple color is rather specific and also quite sensitive for the quantitative determination. The procedure given here is essentially the same as described by Speck and Forist (1954).

Materials and Equipment

Chromotropate reagent: Mix 1 vol of 1% sodium chromotropate (4,5-dihydroxy-2,7-naphthalenedisulfonic acid disodium salt) solution in distilled water and 4 vol of 65% sulfuric acid (2 vol concentrated, analytic grade H_2SO_4 in 1 vol distilled water)

0.1 *M* sodium meta-periodate solution

0.5 m*M* glucose solution, or

0.2 to 1.0 mg/ml oligosaccharide

0.5 m*M* serine solution for the calibration curve, made from analytic grade preparation

0.4 *M* sodium meta-arsenite, analytic grade
Test tubes, size 16 mm × 150 mm or similar size

Procedure

1. To 5 ml each of the carbohydrate and serine solution, add
0.2 ml periodate solution and 4.8 ml distilled water. Keep in
the dark for 4 h at room temperature.
2. Take 0.2, 0.4, 0.6, 0.8 and 1.0 ml of these mixtures and
make the samples up to 1 ml with distilled water. Add to the
tubes 1 ml 0.4 *M* sodium meta-arsenite solution and 5 ml sodium
chromotropate reagent.
3. Prepare the carbohydrate blanks: Mix 5 ml carbohydrate
and 5 ml distilled water. Take 1 ml of this and proceed as above
by adding 1 ml arsenite and 5 ml chromotropate. After mixing all
the samples, heat for 30 min in boiling water bath. Cool and
read against carbohydrate blanks at 570 nm.

Evluation

The serine values obtained serve as a calibration curve. Of the
periodate-oxidized solution 1 ml will contain 0.25 μM (= 7.5 μg)
formaldehyde.

Limitations

A high content of organic materials present in the sample may
form colored products with hot sulfuric acid. These colors may
interfere with the formaldehyde determination.
Another modification of the formaldehyde determination is described by Nash (1953) using acetylacetone-NH_3 reagent. This
quantitative assay is not influenced by the presence of acetaldehyde.

Part B. Formic Acid Determination

In the following experiment, which is based on the procedure of
Flood, Hirst and Jones (1948), the periodate oxidation of a known
carbohydrate will be studied and the amount of liberated formic
acid measured.

Materials and Equipment

0.1 *M* sodium periodate solution
0.01 *N* sodium hydroxide solution
0.1 *M* D-glucose solution (18 mg/ml)
0.1 *M* methyl α-glucoside solution
0.1 *M* lactose solution
10% diethylene glycol, reagent grade, in distilled water,
 pH adjusted to 7.0
Magnetic stirrer and Teflon covered bars
25 ml titration flask (Fig. 17)
Microburet, 10 ml capacity
Electrometric pH meter

Procedure

1. Mix 5 ml of each of the above carbohydrate solutions with
5 ml NaIO$_4$ in test tubes. Keep them in the dark at cold room
temperature. Withdraw 1.0 ml of each mixture immediately after
the addition of periodate and transfer this to the titration flask
which contains 5 ml diethylene glycol solution. Place a magnetic
stirrer bar in each flask, and titrate the solution after 30 min
at room temperature to pH 7.0 with 0.01 N NaOH from the microburet.
2. Withdraw 1.0 ml samples of the carbohydrate-periodate mixtures after 1,2,3,6, and 24 h, and follow the same titration
procedures as described above.

Evaluation

Each milliliter of 0.01 N NaOH corresponds to 10 μM formic acid.
Plot the measured micromols against time in hours. The 1.0 ml
samples taken at different time intervals contained 50 μM carbohydrates. Calculation should express the μM HCOOH liberated by
μM carbohydrate. If the carbohydrate is unknown, express the μM
liberated HCOOH/mg sample.

Use and Limitations

The 0.01 N NaOH must be prepared fresh daily before use from a
standardized 1 N NaOH solution.

References

Flood, A.E., Hirst, E.L., Jones, J.K.N.: J. Chem. Soc. 1679 (1948)
Nash, T.: Biochem. J. *55*, 416 (1953)
Speck, J.C., Jr., Forist, A.A.: Anal. Chem. *26*, 1942 (1954)

Exercise No. 61

Reduction of Carbohydrates With NaBH$_4$

Reduction of aldehyde groups in oligosaccharides by NaBH$_4$ has
been used successfully in different studies investigating the
degree of polymerization of oligosaccharides up to a molecular
weight of 1200-1400. It was also applied to the measurement of
the number of reducing aldehyde groups in higher oligosaccharides
by the gasometric determination of the liberated hydrogen after
sodium borohydride treatment. One of the most popular applications of the sodium borohydride reduction is the identification
of reducing terminal carbohydrates in oligosaccharides.
In *Part A* of the following Exercise, an example is given for
the determination of the degree of polymerization, and in *Part B*
the application of NaBH$_4$ reduction for the identification of
terminal reducing carbohydrates is shown. Lactose will be used
as a model substance with known structure, but an unknown oligosaccharide may also be analyzed in these experiments. *Part C*

illustrates the use of periodate and borohydride treatments for
the analysis of glycosidic linkages.

Part A. Determination of the Degree of Polymerization

The number of carbohydrate components will be determined before
and after the reduction of the reducing terminal carbohydrate to
a sugar alcohol. The ratio of the two values will be used for
the calculation of the degree of polymerization, according to the
method of Peat, Whelan and Roberts (1956).

Materials and Equipment

2.5% sodium borohydride solution dissolved in water. This
  reagent must be prepared fresh every day
Inactivated sodium borohydride solution. This is prepared by
  dissolving 2.5% $NaBH_4$ in 1 N H_2SO_4 instead of water
Concentrated and 4 N sulfuric acid
0.1 mM lactose solution (dissolve 3.60 mg lactose $\cdot$ H_2O in
  100 ml water)
Oligosaccharide solution. The concentration must be adjusted
  to 0.05 mg/ml
Spectrophotometer
5% phenol dissolved in water

Procedure

1. Pipette 1 ml lactose into each of four regular test tubes.
Pipette 1 ml oligosaccharide solution into each of another four
test tubes. Take two tubes containing lactose and two tubes containing oligosaccharide. Add to each 0.2 ml active $NaBH_4$ solution.
Let the tubes stand at room temperature for 4 h. The reduction
of lactose will be complete in 1 h, but some oligosaccharides
require longer treatment. To the remaining lactose and oligosaccharide samples add 0.2 ml inactive $NaBH_4$ solution.
2. Add to all eight tubes 0.8 ml 4 N H_2SO_4. Pipette 1.5 ml
phenol solution into each tube, then add 7 ml concentrated H_2SO_4.
Mix the contents carefully but thoroughly. Let the tubes stand at
room temperature for 60 min, then read the optical density at
490 nm in a spectrophotometer. In this phenol–H_2SO_4 reaction the
reduced hexitols will give a negative reaction.

Evaluation

Q = Reducing power quotient = A/B where A is the nonreduced sample
reading (in optical density scale units) and B is the OD reading of the reduced sample. Degree of polymerization = $Q/Q-1$.
The determination of the absolute values of the carbohydrate
content in the samples is not necessary because the degree of
polymerization is calculated from the relative values of the
readings.

Use and Limitations

As mentioned, some oligosaccharides require longer treatment
than does lactose. Overnight reduction of the preparations may
be necessary. This has to be determined in preliminary experiment
    The phenol-sulfuric acid determination of carbohydrates is
very convenient because of its simplicity, but it must be remembered that hexosamines do not react in this assay. Aldopentoses
have a considerably lower molar extinction than aldohexoses.
Aldoheptoses also differ from aldohexoses. When such carboydrate
components are present in the analyzed polysaccharide, it is
advisable to use one of the methods which measures the number of
reducing aldehydes (see Exercise No. 53).

Part B. Identification of the Reducing Terminal Carbohydrate

The same carbohydrates may be used as in the previous Exercise,
but now in higher concentration. A lactose or any other homogeneous reducing oligosaccharide sample may be reduced by sodium
borohydride. In those tubes which contain active $NaBH_4$, the free
aldehyde groups will be reduced, resulting in a sugar alcohol.
Add inactive sodium borohydride to the control tubes. All samples
will be hydrolyzed and the monosaccharide obtained will be analyzed by paper chromatography. Because the chromatograms will
be sprayed with p-anisidine·HCl, the sugar alcohols will not be
visible. Comparing the chromatographic picture of the $NaBH_4$-re-
duced and of the unreduced samples, it is easy to detect the
partial or complete disappearance of the carbohydrate constituents
which have been at the reducing end of the samples.

Materials and Equipment

    Lactose solution, 1 mg/ml in water, or
    Unknown oligosaccharide solution, 1 mg/ml in water
    Authentic mixture of the following carbohydrates: Water solu-
      tion containing 1% each D-glucose, D-galactose, D-mannose
      and D-ribose
    Active 2.5% sodium borohydride (dissolved in water)
    Inactive 2.5% sodium borohydride (dissolved in 1 *N* sulfuric
      acid)
    4 *N* sulfuric acid
    p-Anisidine·HCl spray reagent (3 g p-anisidine·HCl dissolved
      in 100 ml n-butanol)
    Barium carbonate
    Centrifuge and centrifuge tubes
    Paper chromatographic equipment (Fig. 23 or 24)
    Boiling water bath

Procedure

    1. Take two test tubes, pipette into each 1 ml lactose or
1 ml unknown oligosaccharide solution. To tube No. 1 add 0.2 ml
active $NaBH_4$ and to tube No. 2 add 0.2 ml inactive $NaBH_4$. Let the
tubes stand at room temperature for 4 h.

2. Add to each tube 0.3 ml distilled water and 0.5 ml 4 N H_2SO_4 and cover the tubes with "cold finger" or marble, and hydrolyze the contents in boiling water bath for 4 h. Neutralize the samples by adding approximately 0.3 g of solid $BaCO_3$ to each tube. Transfer the contents of the tubes to a 5 ml centrifuge tube; centrifuge for 30 min at 3000 g and lift the supernate into a clean small test tube.

3. Prepare a 12 cm ×40 cm Whatman No. 1 or any other similar quality filter paper sheet for chromatographic analysis of the hydrolysates. Approximately 10 cm distant from one end of the paper, draw a pencil line. Mark the starting point of the two samples as well as the places where known carbohydrate mixtures will be run simultaneously along this line. Use an authentic mixture of carbohydrates. The spots should be approximately 3 cm apart. Apply 10 µl of the hydrolysates and 2 µl of the known mixture. Develop the chromatogram in a descending chamber using the solvent mixture pyridine:butanol:water = 5:3:2 overnight.

4. Dry the paper and develop with p-anisidine reagent as follows: Spray the paper with the reagent and heat it in an oven at $100°C$ for 10 min.

Evaluation

Observe the location of the developed spots and identify them using the components of the known mixture. If the Exercise is carried out properly, the glucose spot disappears from the lactose sample which was treated with active $NaBH_4$. Record your observations in investigating the paper chromatographic pattern optained for the unknown oligosaccharide.

Use and Limitations

All sorts of homogeneous oligosaccharide preparations ·can be analyzed by the above procedures, but homogeneity of oligosaccharides cannot be easily proven. It is strongly advisable to run those preparations which were obtained in at least three different paper chromatographic solvents to check their homogeneity.

If the oligosaccharide molecule consists of repeating units, the same carbohydrate which was the reducing end of the chain may occur within the chain. These latter units will not be reduced by $NaBH_4$, therefore in the hydrolysate of the reduced oligosaccharide, only a decrease of this carbohydrate will be observed. It is understandable that the larger the molecular weight of the oligosaccharide, the greater the possibility that one cannot determine with certainty which carbohydrate has been the reducing end of the oligosaccharide. Schiffmann et al. (1964) reported that alkaline $NaBH_4$ splits internal glycosidic linkages, thus exposing and reducing carbohydrates not only at the reducing end, but also in the polysaccharide chain.

Part C. Investigation of the Glycosidic Linkages by Periodate and Borohydride Treatment (Smith Degradation)

Smith and associates (1950, 1956) elaborated a procedure which further extends the use of periodate oxidation in the structural studies of polysaccharides. If a carbohydrate chain is treated wi periodate, two aldehyde groups will be formed in each anhydrous sugar unit which has been oxidized by the periodate. The result of this is a polyaldehyde, which usually undergoes profound changes if one attempts to hydrolyze the oxidized polysaccharide in order to liberate and analyze the carbohydrate units after periodate treatment.

As described in *Part A* of this Exercise, aldehydes can be reduced to primary alcohols with $NaBH_4$. This reaction was used by Smith and co-workers to convert the polyaldehyde to a polyalcohol If such a product is hydrolyzed by mineral acids, three different alcohols may be found in the hydrolysate, such as glycerol, erythritol, and glycolaldehyde. The amount of these components depends upon the position of the glycosidic linkages present in the polysaccharide. For example, if a hexose is in a nonreducing terminal position, it will yield glycerol after periodate + sodiu borohydride treatment without hydrolysis. If a hexose is within a polysaccharide chain, linked $1 \rightarrow 3$ to the next carbohydrate, this will not be oxidized by periodate, and after the entire treatment this sugar unit will appear in the hydrolysate as free, unchanged carbohydrate. If the linkage between the hexose units is $1 \rightarrow 6$, this will yield one mol of glycerol after the Smith degradation and subsequent hydrolysis. If the linkage within the chain is $1 \rightarrow 4$, the hexoses will be converted to erythritol after the above treatment.

If, in a poly- or oligosaccharide, the linkages between the hexoses are all $1 \rightarrow 4$, the nonreducing terminal hexose will yield one mol of glycerol and all other hexoses in the straight chain will yield one mol of erythritol. Quantitative determination of the glycerol:erythritol ratio gives exact information about the chain length of such an oligosaccharide. If the polysaccharide has branches, each nonreducing end of the branched polysaccharide will yield one glycerol.

In this exercise two model substances will be used to demonstrate differences in the reaction products of this degradation. The two materials are dextran, which is an α-1:6-D-glucose polyme and glycogen, which is an α-1:4-D-glucose polymer. The reaction products will be analyzed by paper chromatography.

Materials and Equipment

    2% dextran solution in water
    2% glycogen solution in water
    1% glycolaldehyde solution in water
    1% glycol solution in water
    1% glycerol solution in water
    1% erythritol solution in water
    1% glucose solution in water
    0.1 *N* silver nitrate
    10% sodium hydroxide

0.8 *M* sodium periodate solution
Ammonium hydroxide, concentrated
4% sodium borohydride solution
2 *N* sulfuric acid
n-Butanol
Sodium thiosulfate, 5% in water
Ethanol
Dowex 1 ion exchanger in hydrocarbonate form, 100 mesh
Dowex 50 ion exchanger in hydrogen form, 100 mesh
Water bath
Sintered glass filters, coarse grade, diameter 20 mm
Filter tubes, 25 ml capacity, with side arms
Reflux condensers with ⍑ 12/30 joints
Hydrolysis tubes, 18 mm × 150 mm, with 12/30 ⍑ joints
Vacuum distillation equipment (Fig. 12)
Paper chromatography jar (Fig. 23 or 24)
Whatman No. 1 or similar quality paper

Procedure

1. Mix 2 ml dextran with 2 ml $NaIO_4$ solution. In another test
tube add 2 ml $NaIO_4$ to 2 ml glycogen. Keep the tubes in the cold
room in a dark container for 48 h. Add to each tube 2 g freshly
activated Dowex No. 1 (HCO_3^- resin), stir with a glass rod for
10 min, then filter through a coarse glass filter into filter
tubes.

2. Add to each filtrate 2 g freshly activated Dowex No. 50
(H^+ ion exchanger), stir 10 min, then filter into a clean filter
tube. The $NaIO_4$ is removed by these treatments.

3. The reduction of the oxidized polysaccharide takes place
if 1 vol of the above neutral filtrate is mixed with $1/_2$ vol of
4% $NaBH_4$ solution. Let it stand overnight at room temperature.

4. The hydrolysis of the products obtained is best carried
out with sulfuric acid. Take the hydrolysis tubes and add to
1 vol of the above reaction mixture 1 vol of 2 *N* H_2SO_4. Always
use the entire amount of filtrate obtained in the previous steps.
Attach the tubes to a reflux condenser and hydrolyze the preparations for 8 h on a boiling water bath.

5. In order to remove the acid from the sample after hydrolysis,
add to each tube 5 g freshly activated Dowex No. 1 (HCO_3^-) in
small portions, stir for 5 min, then filter the contents of the
tubes through a coarse glass filter as in step No. 1.

6. Concentrate the filtrate to dryness in vacuum.

7. For paper chromatographic analysis of the samples, dissolve
each of the dry residues in 2 ml distilled water. Prepare a paper
chromatogram (Whatman No. 1 or any similar quality paper) size
12 cm × 35 cm. Either descending or ascending chromatography can
be used. Place 5 µl of each of the five authentic samples of
glycolaldehyde, glycol, glycerol, erythritol, and glucose on the
starting line. Take 10 µl of the two degraded polysaccharide
preparations and place them along the known samples on the starting line. Dry the spots with a stream of warm air. The chromatogram has to be developed in n-butanol:ethanol:water = 4:1:5 mixture overnight.

8. Remove the paper chromatogram from the chamber and dry it
at room temperature. Develop it with Tollens reagent which is

prepared as follows: Add to 100 ml 0.1 N AgNO$_3$ 0.5 ml concentrated NH$_4$OH, then 50 ml 10% NaOH. Pull the chromatogram quickly through the solution (do not bathe it), let it drip, and after the color has been well developed (approximately 30 s-60 s), transfer the paper to a tray containing 5% Na$_2$S$_2$O$_3$. Leave it there for 5 min, then wash the paper in large volumes of tap water. Dark spots will appear on a gray background.

Evaluation

Measure the R$_F$ values of the components. Identify the spots obtained from dextran and from glycogen using the R$_F$ values of the authentic samples.

You will see that while dextran yielded mostly glycerol, glycogen produced erythritol after these combined degradation. If the chromatogram showed unaltered glucose present in the product, it indicates either that the periodate oxidation was incomplete, or this may also be due to the presence of glucose units in the structure which had a substituent at C-3 position. The glycogen may show a small amount of glycerol besides the intensive spot of erythritol. This originates from the nonreducing ends of the polysaccharide. The higher the branching in the polysaccharide, the easier the observation of glycerol on the paper chromatogram.

A fairly reliable quantitative procedure for the estimation of glycerol to erythritol ratios is also described by Hamilton and Smith (1956).

Use and Limitations

There are several other procedures which will give us reliable information about the degree of polymerization of an oligosaccharide or methods which will show whether the linkages are 1 → 6 or otherwise in the structure (see Determination of Formaldehyde after Periodate Oxidation, Exercise No. 60). The value of the Smith degradation lies in its capacity to differentiate between most of the glycosidic linkages.

It is very difficult to evaluate the results of the Smith degradation if the oligosaccharide molecule contains different glycosidic linkages. In such cases, other methods (such as total methylation, Exercise No. 46) must also be used to clarify the structure.

Apart from the above mentioned advantages, the Smith degradation is very useful in those studies where the destruction of certain carbohydrates is investigated during periodate oxidation of an unknown oligo- or polysaccharide. As mentioned in the introduction to this Exercise, the resulting polyaldehyde of periodate oxidation cannot be hydrolyzed without unwanted side reactions. For the analysis of the reaction products, such as unchanged monosaccharides, paper chromatography is usually used (see Exercise No. 36 or No. 43). This procedure may give confusing patterns due to the presence of different aldehydes split from the polyaldehyde chain. These problems are eliminated if the reaction product is reduced by NaBH$_4$ after periodate treatment. Watanabe et al. used thin-layer chromatography to analyze the

products obtained after Smith degradation of glycolipids from
human erythocytes (1975).

Methods of structural studies of polysaccharides, using enzymatic degradations, and biosynthesis were edited by Ginsburg
(1972).

<u>References</u>

Abdel-Akher, M., Hamilton, J.K., Montgomery, R., Smith, F.: J. Am. Chem. Soc.
    74, 4970 (1950)
Ginsburg, V. (ed.): Methods in enzymology 28, Complex carbohydrates. London,
    New York: Academic Press 1972
Hamilton, J.K., Smith, F.: J. Am. Chem. Soc. 78, 5907 and 5910 (1956)
Peat, S., Whelan, W.J., Roberts, J.G.: J. Chem. Soc. 2258 (1956)
Schiffmann, G., Kabat, E.A., Thompson, W.: Biochem. 3, 113 (1964)
Smith, F., Montgomery, R.: Meth. Biochem. Anal. 3, 153 (1956). Glick (ed.)
    New York: Interscience Publ.
Watanabe, K., Laine, R.A., Hakomori, S.-I.: Biochem. 14, 2725, 1975

Exercise No. 62

Quantitative Determination of Free Hydroxyl Groups

The simplest procedure for the quantitative measurement of the
number of free hydroxyl groups on a partially substituted polysaccharide molecule employs acetylation and measures the excess
of acetic anhydride added to the unknown sample. The presence
of perchloric acid catalyzes the reaction and leads to quantitative acetylation of the free hydroxyls in the investigated material
at room temperature. After the esterification of the hydroxyl
groups has been accomplished, the unused acetic anhydride will
be hydrolyzed by adding water to the reaction mixture. The amount
of unused acetic acid is determined by titration with standard
base solution. The reaction takes place as follows:

$$
\begin{array}{c}
CH_3 - C = O \\
\hspace{2em} \diagdown \\
\hspace{3em} O \hspace{4em} + \ OH - R \ \rightarrow \ CH_3COOR + CH_3COOH \\
\hspace{2em} \diagup \\
CH_3 - C = O
\end{array}
$$

The procedure described here is based on the publication of Fritz
and Schenk (1959). A micromodification has been described by
Schenk and Santiago (1962).

<u>Materials and Equipment</u>

    Pyridine, reagent grade
    p-Toluenesulfonic acid, reagent grade
    Ethanol
    Acetic anhydride, reagent grade
    1 *N* sodium hydroxide
    50°C water bath

10 mm × 100 mm test tube with No. 9 ℱ stoppers
500 ml volumetric flask
50 ml Erlenmeyer flask with ℱ stopper
Magnetic stirrer
Electrometric pH meter
50 ml titration flask (Fig. 17)
25 ml buret

Procedure

1. With maximum accuracy, weigh between 15 and 20 mg of your
unknown carbohydrate sample into a small ℱ test tube. Be sure
that all material to be analyzed is on the bottom of the tube.
2. Prepare a 2 M acetic anhydride solution in pyridine. Pipett
10 ml acetic anhydride into 30 ml reagent grade pyridine using a
flask equipped with a ℱ stopper. Mix the solvents well and add
1.2 g of p-toluenesulfonic acid. Wait until the reagent dissolves
This acetylating mixture is stable for only a few hours. It turns
yellow-orange very rapidly, therefore it must be freshly prepared
every day before the start of the experiment.
3. Prepare a 0.25 N NaOH solution in ethanol. Pipette 125 ml
of accurately measured standard 1 N NaOH into a 500 ml volumetric
flask and fill up to the mark with ethanol. Shake the contents
well. Determine the titer.
4. If all the weighed carbohydrate sample is on the bottom
of the tube, add exactly 1 ml acetylating reagent. Do not shake
the tube because the sample will then adhere to the wall. Put a
glass stopper loosely into the tube, and place the tube in a 50°C
water bath. In 30 min the carbohydrate sample will be dissolved.
Remove the tube, inspect it, shake it gently, then put it back
in the water bath for another 30 min. If the sample dried in the
tube and therefore forms a film on the tube wall, agitate the
tubes frequently during the first 30 min, trying to remove the
samples from the wall of the tube. It usually goes into solution
in 30 min. Do not exceed a total heating time of 1 h at 50°C. Let
the acetylation proceed for approximately 1 h. Do not tighten the
glass stopper in the test tube.
5. After the reaction is over, cool the tube and pour the
contents into a 50 ml titration flask (see Fig. 17). Use a total
of 10 ml of a pyridine:water = 3:1 mixture to rinse out the test
tube and transfer the sample quantitatively. Put this sample on
the magnetic stirrer and titrate its contents with 0.25 N NaOH
until a pH of 9.0 is reached.
A control sample without hydroxyl-containing material must be
run under identical conditions. The difference between the blank
and the sample titration will be used to calculate the percentage
of hydroxyl compounds in the sample.

Evaluation

The percent hydroxyl content of the dry material can be calculate
by using the following equation:

$$x = \frac{A - B \times F \times 1700}{\text{mg sample analyzed}}$$

where A = NaOH in ethanol used for control, B = NaOH in ethanol
used for sample, F = Normality of NaOH, 1700 = molecular weight
of OH^- × 100.

Use and Limitations

Fritz and Schenk (1959) elaborated the procedure for alcohols,
but it was successfully applied to glycerol, cellobiose and lactose, in addition to aldohexoses. It can be used for hydroxyl-
containing aromatic materials as well. The primary alcohol groups
react in a few minutes even at room temperature; secondary alcohols require longer treatment and elevated temperature. The
p-toluenesulfonic acid serves as a catalyst in this reaction.
For the esterification of hydroxyl groups, the usual catalyst is
perchloric acid, but it cannot be used at elevated temperatures.
     There are certain hydroxyl groups such as those on tertiary
carbon atoms or 2,4,6-trisubstituted phenols where the hydroxyls
do not react in the above system. It has been found that for
certain high molecular weight polysaccharides, in which the secondary alcohols are protected, a longer acetylation time is required. The fact that certain primary and secondary amines will
interfere with this analysis must also be mentioned. Similarly,
low molecular weight aldehydes may also consume acid anhydride
from the acetylating reagent. The mechanism of this latter reaction has not yet been clarified. These nonspecific reactions may
be eliminated by the use of pyromellitic dianhydride instead of
acetic acid anhydride. This method was introduced by Siggia and
co-workers (1961). Aldehydes do not disturb this reaction but
primary amines react as quickly as the primary or secondary
hydroxyl groups. This method has been successfully applied to
different simple alcohols and amines; its usefulness for the
determination of hydroxyl and amino groups in more complex carbohydrates requires further analysis.

References

Fritz, J.S., Schenk, G.H.: Anal. Chem. *31,* 1808 (1959)
Schenk, G.H., Santiago, M.: Microchem. J. *6,* 77 (1962)
Siggia, S., Hanna, J.G., Culmo, R.: Anal. Chem. *33,* 900 (1961)

Exercise No. 63

Lipid Determination With the Hydroxylamine Method

Different classes of lipids which contain ester groups react
with alkaline hydroxylamine to form hydroxamic acid. The hydroxamic acid reacts with acid ferric perchlorate forming a purple-
colored iron chelate complex. Free fatty acids do not react with
alkaline hydroxylamine. This procedure was elaborated for quantitative determination of lipids by Snyder and Stephens (1959).

Materials and Equipment

Ferric perchlorate (Pfalz & Bauer, Flushing, NY 11368)
70% perchloric acid
Glucose pentaacetate, crystalline
Cold ethanol (96%, anal. grade)
Hexane, reagent grade
Hydroxylamine hydrochloride
Sodium hydroxide
Water bath at 67°C
Spectrophotometer
Regular test tubes
Marbles or "cold fingers"

Procedure

1. Dissolve 2 g hydroxylamine hydrochloride in 2.5 ml water
(dissolve completely, use heat if necessary); make up to a volume
of 50 ml with cold ethanol. This is solution A.

2. Dissolve 4 g NaOH in 2.5 ml water; be sure the entire amoun
is dissolved, use heat if necessary. Make up to a volume of 50 ml
with cold absolute ethanol. Mix and warm again to dissolve everything. This is Solution B.

3. Add solution A to solution B. A heavy white precipitate of
NaCl will form. Let this solution stand for at least 10 min before filtering. Pour only the supernate on the filter paper. The
clear filtrate is the alkaline hydroxylamine reagent. It is stabl
only for approximately 1 h, thus it should be prepared fresh before every determination.

4. Prepare a stock solution of ferric perchlorate: A very
pale yellow solution consisting of 5 g of non-yellow ferric perch
lorate dissolved in 10 ml 70% perchloric acid and 10 ml water,
then made up to 100 ml volume with cold ethanol. This keeps
very well under refrigeration. Use only reagent grade ferric
perchlorate.

5. Mix 4 ml of the ferric perchlorate stock solution and 3 ml
of 70% perchloric acid with enough cold ethanol to take the
solution to 100 ml. This yellow-colored reagent mix must be
prepared fresh daily.

6. To the dry lipid samples or hexane extracts add 2 ml hydroxyl
amine reagent and 0.5 ml ethanol. Mix carefully and let stand at
room temperature for 15 min. Immerse the tubes carefully into a
67°C water bath. Because of the vigorous boiling of the hexane-
hydroxylamine mixture during the first minute, the rack containing the tubes has to be lifted out briefly to prevent loss of
the material through boiling over. After the mixture has simmered
down, return the rack to the water bath and let it stand for 15 mi
at 67°C. Since the hexane will evaporate, the final volume will
be 2.5 ml. To this add 5 ml ferric perchlorate reagent, mix well,
and after allowing the mixture to stand for 30 min at room temper.
ture, read in the spectrophotometer at 520 nm against a reagent
blank treated exactly as the unknown sample.

7. As a standard, any reagent grade ester dissolved in hexane
may be used. Glucose pentaacetate, which is commercially available in pure form, can also be used. Dissolve 78.07 mg crystalline glucose pentaacetate in 100 ml ethanol. One ml of this

solution contains 780 µg glucose pentaacetate which contains
10 µEq. ester groups. Take 0.1, 0.2, 0.3, and 0.4 ml of this
solution. Make up to 0.5 ml with ethanol. Treat these four samples in the same way as the unknown extract and read the optical
densities at 520 nm against a blank which contains all the in-
gredients of this reaction except the ester-containing lipids.
Plot the number of microequivalents against the optical density.

Evaluation

For the analysis of unknown natural products, such as lipid extracts obtained from red blood cell membranes or other sources,
it is recommended to express the ester microequivalent content
per milligram dry material. If you wish to express the lipid
content of an extract as percentage of palmitic acid, which obviously cannot be applied to all natural products, you must
multiply the ester microequivalents found per milligram dry
substance by 25.6 (the molecular weight of palmitic acid = 256.42).
The number obtained expresses the lipid content of your sample
as percentage of palmitic acid.

Use and Limitations

Not all esters react with similar rapidity in this assay. Simple
esters such as methyl esters of long chain fatty acids seem to
react very fast; glycerides and phosphatides require slightly
longer time. More complex lipids such as bacterial lipopolysaccharides, which may have - NH_2-bound carboxylic acids require
several hours of alkaline hydroxylamine treatment. Therefore,
this assay seems to be unsuitable for the determination of ester
groups in more complex lipids. For such purposes, the liberation
of carboxylic acids or their transesterification into methyl
esters is more recommended (see Exercise No. 43 or 64).

Reference

Snyder, F., Stephens, N.: Biochim. Biophys. Acta *34*, 244 (1959)

Exercise No. 64

Qualitative and Quantitative Analysis of Carboxylic Acids

Ester-bound carboxylic acids, which give lipophilic, nonpolar
character to certain natural products, may be analyzed qualita-
tively in gas-liquid chromatography or quantitatively by the
hydroxylamine procedure. Glycerides, phosphatides, and simple
esters can be hydrolyzed with acid or alkali and the liberated
carboxylic acids can be measured by different chemical procedures.
The alkaline hydroxylamine procedure of Snyder and Stephens (1959)
may be directly applied to some classes of lipids for the quantitative determination of ester groups. Such an exercise (Exercise
No. 63) is described in this manual. In the case of more complex

glycolipids, such as bacterial polysaccharides, the alkaline hydroxylamine procedure does not give reproducible results, while total hydrolysis and liberation of the carboxylic acid content of these materials usually results in loss of some of the acids. Boron trifluoride in methanol was successfully used by Metcalfe and Schmitz (1961) for the esterification of free fatty acids. It has been shown by Duron and Nowotny (1963) that boron trifluoride in methanol or boron trichloride in methanol are able not only to esterify free fatty acids but also to transesterify certain complex lipids, yielding methyl esters of all bound carboxylic acids. This reaction may be followed by quantiative determination of the carboxylic acid methyl esters or by their gas chromatographic qualitative analysis. The Exercise described here is based on the above experiments.

Materials and Equipment

Bacterial lipopolysaccharide (from Exercise No. 21)
Boron trifluoride-methanol reagent, containing 13%-15% wt/wt BF_3 gas concentration. The reagent can be either purchased from Applied Science Laboratories, Inc., State College, PA., or prepared in the laboratory by dissolving BF_3 gas in methanol and measuring the increase in weight of the methanol solution
Hexane, reagent grade
Methanol, reagent grade
18 mm × 150 mm test tubes equipped with ₸ 12/30 joints
50 cm long reflux condensers with ₸ 12/30 joints
Boiling water bath
Spectrophotometer
Alkaline hydroxylamine reagent and ferric perchlorate reagent as described in Exercise No. 63
10 mm × 120 mm graduated test tubes
Sodium sulfate, anhydrous
15 calibrated centrifuge tubes
Home made centrifuge tubes, 3 mm inside diameter, 100 mm, with rubber stoppers

Procedure

1. Run the determination in duplicate. Accurately measure a lipopolysaccharide sample weighing between 6 and 10 mg into a test tube with ₸ joint. Add 1 ml BF_3-methanol reagent, Alundum crystals or other type of pure boilingstone chips and connect the tube with the condenser and reflux it on a hot water bath. Be sure that the condenser is at least 50 cm long and that low molecular weight methyl esters will not escape because of improper cooling. If the tap water in the laboratory is not cold enough, circulate ice water in the condenser. Reflux the samples for 6 h. Before disconnecting the tube from the condenser, immerse the tube up to its neck in ice water. When the suspension is completely cooled, add approximately 3-4 ml methanol to the tube through the condenser from a pipette. Try to rinse the inner walls of the condenser with this methanol. Wait a few minutes, then disconnect the tube from the condenser. Add 4 ml water and

4 ml reagent grade hexane to the tube. Close the tube tightly with a $\text{\P}$ 12/30 stopper and shake it vigorously for 30 s. Let it stand at room temperature for approximately 15 min. The reaction mixture will be separated and the upper hexane phase will contain the transesterified methyl esters.

2. After the transesterification, the methyl esters may be determined quantitatively in the hexane extract by the Snyder and Stephens procedure (1959). It is recommended to take different aliquots of the hexane extract such as 0.5 ml and 2.0 ml. Add to the 0.5 ml sample 1.5 ml hexane. Add to both tubes 2.0 ml alkaline hydroxylamine reagent, mix the contents of the tubes carefully, and let the tubes stand at room temperature for 15 min. Thereafter, immerse the tubes very briefly in a 67°C water bath, as described in Exercise No. 63. When the mixture simmers down, the rack with the test tubes is left in the bath for 15 min. During this time the hexane evaporates completely without any loss of the methyl esters, which react very rapidly with the hydroxylamine reagent. The final volume after 15 min at 67°C is 2.0 ml. After the tubes have cooled, pipette 5,0 ml ferric perchlorate reagent into each sample. The purple color develops in 30 min at room temperature and can be read in the spectrophotometer at 520 nm against a reagent blank treated exactly as the unknown sample, using 2.0 ml reagent grade hexane instead of the ester-containing hexane extract.

3. The volatility of the low molecular weight methyl esters may be used to estimate the short chain and long chain fatty acid content in the hexane extracts. Pipette 2.0 ml hexane extract into open test tubes. Use graduated test tubes for this purpose. Place the 2.0 ml hexane into a 67°C water bath. Proceed carefully, as above, then leave the open, graduated test tubes in a water bath at 67°C for 60 minutes. During this period the methyl acetate evaporates. If necessary, replace the hexane lost by evaporation. Add to the tube pure hexane until the volume reaches the volume con had before evaporation. Add now 2.0 ml hydroxylamine to the cooled tubes, and complete the procedure as described in the previous paragraph.

4. The hexane extract containing methyl esters can be used directly for gas chromatographic analysis. In case the methyl ester concentration is too low to be easily detectable by gas chromatography, repeat the experiments and use only 0.5 ml hexane instead of 4 ml for the extraction of the reaction mixture after transesterification. Lift as much as you can with the help of a fine capillary into a home-made test tube with 2 or 3 mm inside diameter and 100 mm length. Some water will also be transferred with the capillary pipette, therefore add approximately 20 mg to 50 mg anhydrous Na_2SO_4 to the tubes; close the tubes with a rubber cap, shake well, and centrifuge them at 1000 g for 10 min. Take samples with a Hamilton syringe for gas chromatography from the upper phase. For the gas chromatographic procedure, use the instructions given in Exercise No. 39.

Evaluation

In the quantitative determination of methyl esters by the hydroxylamine procedure, proceed as described in Exercise No. 63. Express the ester content of the analyzed natural product as number of ester microequivalents per milligram material.

For the identification of fatty acid methyl esters, gas-liquid chromatography may be used (see Exercise No.39). The use of internal standards is as helpful in this assay as in any other chromatography procedures.

Use and Limitations

The procedure described here can be used without modification for other lipids or fatty acid-containing natural products. A remarkable difference can be observed in the case of transesterification of certain carboxylic acid esters. Neutral fats, phosphatides, and simple esters can be transesterified by refluxing the samples for one-half hour as described above. Amine-bound long chain carboxylic acids which are known to be present in certain sphingolipids and also in bacterial lipopolysaccharides require a much longer treatment. This property of the transesterification procedure with boron trifluoride-methanol may be utilized for differentiation between O-acyl and N-acyl linkages. The difference in the transesterified methyl ester content of the hexane extracts after 30 min and 6 h refluxing indicates approximate distribution between O-acyl and N-acyl functional groups. An ultramicroprocedure for the quantitative determination and identification of fatty acids in lipids has been described by Archibald and Skipski (1966).

References

Archibald, F.M., Skipski, V.: J. Lip. Res. 7, 442 (1966)
Duron, O.S., Nowotny, A.: Anal. Chem. 35, 370 (1963)
Metcalfe, L.D., Schmitz, A.A.: Anal. Chem. 33, 364 (1961)
Snyder, F., Stephens, N.: Biochim. Biophys. Acta 34, 244 (1959)

3 Immunologic and Other Biologic Assays

3.1 Antibody Production

Exercise No. 65

Immunization and Adjuvant Effect

It is not possible to describe a general procedure of immunization which can be applied to many different antigens and different experimental animals. One reason for this is the wide variety of antigen materials. Another is the fact that different animal species respond differently to the same antigen and that some apparently healthy individuals of a species hardly produce antibody to a certain antigen, while other individuals of the same species respond with a high output.

The antibody-producing ability (immunogenicity) of the materials is usually determined by the following factors: (a) It must have one or more characteristic features in its structure which makes the substance foreign to the host; (b) It must have a certain molecular weight; (c) It has to be present in the host organism for a certain length of time; and (d) It must be able to reach the antibody-producing immunocompetent cells. Some of the above points can be influenced artificially. For example, a small molecule, which alone would not elicit antibody production, can be coubled with a large inert carrier. This complex will act as an antigen, and the immunologic specificity of this preparation will still be determined by the relatively small molecule (see Exercise No. 15). The introduction of new groups into the serum albumin molecule will result in a preparation which acts as an antigen if injected into the parent animal. If we mix an easily soluble weak antigen with mineral oils, it will not be eliminated by the host as easily as it would be without the added oil. The result is an increased antibody production and this "adjuvant effect" has been shown using soluble antigens adsorbed to inorganic precipitates such as alum. For details and additional information, students should read any of the numerous textbooks on Immunology.

In *Part A* of this Exercise, high titer of antihuman serum albumin antibodies will be produced in rabbits for use in several forthcoming Exercises. At the same time the adjuvant effect of alum precipitate and "complete Freund's adjuvant" (a mineral oil plus killed mycobacterial suspension) will be demonstrated. In *Part B* of this Exercise, particulate antigen will be used for immunization.

Part A. Soluble Immunogen

Materials and Equipment

    White rabbits approximately 2.5 kg weight
    Sterile human serum albumin (HSA)
    DNP-human serum albumin conjugate (from Exercise No. 15)
    Commercial "complete Freund's adjuvant"
    Potassium aluminum sulfate, 10% sterile solution in water
    N sodium hydroxide
    Sterile physiologic saline
    2 ml and 10 ml sterile syringes
    Needles, 20 and 26 gauge
    Sterile 50 ml centrifuge tubes
    Sterile 20 ml serum bottles with stoppers
    Sterile 10 ml pipette with cotton plug
    Pipetting balloon
    Rabbit boxes

Procedure

1. The 1% HSA solution in saline has been filtered through a sterile Millipore filter. It is supplied for the exercise in sterile containers.

2. To 5 ml of the above HSA, add 15 ml sterile NaCl. This dilution can be used for immunization according to the schedule given below.

3. Preparation of alum precipitated HSA. Mix 5 ml HSA under sterile conditions with 10 ml sterile NaCl. Add 1 ml sterile 10% alum solution and mix thoroughly. The solution must be neutralized with N NaOH. A fine white precipitate will form; make up the volume to 15 ml with saline. This suspension is ready for injection.

4. The HSA used with complete Freund's adjuvant is prepared as follows: Mix 10 ml HSA and 10 ml NaCl under sterile conditions. Before injection, mix 1 vol of the above dilution in a small sterile flask or test tube with 1 vol of complete Freund's adjuvant. For mixing, a sterile syringe should be used. Care must be taken to avoid excess air bubble formation. For easier handling, always prepare twice the volume needed for the injection.

5. Prepare a 0.25% DNP-HSA solution in saline from a freeze-dried preparation.

Table 4. Immunization schedule

Weeks	HSA	HSA-Alum ppt	HSA + Freund's adjuvant*	DNP-HSA conjugate
First	0.2 ml I.P.	0.2 ml I.P.	0.2 ml I.P.	0.2 ml I.P.
Second	1.0 ml I.V.	1.0 ml I.M.	1.0 ml S.C.	1.0 ml I.V.
Third			Rest, no injection	
Fourth	1.0 ml I.V.	1.0 ml I.M.	1.0 ml S.C.	1.0 ml I.V.
Fifth	Bleeding	Bleeding	Bleeding	Bleeding

* Only the last injection should contain Freund's complete adjuvant. Inject only HSA in the first two weeks.

6. All four preparations have practically the same HSA concentration. Inject rabbits with each preparation. The immunization schedule, given in Table 4, is not an optimal one to obtain
high titer antisera, but is adapted to the conditions of teaching
courses. A booster injection is given in the fourth week.

7. Bleed the rabbits through the central artery of the ear
five days after the last injection has been given. Another way
of taking blood from an experimental animal is by heart puncture.
Blood must be collected in properly marked sterile and dry 50 ml
centrifuge tubes. After allowing it to stand for 60 min at room
temperature, separate the serum by centrifugation at 1000 g for
30 min. Remove the serum with a sterile syringe or pipette and
transfer it to sterile serum bottles supplied with rubber stoppers.

Evaluation see *Part B.*

Use and Limitations see *Part B.*

Part B. Particulate Immunogen

The same basic principles are applicable to the immunization with
cells or other nonsoluble antigens as to the immunization with
soluble antigens.

In this Exercise, the adjuvant effect of bacterial endotoxic
lipopolysaccharides will also be demonstrated. Sheep red blood
cells (SRBC) will be injected with and without endotoxin. The
adjuvant effect of endotoxins was first described by Landy and
Johnson (1955).

Materials and Equipment

Sheep red blood cells, 10% suspension of washed erythrocytes,
  made in physiologic saline
Serratia marcescens cell suspension, autoclaved, containing
  2.5 mg/ml dry cells in saline
Endotoxic lipopolysaccharide solution from *Serratia marcescens,*
  containing 10 µg/ml saline
Syringes, rabbit boxes, centrifuge tubes, etc., as in *Part A*

Procedure

1. The schedule of immunization is given in Table 5 for *Serratia
marcescens* cells, and in Table 6 for SRBC antigens.

Table 5. Immunization with
Serratia marcescens cells

Weeks	Dose, ml	Route
First	0.1	I.V.
Second	0.3	I.V.
Third	Rest, no injection	
Fourth	1.0	I.V.
Fifth	Bleeding	

Table 6. Immunization with
sheep red blood cells

Weeks	Dose, ml	Route
First	0.2	I.P.
Second	0.2	I.M.
Third	Rest, no injection	
Fourth	1.0	I.P.
Fifth	Bleeding	

2. To demonstrate the adjuvant effect, in another group of
rabbits the first, third, and fifth unjections should be accom-
panied by the injection of 1 μg (0.1 ml) of toxic endotoxin.
Use exactly the same schedule for this group of animals as
above.
3. Bleed the animals as described in *Part A*.

Evaluation

The antibody titer can be measured by any of the different procedures described in this Manual (Exercise Nos. 69,70,76,77, 78 or 8
The immunoplaque method (Exercise No. 66) is especially useful for
the demonstration of antibody-producing activity in mice, using
SRBC with or without adjuvant. Observe the higher antigenicity
of particulate material compared to soluble antigens on a weight
basis.

Use and Limitations

The procedures above have been described for the immunization of
rabbits, and they will yield, in case of final bleeding by heart
puncture, approximately 30 ml - 45 ml blood per animal. It is much
more convenient to handle mice, and for several other reasons it
is frequently preferred. A serious disadvantage of this is the
small volume of serum one can obtain, which is usually less than
0.5 ml per mouse. A very convenient method has been developed by
Muñoz (1957) to harvest mouse ascites fluid in several milliliter
quantities. The ascites fluid is quite similar in its composition
to serum and it has approximately the same IgG content. The amount
of other immunoglobulin classes may be different from serum, but
it still serves as an excellent source for various immunoglobulin
isolations. The procedure as described by Muñoz is as follows:
   Mix under sterile conditions 10 mg bovine serum albumin (or
any other immunogen) dissolved in 1 ml saline and 2.0 ml complete
Freund's adjuvant by using a glass tissue grinder. Inject 0.25 ml
emulsion intraperitoneally. The second injection of 0.25 ml should
be given two weeks later. The ascites fluid will accumulate in
the peritoneal cavity within two to three weeks after the last
injection. The ascites fluid can be obtained by tapping the
peritoneal cavity with an 18 gauge needle, allowing the fluid to
flow directly into a centrifuge tube. The mice will survive and
tapping can be repeated at two week intervals several times.
Muñoz reported the collection of 492 ml ascites fluid high in
anti-BSA content from a total of 50 mice.

References

Landy, M., Johnson, A.G.: Proc. Soc. Exp. Biol. Med. *90*, 57 (1955)
Muñoz, J.: Proc. Soc. Exp. Biol. Med. *95*, 757-759 (1957)

Exercise No. 66

Demonstration of Antibody Production at Cellular Level (Immunoplaque Method)

Jerne (1963) described a procedure by which it is possible to
determine how many of the total cells present in an organ, such
as spleen, lymph nodes, etc., are producing and secreting antibodies. Friedman (1964) made some modifications of this basic
technique and applied it in a series of different experiments.
The description of the immunoplaque technique given here is based
on the above references.

Materials and Equipment

      10 Swiss albino mice weighing between 18 and 20 g each
      Sheep erythrocytes
      Guinea pig complement
      0.1 M sodium bicarbonate
      Hanks balanced salt solution
      Physiologic saline
      Agar, Noble (Difco)
      Hemocytometer
      Scissors
      Dissecting needles
      Forceps
      Plastic Petri dishes, 90 mm diam
      Nylon filters
      Funnel
      Incubator at 37°C
      Microscope

Procedure

   1. Prepare the balanced salt solution according to the description of Hanks and Wallace (1949). This solution is also
commercially available. Adjust the pH to 7.2 with 0.1 M NaHCO$_3$.
   2. Wash sheep red blood cells with saline three times by serial
centrifugation at 600 g and prepare a 5% suspension by pipetting
0.5 ml packed cell sediment into 10 ml saline. Rinse the pipette
with the saline. Immunize five mice by intraperitoneal injection
of 1.0 ml of the red blood cell suspension. The other five mice
will serve as uninoculated controls.
   3. After four days, kill the animals. Remove the spleens of
the immunized mice and place them in ice-cold balanced salt solution. Do the same with the spleens of the nonimmunized mice, but
keep them separate. Tease the spleens with dissecting needles.
Place two layers of a Nylon stocking in a funnel and filter the
spleen homogenate through the Nylon. Prepare a similar suspension
from the spleens of the control mice. Form a bag from the Nylon
containing the homogenate. Gentle squeezing will yield additional
cells in the filtrate. Wash the filtrate twice with the balanced
salt solution by suspending and centrifuging at 1000 g for 10 min
in a refrigerated centrifuge. Discard the washings and resuspend
the cells in 10 ml salt solution. Keep in the refrigerator.

4. Prepare agar plates: Pour 10 ml of 1.5% Noble agar (Difco)
in salt solution into a Petri dish. Let it solidify. Carefully
mix 0.1 ml of a 10% suspension of freshly washed sheep red blood
cells and 0.1 ml of spleen cell suspension with 2.0 ml of 0.7%
melted agar at $40^{0}-45^{0}C$. Carefully and quickly pour this mixture
on the prepared agar base layer in a Petri dish so as to form a
thin, even upper layer. Incubate the plates without inverting at
$37^{0}C$ for 60 min.

5. While the incubation goes on, make a cell count from the
spleen cell suspension with a hemocytometer and determine the
number of nucleated cells per milliliter. (See Exercise No. 82).

6. Make a ten-fold dilution of guinea pig complement with the
balanced salt solution, and pipette 2.5 ml on each incubated plate.
Return the plates to $37^{0}C$ for an additional half hour.

7. Plaques are visible on the plates where antisheep hemolysin-
producing spleen cells were plated. A clear zone indicates these
plaques, and with the microscope a spleen cell can usually be
found in the center. In order to make the lysed zones more visible,
the hemoglobin of the unlysed red blood cells may be stained with
benzidine-H_2O_2 solution, according to Jerne (1963).

Evaluation

Count the plaques per plate. Compute the number of antibody-
producing cells per 10^{6} spleen cells and also per spleen. Compare
the findings with the control group.

Use and Limitations

Several modifications make this simple technique applicable to
a variety of experiments. It could be shown that human O red
blood cells, coated with bacterial lipopolysaccharides (see passive hemagglutination, Exercise No. 69), or conjugated with simple
hapten may be used to indicate the presence of those spleen or
lymph node cells which produced anti-lipopolysaccharide or anti-
hapten antibodies. This is an application of passive hemolysis
to the immunoplaque method (Merchant and Hraba, 1966).

An elegant modification of the immunoplaque procedure has been
developed by Sterzl and Riha (1965), and Dresser and Wortis (1965)
to distinguish between IgM and IgG antibody production. In the
procedure described above the IgM will result in lysis because
it is highly "efficient" in fixing complement. In the same test
the IgG will not be visible unless present in very high quantity.
The modification is as follows: Mice receiving two or more injections of sheep or other erythrocytes, at least two weeks apart
are to be used rather than animals which received a single, primary immunization. Otherwise, steps 1 to 7 are followed exactly as
above. Then rinse the surface of the lysed (but unstained) plates
three times with approximately 10 ml balanced salt solution to
remove the complement. Add 1 ml 1:10 diluted antimouse IgG serum
(produced in rabbits or goats). Incubate at $37^{0}C$ for 60 min. Add
1 ml 1:10 diluted fresh complement to the plates and incubate for
30 min. Repeat the plaque count on the plates and compare the
findings with the previous ones. The increase in plaque count is
due to the complex formation of antisheep red blood cell mouse
IgG antibodies and antimouse IgG rabbit antibodies. This complex

will fix complement and therefore it will readily produce lysis,
whereas the low efficiency 7 S hemolysins will not do so.
     The immunoplaque method has also been successfully used in
quantitative studies of antibody production due to the application of different adjuvants both *in* vivo and *in* vitro (Behling
and Nowotny, 1977 and Frank et al., 1977).
     The fact that only 1% or less of the total cell count in a
spleen can be demonstrated to be antibody-producing by the immunoplaque technique makes it clear that not all antibody-
producing cells will be detected by this method. An inherent
limitation of the process lies in the fact that only secreted
antibodies will be demonstrated. Microagglutination techniques
with lymphoid cells, rosette or cluster formation methods have
been developed by Biozzi, Stiffel and Mouton (1967). Exercise
No. 87 describes such procedure.

<u>References</u>

Behling, U.H., Nowotny, A.: J. Immunol. *118*, 1905, 1977
Biozzi, G., Stiffel, C., Mouton, D.: In: Immunity, cancer and chemotherapy.
   Mihich, E. (ed.), p. 103. New York: Academic Press 1967
Dresser, D.W., Wortis, H.H.: Nature London *208*, 859 (1965)
Frank, S., Specter, S., Nowotny, A., Friedman, H.: J. Immunol. *119*, 855, 1977
Friedman, H.: Proc. Soc. Exp. Biol. Med. *117*, 526 (1964)
Hanks, J.H., Wallace, R.E.: Proc. Soc. Exp. Biol. Med. *71*, 196 (1949)
Jerne, N.K., Nordin, A.A., Henry, C.: Cell-bound antibodies. Philadelphia,
   PA.: Wistar Institute Press 1963, p. 109
Jerne, N.K., Nordin, A.A.: Science *140*, 405 (1963)
Merchant, B., Hraba, T.: Science *152*, 1378 (1966)
Sterzl, J., Riha, I.: Nature London *208*, 858 (1965)

3.2   Antigen-Antibody Interactions

3.2.1   Methods of Agglutination

Exercise No. 67

Bacterial Agglutination

Particulate antigens, such as bacteria, combine with their specific
antibodies to form complexes that usually aggregate as visible
clumps. This is called bacterial agglutination.
     The procedure is used both to demonstrate the presence of
antibodies in serum and to identify antigens on microbial cell
surfaces. The principle is the same in both applications. If the
serum contains antibodies against a surface antigen, they will
agglutinate the bacterial cells. Using a variety of typing sera,
bacteria can be identified and classified. If a constant amount
of a bacterial cell suspension is mixed with graded volumes of
dilutions of homologous antiserum, one obtains a measurement of
the concentration of antibodies in the serum. The term used to

describe the concentration of antibody in serum is "titer",
which is the reciprocal of the highest dilution producing a definite reaction.

Materials and Equipment

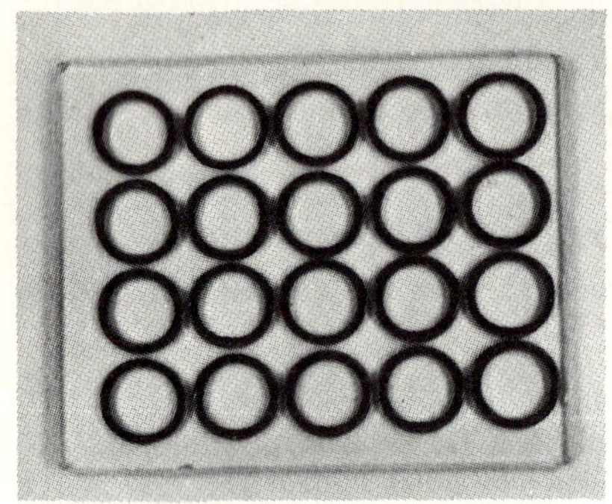

Heat killed suspension of bacterial
  cells, containing 1 mg/ml dry
  weight
Rabbit antiserum against the above
  bacteria
Normal rabbit serum
Disposable plastic plates with cup-
  form depressions, or microtest glas
  plates with raised ceramic rings.
  (See Fig. 39)
Physiologic saline, pH 7.4
Test tubes, 10 × 100 mm
Pipettes, 1 ml capacity
Glass rods

Fig. 39. Microtest slide with
ceramic rings, available from
Arthur H. Thomas, Philadelphia,
PA.

Procedure

   1. If no plastic plates or microtest slides are available,
take a glass plate which has been well cleaned with chrome-sulfur
acid, rinsed and dried. By using a wax marking pencil, draw lines
on the glass plate dividing its surface into squares approximately 2 × 2 cm. Do not touch the glass with your fingers.
   2. Prepare 10 twofold dilutions of the rabbit antiserum in the
small test tubes, making 0.2 ml of each dilution with pH adjusted
saline. Make 10 similar dilutions of the normal rabbit serum as
well. These will serve as controls.
   3. Starting from the highest serum dilution, transfer one drop
from the test tube into the last cup of the plastic plate (or last
circle of the glass plate). Proceed towards the lowest dilution.
Add one drop of bacterial cell suspension to each drop of serial
dilution. Mix the two drops together with a glass rod, starting
with the highest dilution. Between each dilution, rinse and wipe
the glass rod.

Evaluation

Tilt the slide back and forth slowly for a few minutes, then
watch closely for developing agglutination. Record the highest
dilution which produced bacterial agglutination. The reciprocal
of this dilution is the "titer" of the serum.

Use and Limitations

The bacterial agglutination method is generally not very sensitive. It is also known that certain antigens of the bacterial
cell may not be exposed and therefore are not available for reaction with antibodies. Capsular antigens, for example, may mask
O-antigens in the bacterial cell wall, as with *Serratia marcescens*
(Tripodi, 1966). In such cases, it is essential to use bacterial
cells which were boiled in a water bath for 60 min. This treatment
sets free O-antigen receptors on the bacterial wall.
    The various serodiagnostic procedures used in routine bacterio-
logy were summarized in two volumes edited by Kwapinski (1969).
    A more sensitive version of bacterial agglutination uses
microscopic observation. A loopful of serum can be mixed on a
microscopic slide with the bacteria taken from a simple colony
grown on a nutrient agar plate. Using a low-power resolution,
the agglutination can be observed under a microscope at room
temperature or, if available, on a 37°C heated microscopic stage.
If the droplet on the microslide is encircled with vaseline, a
cover slide may be placed on the drop. This will prevent evaporation and thus will make prolonged incubation at 37°C possible.

References

Kwapinski, J.B.G., (ed.): Analytical serology of microorganisms. Vols. I and
    II, New York: Wiley Interscience 1969
Tripodi, D.: P.D. Thesis. Temple University School of Medicine, Philadelphia,
    PA., 1966

Exercise No. 68

Hemagglutination and Its Inhibition

If antigens on the surface of erythrocytes react with their
homologous antibodies, agglutination of the red blood cells will
be observed. This hemagglutination may be inhibited if isolated
antigens of the erythrocyte membrane are first incubated with the
antiserum. This will block the receptor sites of the antibodies,
thus inhibiting their capacity to cause hemagglutination.
    The hemagglutination inhibition test will be used in this
Exercise for the semiquantitative determination of blood group
antigen activity of some preparations. To a certain amount of
group A (or B) material, an excess of anti-A (or B) serum is
added. After incubation, the group A antigens will absorb a certain amount of the anti-A agglutinins. The amount of absorbed
agglutinins is related to the activity or amount of the group A
antigens in the preparation. In a control setup, the same amount
of anti-A serum is incubated at room temperature with saline.
The amount of nonabsorbed agglutinins will be determined.

Materials and Equipment

Hog gastric mucin, or pepsin from hog stomach (available from
  Worthington Biochemical Corp., Freehold, NJ 07728)
Group A or B RBC membrane preparations (from Exercise No. 27)
Anti-A or anti-B human serum (high titer sera are usually
  available from blood banks)
Saline, physiologic
Takatsy "Microtitrator" kit (Fig. 40). An American model of
  the microtitrator, based on the same principle as the Takatsy
  model is available from Flow Laboratories, Inc. Rockville,
  MD 20852
Magnifying glass

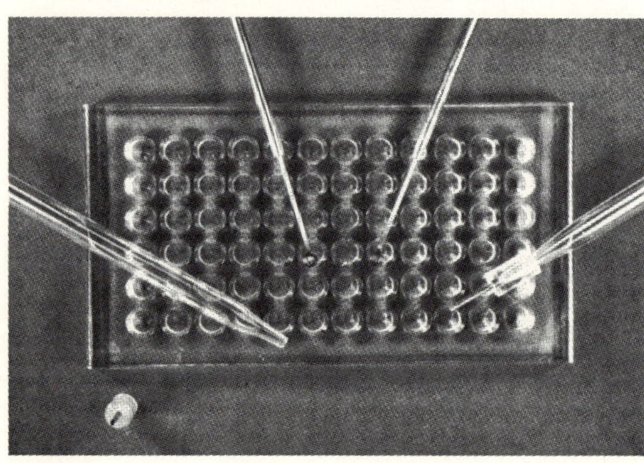

Fig. 40. Takatsy micro-
titrator kit

Procedure

1. In Exercise No. 27, stroma preparations from human blood
were prepared. This stroma was solubilized and its group A antigen
content will be measured. In this exercise we will prepare a solution of commercially available pepsin or mucin preparations and
we will compare the group A antigen content thereof with the
solubilized group A human RBC.
2. 1 mg mucin or pepsin powder will be dissolved in 100 ml
saline. Keep the solution refrigerated.
3. Serial dilutions will be made from both the antigen + isoagglutinin and from the saline + isoagglutinin mixtures. After
adding a constant amount of a 2% group A RBC suspension to each
dilution, we will observe and quantitate the degree of hemagglutination in the different dilutions. Detailed descriptions
are given below.
4. For the rapid preparation of serial double dilutions and
for the convenient observation of hemagglutination, the Takatsy
microtitrator will be used. This kit consists of the following:
Loops which are made of thermoresistant wire individually calibrated to pick up and deliver 0.025 ml of a standard 0.85% saline
solution. They have tapered handles to facilitate accurate simultaneous titrations. Larger size loops are calibrated to pick up
and deliver 0.05 ml of a normal saline solution. Calibrated

droppers are also included in the kit. They are stainless steel
tube sections mounted in Plexiglas hoods. The tube end has been
designed to form exactly 0.025 ml drops of saline solution. The
calibrated dropper is fitted with a conical bore so that it fits
a glass pipette with a Luer tip. Instead of test tubes, the
Takatsy kit uses plates made of transparent Plexiglas with 6 rows
of 12 wells. The wells have a diameter of approximately 5 mm.
For better observation of the degree of agglutination, plates
with larger and shallower depressisons are more suitable. Such
plates, with well diameters of 12 mm, are available from Instrumentation Associates, Inc., New York, N. Y. 10023. Fig. 41
shows the cross section of such a well with the Takatsy loop in
it.

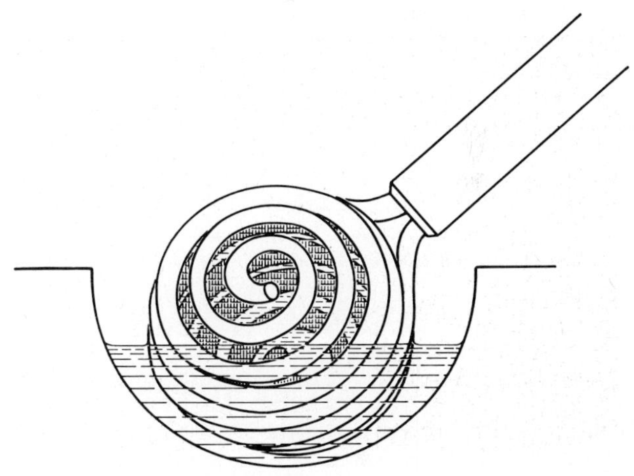

Fig. 41. Closeup illus-
trating Takatsy loop in a
well of the hemagglutina-
tion plate

   The technique is as follows: The Takatsy loops are flamed
with a Bunsen burner to remove contaminants. After flaming, the
loops are allowed to cool. Meanwhile, the diluent is picked up
by the calibrated dropper and measured into the wells; the
quantity of the diluent should be equal to the volume of the
Takatsy loops, or double, according to the desired ratio of
dilution (1:1 or 1:2). The cooled loop fills when it touches
the liquid surface. The liquid is retained by the Takatsy loop
through surface tension. The loop with the liquid is submerged
in the first well and vigorously twirled. The diluted liquid is
transferred from the first well to the next well by the loop.
Serial dilutions are prepared by repeating the above process as
many times as necessary.
   5. Group A RBC will indicate the presence of nonabsorbed
isoagglutinins in the incubated antigen + isoagglutinin mixture.
For this purpose, a fresh suspension of washed group A RBC must
be prepared every day. The procedure is as follows: 2 ml of group
A blood is washed with 8 ml saline in a centrifuge tube and
sedimented. This washing should be repeated with fresh saline.
Pipette 0.5 ml of the washed and packed cell sediment into 25 ml
saline and suspend the cells by stirring with the pipette.
   6. The incubation of the group A antigen with the isoagglutinin
takes place in the first well of the Plexiglas plate described in

the introduction. With the calibrated control dropper, put two
drops of the antigen preparation (0.5 ml) and two drops of the
anti-A serum into the first well of the rows. In the first well
of the control row, mix two drops of saline with two drops of
the above serum. Add two drops of saline to the remaining wells
in the first and second rows. The contents of the first wells
must be thoroughly mixed with the small glass rods.

7. With the 0.05 ml capacity loop, pick up a loopful of mixture
from the first well and transfer it to the second, which contains
two drops of saline. Rotating the loop between the fingers for
approximately 10 s, stir the transferred mixture thoroughly with
the saline in the well. Lifting the loop and transferring its
contents to the third well, proceed as before. Repeat this procedure for each well.

8. After the serial double dilutions have been completed, add
two drops of the washed 2% suspension of group A human RBC to
each well, using the calibrated dropper.

The contents of all the wells must be mixed by vigorous shaking of the plates in a horizontal plane with short strokes. Care
should be taken to avoid spilling. The Plexiglas plates must be
covered with a glass plate and allowed to react at room temperature for 60 min. Shaking of the plates has to be repeated every
20 min during this hour, and once more before evaluation, as
described below.

Evaluation

To obtain reproducible reading of the degrees of hemagglutination
using the diagram (Fig. 42) requires experience. However, some
useful information can be obtained from the first trial. The
maximal agglutination is considered as No. 10 and the least agglutination is considered as No. 1. Agglutination No. 1 is termed
the case where, under an eight- to tenfold magnifying glass,
clumps of erythrocytes are observed to float in the drop. In

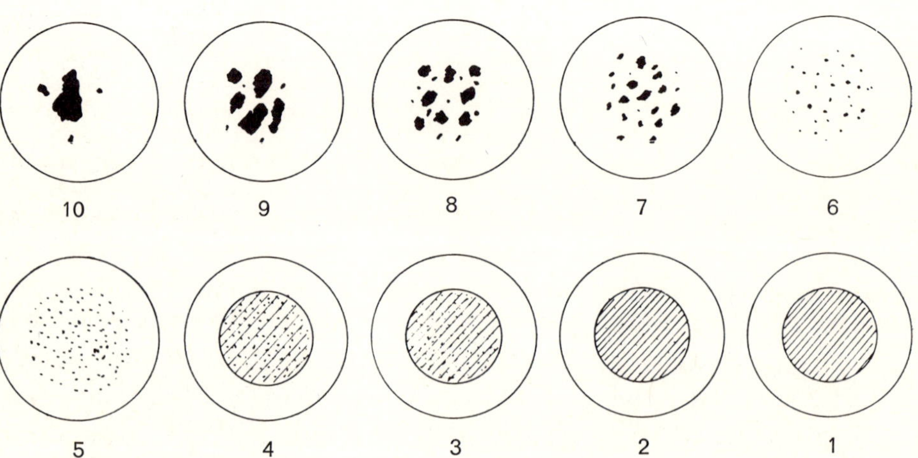

Fig. 42. Schematic illustration of the different degrees of hemagglutination

agglutination No. 2, a larger number of erythrocytes is clustered,
therefore they are readily discernable through the lens. Agglutination No. 3 is one visible to the normal naked eye. No. 4 is dis-
tinguished from No. 5 by the circumstance that while at No. 5
the fluid surrounding the agglutinated erythrocyte is colorless,
with No. 4 it appears stained due to the presence of unaggluti-
nated red cells. The difference between the higher grades is
visible in the diagram.

    For semiquantitative evaluation, the calculation is as follows:
The degrees of agglutination read in the various wells are written
down and totaled. The total represents the agglutination titer
of the serum. Where antigen is added to the serum, depending on
the activity of the antigen solution, more or less serum agglutinin was neutralized; in other words, the serum titer decreased.
Deducting the titer of the serum + antigen solution from the
titer of serum + saline, the number of antigen units (AU) contained in the measured antigen solution is obtained. Antigenic
activity (AA) is expressed by the number of antigen units per
1 mg of the dry antigen substance. This procedure has been developed by Nowotny and Backhausz (1957).

    A considerably simpler but less quantitative method of evalua-
tion is to read the end point. This means to identify the highest
serum dilution which still gives hemagglutination. In a narrow
test tube with about 5 mm i. d. the erythrocytes which did not
agglutinate will settle at the deepest point of the tube, forming
a compact small red button. If microtiter plates are used where
the wells have a conical (v-shaped) bottom, the erythrocytes will
form an even smaller, point-like sediment. The cells on the other
hand, which reacted with sufficient number of antibodies, will
form a network of agglutinated cells, which will cover the inside
of the tube, forming a spread-out, film-like deposit in the bottom
of the tube. The two patterns are clearly distinguishable. The
highest dilution, which still shows agglutination, is the end
point. Figure 43 shows the patterns of positive and negative agglutination.

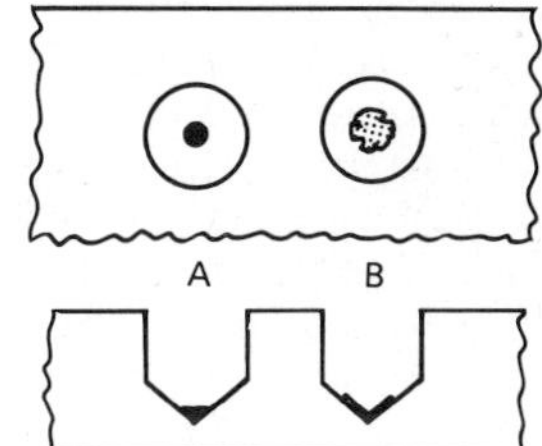

Fig. 43. Microtitration pattern of hemagglutination
A negative, *B* positive

Use and Limitations

With some experience, this method of determination yields unex-
pectedly useful comparative values for antigen solutions of
diverse activity. In the above process, care must be taken to
dilute the antigen solution to such an extent that it does not
reduce the agglutination titer of the serum by more than 50%. At
higher concentrations the relationship between concentration and
diminution of agglutination is no longer linear.

References

Nowotny, A., Backhausz, E.: Acta Phys. Acad. Sci. Hung. *12,* 53 (1957)
Takatsy, Gy.: Acta Microbiol. Acad. Sci. Hung. *3,* 191 (1955)

Exercise No. 69

Passive Hemagglutination and Its Inhibition

The binding of antigens to red blood cells may render them agglutinable with the corresponding antiserum. For example, human
group O erythrocytes will be agglutinable by antiegg albumin
antibodies if the red blood cell is coated with egg albumin.
In these reactions the red blood cell serves as a carrier of the
antigen and also as an easily visible indicator of the antigen-
antibody reaction. Because of the passive role of erythrocytes
in these immunologic reactions, the method is called "passive
hemagglutination".
   If the antibodies have been incubated with antigens or their
hapten breakdown products before this reaction, hemagglutination
will not take place because the antigen receptor sites of the
antibody molecules are already occupied. This phenomenon is called
inhibition of passive hemagglutination. Passive hemolysis may
also be inhibited in the same way. The great importance of these
two assays can be found in their sensitivity in the detection of
nonprecipitating antibody + hapten reactions. The same inhibition
may be used for demonstration of the formation of soluble antigen
+ antibody complexes.
   Several methods are known for the fixation of antigens to
erythrocytes. Depending upon the chemical nature of the antigen,
different procedures are recommended. Bacterial lipopolysaccharide
and polysaccharide antigens can be attached relatively easily.
Neter and associates (1952) found that often heat treatment of
the antigen is sufficient to achieve its adsorption. Such a
method is given in this Exercise under *Part A.* Protein antigens
can usually be fixed to erythrocytes if the surface of the cells
has been modified by chemicals such as tannic acid. The method
in *Part B* uses this treatment, according to the original description of Boyden (1951). In *Part C* the more recently developed
cabodiimide coupling procedure is described, which has been introduced by Johnson et al. (1966). In *Part D* of this Exercise,
inhibition of passive hemagglutination will be demonstrated using
partial hydrolysis products of bacterial lipopolysaccharides.

Part A. Neter's Method

Materials and Equipment

   Human group O blood, or sheep red blood cells (SRBC)
   Bacterial lipopolysaccharides (obtained from Exercise No. 21)
   Anti-bacterial O-antigen serum produced in rabbits, inactivated
     at 56°C for 60 min (Exercise No. 65)

Normal inactivated rabbit serum
Physiologic saline, pH adjusted to 7.2
Small test tubes, 8 × 100 mm
0.2 ml pipettes
Centrifuge
Boiling water bath

Procedure

1. Dissolve 5 mg bacterial lipopolysaccharide in 5 ml saline, immerse it in a boiling water bath for 2 h.

2. Centrifuge 1 ml group O human blood (or SRBC) at 1000 g for 10 min and discard the supernate. Wash the cells three times with physiologic saline.

3. Suspend the red blood cells in the lipopolysaccharide solution at room temperature, transfer the mixture to 37°C and incubate 60 min. Occasionally shake the cells. Centrifuge the suspension and wash three times as above with saline. Pipette 0.25 ml of the packed cell sediment into 10 ml saline. Keep in cold room.

These cells are now coated with lipopolysaccharide and can be used only for a few days. To prolong the usefulness of antigen-coated erythrocytes, several methods have been elaborated, the most useful being formalinization of the cell. Such treatment of the red blood cell may be carried out before combining it with antigens. The formalinized cells may be kept in the refrigerator for several weeks, thus yielding a stable preparation with very reproducible results in antibody titration. Such a method was described by Ingraham (1958) and by Csizmas (1960).

4. For titration of antiserum, first prepare double dilutions. Pipette 0.2 ml saline into ten small test tubes. Add 0.2 ml 1:10 diluted serum to the first tube, mix with a 0.2 ml pipette, transfer 0.2 ml into the next tube. The dilution in the last tube is 1:5120. If the antibody titer of the rabbit sera is high, this dilution will not give the hemagglutination end point, therefore it may be necessary to use 15 tubes in the serial double dilution. Needless to say, the Takatsy microtitrator can be used as well for preparing the dilutions. (See Exercise No. 68.)

5. Add 0.2 ml of the antigen-coated red blood cells to each tube, stir well. In a control series, make dilutions of normal rabbit serum with saline, add coated red blood cells to the tubes. Incubate all tubes at 37°C for 30 min, stir again, and let them stand at room temperature for another 30 min.

Evaluation

Read the agglutination. The titer of a serum is the reciprocal of the highest dilution which gave visible agglutination.
Use and Limitations see *Part C*.

Part B. Boyden's Method

Materials and Equipment

Crystalline egg albumin (commercially available)
Antiegg albumin rabbit serum (see Exercise No. 65) inactivated
  at 56°C for 60 min
0.01% tannic acid in physiologic saline
Human group O blood, tubes, pipettes as in *Part A*

Procedure

1. With another 1 ml human group O blood (or SBBC) prepare a
washed erythrocyte sediment as above.
2. Add 0.25 ml of the packed cell sediment to 10 ml saline.
Mix this cell suspension with 10 ml 0.01% tannic acid which has
also been prepared in saline. Mix the suspension gently and place
in a refrigerator for 15 min.
3. Wash the cells with saline three times. Discard the supernate. Resuspend the cells in 2 ml saline.
4. Prepare an egg albumin solution by dissolving 3 mg egg
albumin in 10 ml saline. After the egg albumin dissolves, mix
with 2 ml tannic acid-treated erythrocytes. Let the mixture
stand for approximately 30 min at room temperature. Sediment
the cells in a centrifuge at 1000 g for 10 min and discard the
supernate.
5. Dilute 4 ml inactivated normal rabbit serum to 100 ml with
the above saline. Use this diluted rabbit serum to wash the red
blood cell suspension. Suspend the red blood cells in 10 ml serum,
stir gently, and centrifuge as above. Repeat the washing, discard
the supernates. Finally, resuspend the cell sediment in 10 ml
diluted rabbit serum. The cells are now ready for the titration
of antiegg albumin serum. The procedure is described in *Part A*.
Evaluation see *Part A*.
Use and Limitations see *Part C*.

Part C. Carbodiimide Coupling Method

Materials and Equipment

Sheep red blood cells, washed as described in Exercise No. 27
pH 7.0 Phosphate buffered saline (PBS). See Exercise No.1.
1-Ethyl-3-(3-dimethylaminopropyl) carbodiimide.HCl, available
  from Pfaltz & Bauer, Inc., Flushing, NY 11368. Prepare a 2%
  solution in PBS
Washing fluid: 5 g BSA + 10.2 g dextrose + 10 g EDTA + 2.1 g
  NaCl dissolved in a total of 500 ml distilled water

Procedure

1. Dissolve the protein antigen to be used for coating the
erythrocytes in PBS, making a 10 mg/ml solution. A total of 2 ml
solution will be sufficient.
2. Mix 0.6 ml washed and packed erythrocyte sediment with 0.5 ml
carbodiimide and 2 ml antigen solutions. Let it stand at room
temperature for 60 min with occasional gentle shaking.

3. Centrifuge the mixture, discard the supernatant. Wash the sedimented cells three times with the "washing fluid" given in the Materials and Equipment list.

Discard the supernatants. Resuspend the conjugated RBC sediment in 25 ml PBS. The suspension is ready for passive hemagglutination. Proceed as described above in this exercise, *Part A.*

Use and Limitations

The carbodiimide coupling is a simple procedure, it uses a well-defined chemical which can be used to couple diverse proteins to red blood cells. The complex is stable and can be stored in the cold room for several days. The coupled proteins will not be altered in their structure to the effect that their reactivity with the specific antibodies would be changed noticeably.

Carbodiimide can be used very efficiently to establish covalent linkage between two soluble proteins. Goodfriend et al. (1964) used this method to couple the small molecular weight peptide bradykinin to albumin and used this complex to produce antibodies specific to bradykinin.

Part D. Inhibition of Passive Hemagglutination

In this exercise, use the product of partially hydrolyzed bacterial lipopolysaccharides which could be separated and isolated either by preparative paper chromatography or by preparative high-voltage paper electrophoresis (see Exercise No. 36 or No. 37). The fractions obtained by hydrolysis with ion exchangers in continuous dialysis equipment (Exercise No. 35) can also be analyzed in this assay.

Materials and Equipment

   2% suspension of lipopolysaccharide-coated human group O red
    blood cells or SRBC (prepared as described in *Part A)*
   1:10 diluted anti-lipopolysaccharide rabbit serum
   Partially hydrolyzed fractions containing fragments of lipopoly-
    saccharides. Mix 0.9 ml of these haptens with 0.1 ml 9% sodium
    chloride
   9% sodium chloride
   8 × 100 mm test tubes (or Takatsy microtitrator)
   0.2 ml pipettes
   37°C incubator

Procedure

   1. In the first tube of a series of 12 tubes, pipette 0.2 ml of a diluted antiserum, add 0.2 ml of the hapten-containing solution.

   Simultaneously prepare the controls: In the first tube of a second series of small tubes pipette 0.2 ml diluted serum and 0.2 ml saline. In the third row of tubes, mix 0.2 ml hapten with 0.2 ml saline. Pipette into the rest of the tubes in all three series 0.2 ml saline. Mix the first tubes in the three series

and incubate all tubes for 30 min at 37°C. Remove the tubes,
mix, and incubate for 30 min at room temperature.
    2. Remove 0.2 ml from the first tube and transfer this into
the second tube. Mix the contents of the tubes and continue
making serial double dilutions in all 12 tubes. Do the same with
the control series.
    Add 0.2 ml 2% erythrocyte + antigen conjugate suspension to
each of the 36 tubes. Mix and incubate as above. After incubation
read the degree of hemagglutination.
    The Takatsy microtitrator may also be used for the preparation
of serial twofold dilution, as described in Exercise No. 68.
Proceed according to the instructions given there, but use a 2%
erythrocyte suspension, which has been coated with the corresponding antigen, as described here.

Evaluation

Determine the highest dilution of antiserum and of the antiserum
+ hapten mixtures which still give visible hemagglutination. The
hapten alone should not agglutinate the red blood cells. If more
than two dilution tubes difference can be seen between the titer
of the antiserum (first row) and antiserum + hapten mixture (second row), this is due to the hemagglutination inhibitory effect
of the hapten. One tube difference is considered to be insig-
nificant.
    Those breakdown products of the lipopolysaccharide which in-
hibited the hemagglutination contain the receptor sites for the
antibodies.
    If the antibody + hapten mixture does not agglutinate the
antigen + red blood cell conjugate at all, this may mean that
the amount of hapten added completely absorbed the antibodies.
In such case the reaction has to be repeated with ten-fold or
higher dilution of the hapten.

Use and Limitations

Coating of the red blood cells may be achieved not only by the
procedures described here, but also by periodate treatment of
the cells or by coupling protein antigens to the erythrocyte
membrane by the diazo reaction. The use of bis-diazotized benzidine for this reaction was described by Pressman, Campbell
and Pauling (1942). Avrameas et al. (1969) used tetrazotized
O-dianisidine, cyanuric chloride and glutaraldehyde to couple
proteins to erythrocytes. Gold and Fuedenberg (1967) applied
$CrCl_3$ for similar purposes.
    An important disadvantage of the reaction which links antigen
proteins to red blood cells through a diazo bond is that it may
be more influenced by the presence of nonspecific proteins than
the coating of tanned red cells. While the tanned cell method
adsorbs equally all proteins present in a crude antigen extract
(such as tissue antigens), the BDB reaction may prefer certain
types of protein for the diazo linkage. The preferred protein
may be present as a contaminant. This may result in enhanced nonspecific reactions suppressing and inhibiting the reaction between
the investigated antigen and its homologous antibody (Richter and
Cohen, 1965).

A great advantage of these methods lies in their high sensitivity. The erythrocyte + antigen conjugates, especially in the BDB complexes, have been used to convert passive hemagglutination reaction into a passive hemolytic reaction. Addition of complement to the conjugate will result in lysis of the erythrocyte if the corresponding antibodies are present. This modification does not result in an increase of sensitivity but the important feature of this is the more accurate quantitative evaluation using photometric measurement of the liberated hemoglobin.

Reviews of the procedure for the fixation of antigens to red blood cells have been written by Neter (1956) and by Stavitsky (1977).

The inhibition of passive hemagglutination or passive hemolysis can be used in a large number of immunochemical experiments. It has been used extensively in studies investigating the serologic determinants of oligosaccharides in blood group antigens, in bacterial antigens, and in a variety of different natural products. As a diagnostic application of this assay, the detection of chorionic gonadotropin in pregnancy tests may be mentioned.

Limitations of the procedure are reflected by the occasionally observed nonspecific reactions. All materials which inhibit antigen + antibody complex formation will interfere. Several colloidal or surface active materials are known to influence the hemagglutination nonspecifically. If the presence of such substances in the preparations is known (for example, detergents, soaps, lipid emulsions, etc.), they must be included in the controls, applying them in the corresponding concentrations.

Another disadvantage of the passive hemagglutination assays was the lack of required reproducibility, but this may be overcome by the application of formalinized red blood cells. If this latter preparation is used, controls must be included in which formalinized but not antigen-coated cells will be tested under identical conditions for their possible reactivity with antiserum.

It is important to keep a known antiserum sample in the freezer as a standard and run this as a control together with the unknown antiserum. Faulty erythrocyte or antisera preparations can easily be detected by this means.

References

Avrameas, S., Taudou, B., Chuilon, S.: Immunochem. *6,* 67 (1969)
Boyden, S.V.: J. Exp. Med. *93,* 107 (1951)
Csizmas, L.: Proc. Soc. Exp. Biol. Med. *103,* 157 (1960)
Gold, E.R., Fuedenberg, H.H.: J. Immunol. *99,* 859 (1967)
Goodfriend, T.L., Levine, L., Fasman, G.D.: Science *144,* 1344 (1964)
Ingraham, J.S.: Proc. Soc. Exp. Biol. Med. *99,* 452 (1958)
Johnson, H.M., Brenner, K., Hall, H.E.: J. Immunol. *97,* 791 (1966)
Neter, E.: Bact. Rev. *20,* 166 (1956)
Neter, E., Bertram, L.F., Zak, D.A., Murdock, M.R., Arbesman, C.E.: J. Exp. Med. *96,* 1 (1952)
Pressman, D., Campbell, D.H., Pauling, L.: J. Immunol. *44,* 101 (1942)
Richter, M., Cohen, J.: Nature London *205,* 610 (1965)
Stavitsky, A.: In: Methods in immunology and immunochemistry, Williams, Chase (eds.) Vol. IV, pp. 30, 47. New York: Academic Press 1977

Exercise No. 70

Charcoal Agglutination

Very fine charcoal particles may serve as highly visible inert
carriers for viral antigens, such as in the method of Klein and
associates (1966). A modification is also given here in which
the charcoal has been coated with a bacterial O-antigenic preparation (Radvany et al., 1966). Agglutination of the antigen-
coated charcoal granules takes place in the presence of homologous
O-antiserum.

Materials and Equipment

    Endotoxic bacterial O-antigen solution, 1 mg/ml concentration
    Virus preparation (see Procedure)
    Rabbit antiserum against above endotoxin
    Antiserum to above virus
    Colloidal suspension of charcoal (available from Hynson,
     Westcott & Dunning, Baltimore, MD)
    Physiologic saline, pH 7.4
    White Brewer diagnostic cards (available from above Company).
     If such a card is not available, chromic acid cleaned glass
     plates divided into squares or circles as in Exercise No. 67
     will be satisfactory (Fig. 44)
    Calibrated standard droppers (available from above Company)
    Calibrated microcapillary pipettes with small rubber bulbs
     (available from above Company)
    Horizontal shaker

Procedure

    1. Preparation of virus + charcoal complex. The method described
here has been elaborated by Klein and co-workers (1966) for the
detection of viral antibodies in human serum to adenoviruses
types 1,2,5, and 6. The preparation of the antigen + charcoal
complex is carried out by simple adsorption of antigens from the
supernate of infected tissue culture cells. 5 mg of charcoal
suspension is added to each 2 ml of tissue culture supernate,
mixed and centrifuged at 1000 g for 15 min at room temperature.
The sediment is resuspended in saline to obtain a final charcoal
concentration of 2.5 to 1 mg/ml. The preparation is stable at
+5°C for at least six months.
    2. Preparation of bacterial O-antigen + charcoal complex.
Dissolve 2 mg bacterial O-antigen in 2 ml pH 7.4 saline. Add
2 ml charcoal suspension which contains 5 mg colloidal charcoal.
Rotate in a shaker overnight in the cold room. Centrifuge the
suspension and discard the supernate. Wash the charcoal + antigen
complex three times with 10 ml saline. To make the final suspension, add 5 ml saline. This antigen-coated charcoal is stable for
several years in the cold room.
    3. The procedures described from here on can be applied equally
well to detect viral antibodies or bacterial antibodies. Prepare
serial double dilutions of the antisera on glass plates or on the

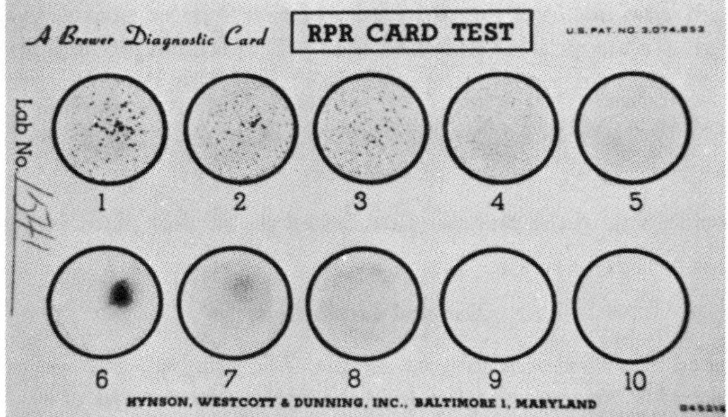

a

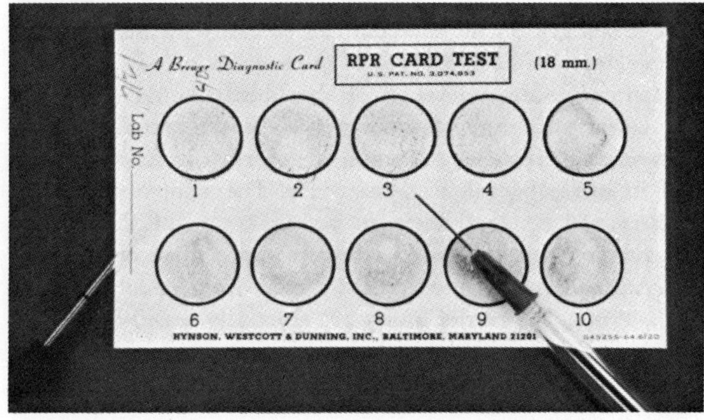

b

Fig. 44 a and b. Diagnostic cards and tools for charcoal agglutination,
available from Hynson, Westcott & Dunning, Baltimore, MD.

RPR cards as follows: Place one drop of saline in every sector.
Use the calibrated dropper to deliver the drops. Add one drop of
undiluted serum to the first sector. Take the capillary micropipette which is calibrated to deliver one and two drops (see
Fig. 44) and mix the first dilution in the first sector with it.
Pick up one drop from this circle and transfer it into the next
sector, mix it well with the saline which was placed in it before.
Proceed to make ten dilutions, and discard one drop from the last
circle. Naturally, the serum dilutions can also be prepared in
test tubes, as described before.

As a control, prepare ten dilutions from a normal serum in the
same manner.

4. Add one drop of antigen-coated charcoal to each serum
dilution, and starting from the highest dilution, mix the serum
and antigen carefully with a toothpick, spreading each drop out
over the entire area of the circle or square divisions. Place
the card or glass plate on a horizontal shaker (100 rotations per
minute) and let the drops dry out at room temperature while rotating. This occurs in approximately 30 min.

Evaluation

Observe clumping of the charcoal particles on the card or plate.
Read the titer of the serum.

Use and Limitations

The method described here is very simple. The antigen + charcoal
suspension is stable at cold room temperature for several years
without loss of reactivity with antiserum to any measurable degree.
   Other viral antigens may require different handling to be
absorbed to charcoal. This has to be elaborated individually.
Klein and associates (1966) have described some additional procedures used in their laboratory.
   The amount of antigen on the charcoal can be easily measured
by chemical reactions. For example, total acidic hydrolysis of
the endotoxin-coated charcoal suspension followed by quantitative
determination of the liberated monosaccharides will give the
amount of O-antigenic lipopolysaccharide on the charcoal. Such
measurements showed that *Serratia marcescens* endotoxin, if adsorbed
to charcoal, shows approximately 10% of its original serologic
reactivity. The amount of adsorbed antibody can also be determined by total nitrogen determination of the charcoal and antibody + antigen complex after subtracting the nitrogen content
of the antigen from the total nitrogen.

References

Klein, M., Chaefsky, S., Muller, M.: J. Immunol. *97*, 131 (1966)
Radvany, R., Neale, N.L., Nowotny, A.: Ann. N.Y. Acad. Sci. *133*, 763 (1966)

3.2.2   Precipitation Methods

Exercise No. 71

Double Gel Diffusion/Ouchterlony Method

In the gel diffusion techniques, gels, usually clarified agar,
are used as matrices for combining diffusion with precipitation.
The reactants simply diffuse through the gel towards each other
and precipitation results where the optimal antibody/antigen
ratios have been reached. A single antigen will give rise to a
single line of precipitation in the presence of its homologous
antibody. When two antigens are present in a system, each behaves
independently of the other. Thus, if several bands of precipitation are evident, there are at least that many antigen-antibody
combinations present.
   In this exercise antihuman rabbit sera will be investigated
in Ouchterlony gel diffusion test against human serum. The antibodies were obtained in Exercise No. 65.

Materials and Equipment

Human serum
Antihuman rabbit sera (see Exercise No. 65)
Ion agar, 1%, preserved with 0.01% merthiolate
Regular Petri dish
Feinberg gel cutter (Fig. 45). Various other types of cutters
  are commercially available (Gelman Co., Ann Arbor, MI 48106
  or LKB Instruments, Rockville, MD 20852). Precut agar plates
  with different patterns, ready for double diffusion experi-
  ments are available from several vendors. One of them is
  Hyland, Costa Mesa, CA 92626
Sterile 1 ml syringe
Moist chamber

Fig. 45. Feinberg type gel cutter, manufactured by Shandon Scientific Co., London, England

Procedure

1. The easiest way to prepare a good agar layer for gel diffusion is to use the commercially available "Ion agar" or "Noble
Agar". 1 g of this powder should be autoclaved for 15 min in
90 ml water or saline. Before it solidifies, 10 ml merthiolate
(1:1000 dil.) should be added. It is convenient to have this
agar distributed in 10 ml portions in test tubes for storage.
The tubes must be closed to prevent drying.
2. Pour 10 ml "Ion agar" into a Petri dish; wait until it
solidifies.
3. Take the standard Feinberg cutter (product of Colab Laboratories, Inc., Chicago Heights IL 60411), press it into the
agar. Then remove the agar from the inside of the cutter tubes
with a disposable pipette attached to a water aspirator pump.
Lift the cutter, observe whether the wells are clear and the agar
is removed. With the disposable pipette, punch a small additional
hole above one of the six outer wells. This well will be number
1 and the others will be numbered clockwise. No writing on the
Petri dish is necessary.
4. Fill the center well with human serum and the six wells
with antisera obtained from the different rabbits according to

Exercise No. 65. Cover the Petri dish and place it in a humid
atmosphere. For this purpose, a desiccator with water in the
bottom is suitable. Let it stand at room temperature for three
to four days.

Evaluation

Observe (a) the number of precipitation lines, (b) the differences
in intensity of the lines formed from the sera obtained from
different rabbits. It will be easy to see that antisera obtained
by antigens injected with complete Freund's adjuvant or with alum
precipitate will show much more intense bands than the antiserum
obtained by using human serum alone.

Use and Limitations

This technique has infinite applications, but in general they
are: 1. To determine the homogeneity of antigen-antibody systems.
2. To enumerate the minimum number of systems present. 3. To
follow the purification of an antigenic mixture. 4. To elucidate
the reactions among serologically related antigens.
    As a tool to follow the purification of certain antigenic
natural products, gel diffusion has the disadvantage of not being
able to detect impurities which are not antigenic. Similarly,
those antigen-antibody complexes which do not form a precipitate
will remain invisible. The relatively short distance between the
wells does not allow the clear separation of all precipitin bands
Some will merge into a wider zone. Short incubation will not
develop all precipitin bands, giving the false impression of
homogeneity. If the gel diffusion plate is transferred to a cold
room after three to four days at room temperature and is observed
again in one or two weeks, frequently additional bands can be
observed. If too high antigen or antibody concentrations are used
the entire area between the two wells will be in antigen or antibody excess, respectively. This means that weak bands or no bands
will be formed. It is necessary to use several antigen dilutions
and select the one which gave the clearest separation and largest
number of visible bands for further studies.
    Several applications of the method are known. A micromodifica-
tion can be made on microscopic slides, using the gel cutter tool
shown in Exercise No. 76. Good monographs covering gel diffusion
systems were written by Crowle (1961) and Backhausz (1967).

References

Backhausz, R.: Immunodiffusion und Immunoelektrophorese, Jena: Fischer 1967
Crowle, A.J.: Immunodiffusion, New York: Academic Press 1961
Ouchterlony, O.: Acta Pathol. Microbiol. Scand. *26,* 507 (1949)

Exercise No. 72

Immunoelectrophoresis

Immunoelectrophoresis, developed by Williams and Grabar (1955), has a much better resolution than gel diffusion. In this system the components of the antigen will first be separated by electrophoresis. The antigen components will move through the agar gel to a distance characteristic for the component in the given system. After the antigen has been separated into its components, antiserum will be put into a channel cut parallel to the direction of the electrophoresis. From this channel, the antibodies will diffuse towards the electrophoretically separated antigen components, and vice versa. Specific precipitate will be formed where an optimal ratio between the corresponding antigen and antibody components has been reached.

In this Exercise the antigen-antibody system, which was investigated in the previous Exercise by gel diffusion, will be analyzed by immunoelectrophoresis.

Materials and Equipment

    Antihuman rabbit serum, commercially available
    Human serum
    Ion agar
    Veronal buffer pH 8.6 with 0.01% merthiolate (see Exercise No. 9)
    Microscope slides
    Gel cutter for immunoelectrophoresis (see Procedure)
    Immunoelectrophoresis apparatus (see Fig. 46)
    Power supply
    Moist chamber

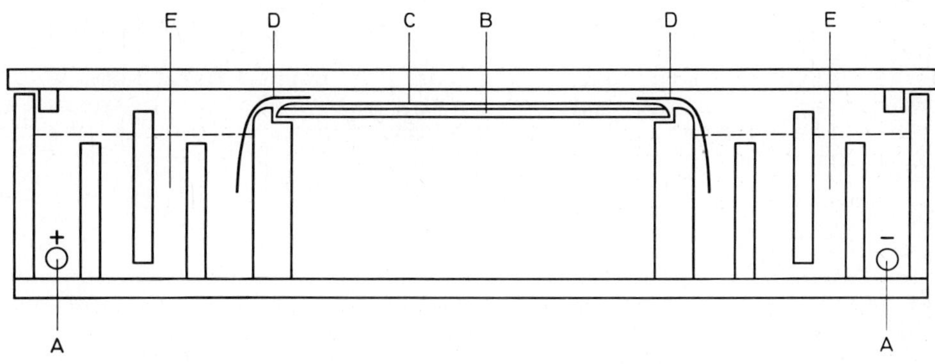

Fig. 46. Cross section of laboratory made equipment for immune electrophoresis

Procedure

1. Pour 2 ml melted ion agar (1% in pH 8.6 Veronal buffer) on the leveled, dry, clean microscope slides (B). Let it solidify.

2. Place a microscope slide between two razor blades so that
the parallel edges of the blades will extend approximately 5 mm
beyond the edge of the slide. Fix the blades to the slide in this
position with adhesive tape. Press the cutter in the gel, making
a channel in the middle of the slide parallel to its longer edge.
Remove the cut material with a 15 gauge or similar needle attached
to vacuum. Slip the needle under the cut surface, carefully lift
the needle, and it will remove the cut strip from the slide.
Figure 47 shows the pattern for immunoelectrophoresis.

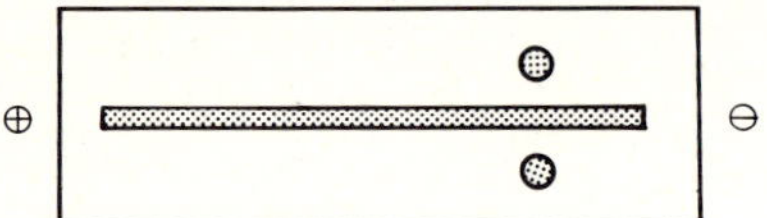

Fig. 47. Pattern of gel cut for immunoelectrophoresis on microscope slide

To punch the antigen wells, cut off the tip of a 13 gauge
syringe needle and clean its edge with a fine file. Attach a
rubber tube to this needle and apply a slight suction to it.
Press this blunt needle into the agar, 3 mm from the central
channel and approximately 20 mm from the negative end of the
slide. Lifting the needle while applying slight suction will
leave a hole in the agar layer (C). Immediately fill this well
with antigen, using a 1 ml syringe and a 23 gauge needle or a
very fine glass capillary. Place the slide in the electrophoresis
chamber.
    3. Make contact between the slide and the electrode vessels
(E), using approximately 1 cm wide filter paper strips (D).
Cover 5 mm of both ends of the slide with the paper strips
soaked in Veronal buffer. Immerse the other end of the papers
in the electrode vessel. Cover the electrophoresis chamber.
    4. Turn on the power supply. If the contact between the slide
and electrode vessels is good, approximately 2 mA current should
go through per slide. Adjust the voltage to 150 V between the
two electrodes (A).
    5. After 60 min, the power must be turned off. Remove the
slides and fill the central trough with rabbit antiserum. Place
the slide carefully in a moist atmosphere. Allow the precipitation lines to form at room temperature.

Evaluation

The lines will be visible in 24 h, but will be more intense after
48 h. Observe the number of lines. Compare the resolution of
immunoelectrophoresis with gel diffusion.

Use and Limitations

The precipitin zones can be stained in the gel layer with protein
stains, such as acidic fuchsin, amidoblack, etc. First the nonreacted proteins must be removed from the slide by washing it
in a saline-filled Petri dish for a few days, changing the saline
twice daily. Then the saline must be washed out, using distilled

water. This takes another two days. Now cut a filter paper strip
approximately the size of the microscope slide. Moisten it and
place it on the surface of the agar. Be sure that no air bubbles
remain under the paper. Let them dry together at room temperature.
Now the agar will form a solid film, strongly adhesive to the
glass. Lift and discard the paper strip. The protein staining of
the precipitin bands can be carried out in the same way as the
treatment of the paper strip after electrophoresis of serum samples
in Exercise No. 9.

By using this method, human serum can be resolved into more
than 20 different antigenic components. It is very useful in
following the purification of serum proteins. It is widely used
today not only in research laboratories but also in clinical
diagnoses. Nonantigenic or nonprecipitating components cannot
be detected with this procedure. The same limitation is valid
here as was mentioned briefly in the gel diffusion assay. A considerably better resolution can be obtained by using the crossed
immunoelectrophoresis as described in Exercise No. 74.

Uriel and Grabar (1956) described a method which makes possible the use of specific stains to differentiate between poly-
saccharide-containing and pure protein precipitates. Skarnes (1966)
used enzymatic assays to detect esterases in immunoelectrophoretic
plates.

A detailed discussion of the application, methodology, and
limitations of the method can be found in Wieme (1965) and in
Crowle (1961).

<u>References</u>

Crowle, A.J.: Immunodiffusion. New York: Academic Press 1961
Skarnes, R.C.: Ann. N.Y. Acad. Sci. *133,* 644 (1966)
Uriel, J., Grabar,P.: Bull. Soc. Chim. Biol. *38,* 1253 (1956)
Wieme, R.J.: Agar gel electrophoresis. New York, N.Y.: Elsevier 1965
Williams, C.A., Jr., Grabar, P.: J. Immunol. *74,* 158 and 397 (1955)

Exercise No. 73

Quantitative Immunoelectrophoresis/Rocket Method

There are numerous new developments in immunoelectrophoresis.
One of them is the so-called rocket immunoelectrophoresis, which
is a derivative of quantitative radial immunodiffusion, also
called the Mancini technique, which is described in Exercise
No. 78. It can be used to determine the amount of a certain antigen in any mixture, provided a monospecific antiserum to this
antigen is available. While in the Mancini technique the antigen
will diffuse out of the wells punched in the antibody-containing
gel, in rocket immunoelectrophoresis, the antigen will be elec-
trophoresed into a similar gel. This migration of the antigen
into antibody-containing gel will produce rocket-shaped precipitates. The distance the antigen can reach during this electrophoretic migration out of its well is directly proportional to
the amount of antigen present in the sample applied. This procedure was developed by Laurell (1966).

Materials and Equipment

U-shaped mold to prepare the gel plate (available from BioRad
Laboratories, Richmond, CA 94804, or P.O. Box 519, Rockville
Center, NY 11570). The same mold can also be prepared in the
laboratory (See Fig. 48)
0.5 *M* Veronal buffer, pH 8.6. This buffer is the same as that
used in Exercise No. 72
Agarose (available from various commercial sources)
Human serum
Human serum albumin (crystallized)
Rabbit antihuman albumin (available from numerous suppliers)
Electrophoresis equipment (available from BioRad Laboratories
or from LKB Instruments, Rockville, MD., 20852). It can also
be constructed easily in the laboratory

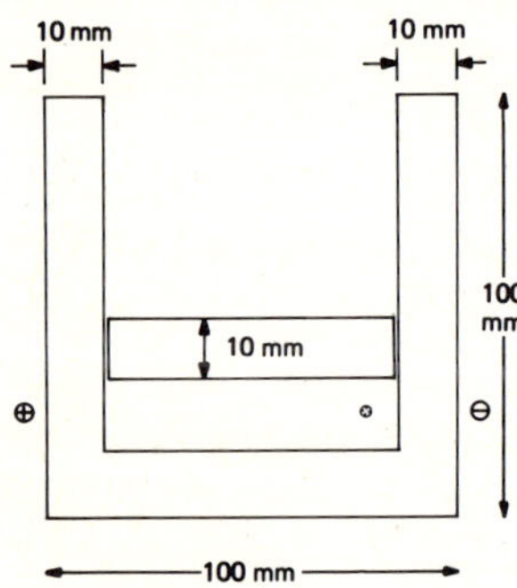

Fig. 48. Mold for pouring gel to run crossed immunoelectrophoresis in the first dimension

Procedure

1. Place the mold on a glass plate and close its open end with
a straight piece cut from the same material used to prepare the
mold.
2. Dissolve 1.5% agarose in 100 ml of the above Veronal buffer
at boiling water-bath temperature. Cool it to 45°C and at that
temperature add rabbit antihuman albumin antiserum, so that the
final concentration of antiserum in the agarose is 2.5%. Stir
well. The antiserum must have a titer between 128 and 1024, as
determined by passive hemagglutination (see Exercise No. 69).
The higher the titer, the shorter the rockets will be. In case
of high titers the serum should be diluted with physiologic
saline. Low titers will give long rockets but sometimes diffuse
contours.
3. Before the gel solidifies, pour it into the mold and let it
cool to room temperature. Remove the mold, but keep the agarose
gel on the glass plate. Punch 3 mm diam holes, approximately
10 mm apart, along one edge of the gel as shown in Figure 49.
4. In this Exercise we will determine the amount of human
serum albumin in various samples. The normal human serum to be
analyzed should be diluted 1:100 using physiologic saline. In
the same experiment we will prepare our calibration curve, necessary for the quantitative determination, by dissolving crystalline human serum albumin in saline. The concentration of the

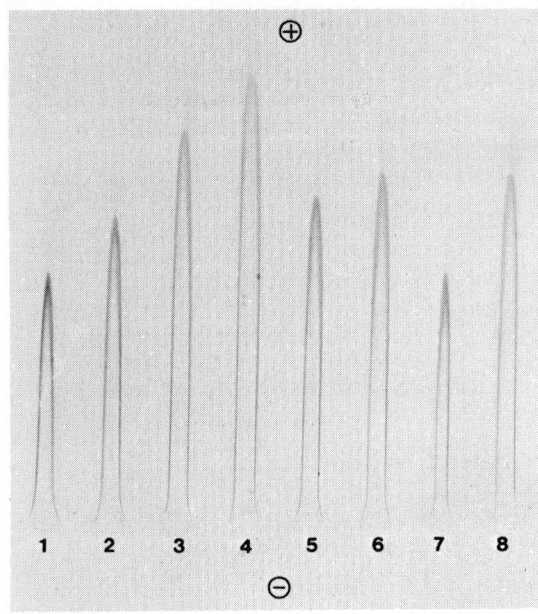

Fig. 49. Pattern of "rocket" immunoelectrophoresis. (Courtesy of LKB-Produkter AB, Bromma, Sweden)

stock solution should be 1 mg/ml. From this standard make the following dilutions: 0.8, 0.6, 0.4, and 0.2 mg/ml. Fill the first five wells, starting with the lowest concentration in the first well.

5. Place the gel plate into the electrophoresis equipment, connect the ends of the gel with the electrode chambers using paper wicks. These chambers are also filled with the same Veronal buffer which was used to prepare the agarose gel. The albumin will migrate to the positive end, therefore apply the polarity accordingly, as indicated in Figure 49. Close the electrophoresis apparatus, turn on the cooling system which should provide a constant temperature under the glass plate on which the agarose gel rests. Wait for 30 min to cool the system and now connect it to the power supply. If no cooling system is available, transfer the entire apparatus into the cold room. The potential gradient during the electrophoresis should be 10 V/cm. Carry out the electrophoresis for 2 h.

6. Disconnect the electrophoresis equipment from the power supply and, holding the gel plates against a black background, observe the rockets.

Evaluation

Measure the height of the rockets in millimeters. Plot these values against the concentration of known human albumin solutions used as standard to prepare the calibration curve. Now read the height of the 1:100 diluted serum sample rockets. To determine the albumin content in these samples, use the calibration curve. Calculate the mg/ml albumin content of undiluted human serum samples under investigation.

Use and Limitations

This Exercise used human serum albumin as an antigen, but any
soluble antigen preparation which will migrate to the anode at
this pH can be detected and quantitatively determined, provided
that a good antiserum to the antigen is available. For example,
rocket immunoelectrophoresis can be applied for the quantitative
determination of human IgG or IgM content in whole serum, if
monospecific rabbit antihuman IgG or IgM is available. Either
the antigen or the immunoglobulin has to be pure, but it is not
necessary to purify both. If neither of the two are homogeneous,
multiple rockets will start out of each well, overlapping each
other. The heterogeneity of such systems can be easily detected
by either regular or crossed immunoelectrophoretic procedures,
as described in Exercise No. 72 or No. 74.

Reference

Laurell,C.-B.: Anal. Biochem. *15,* 45 (1966)

Exercise No. 74

Crossed Immunoelectrophoresis

This procedure further enhances the resolution of immunoelectrophoresis. The principle is that electrophoresis of the components
in one dimension is followed by a second electrophoresis into an
antibody-containing gel. The electrophoresis into the second
dimension is carried out in a fashion identical to the "rocket"
technique, as described in Exercise No. 73. Figure 50 clearly
shows the resolution by which even very minor antigenic component
of human serum can be detected. The area under the peaks is proportional to the amount of component represented in the mixture.
   The principle of crossed immunoelectrophoresis was described
first by Ressler (1960), later elaborated in detail by Laurell,
(1965 and 1966). The procedure here is based on these description

Materials and Equipment

   The same glass plates, molds, electrophoresis equipment,
    buffers, and agarose are used as in Exercise No. 73
   Goat antihuman antiserum

Procedure

   1. Prepare agarose for the electrophoresis in the first dimension. Dissolve 0.3 g agarose in 20 ml veronal electrophoresis
buffer at boiling water bath temperature. The buffer is identical
to that used in Exercise No. 72.
   2. Place the mold on a 10 × 10 cm glass plate so that an
8 × 1.5 cm large and 0.2 cm thick gel slab can be formed on the
plate, as shown in Figure 48. Pour the warm agarose solution into

Fig. 50: Crossed-immunoelec-
trophoretic separation of
serum components. (Courtesy
of LKB-Produkter AB, Bromma,
Sweden)

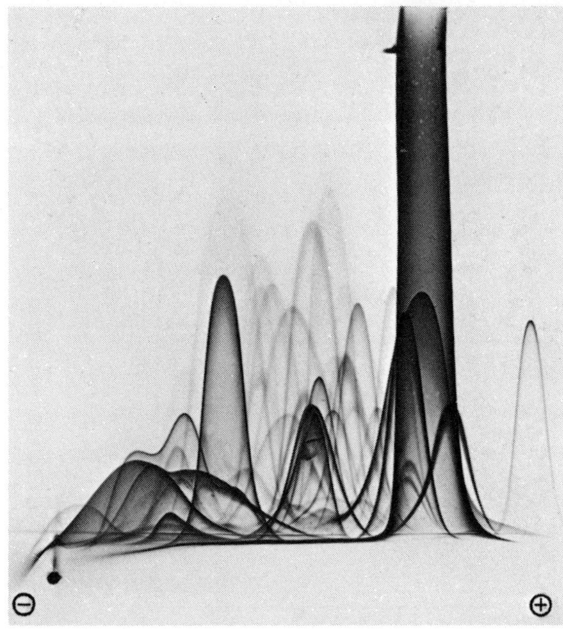

the mold and let it cool at room temperature until it solidifies
completely. Transfer the plate to the electrophoresis equipment,
which is kept in a cold room.

3. Punch a 3 mm diam hole close to the negative end of the
plate (as indicated in Fig. 48), connect the two edges of the
gel with paper wicks to the positive and negative compartments
of the elctrophoresis chamber. Fill the hole with normal human
serum. Apply 150 V potential to the two electrodes and let the
electrophoresis proceed for 1 h in the cold room. The current
going through the system should not be more than a few milli-
amperes/slide.

4. Now place the mold so that the rest of the 10 × 10 cm glass
plate can be filled with melted agarose containing 2.5% (vol/vol)
goat antihuman serum. See Figure 51 for this step. Prepare the
antibody-containing agarose as described for "rocket" electrophoresis in Exercise No. 73. Pour the gel into the frame at room
temperature so that it makes bubblefree contact with the gel slab
of the first electrophoresis. After this new gel solidifies,
carry out the electrophoresis in a cold room in the second dimension (at right angles to the direction of the first electrophoresis).

Evaluation

The pattern is visible after electrophoresis if the plate is
held against a black background. A home-made viewing box is shown
on Figure 52. This can be photographed and the print enlarged for
detailed study. The gel can also be stained, but only after
complete removal of the nonreacted proteins from the gel. This
can be achieved by placing the gel in a larger tray and soaking

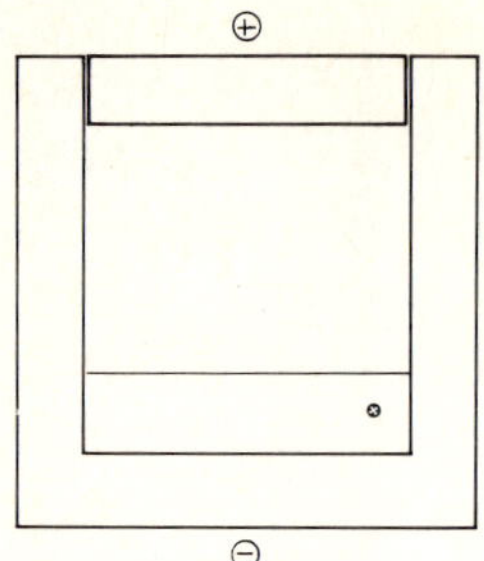

Fig. 51. Mold for pouring gel to run crossed immunoelectrophoresis in the second dimension.

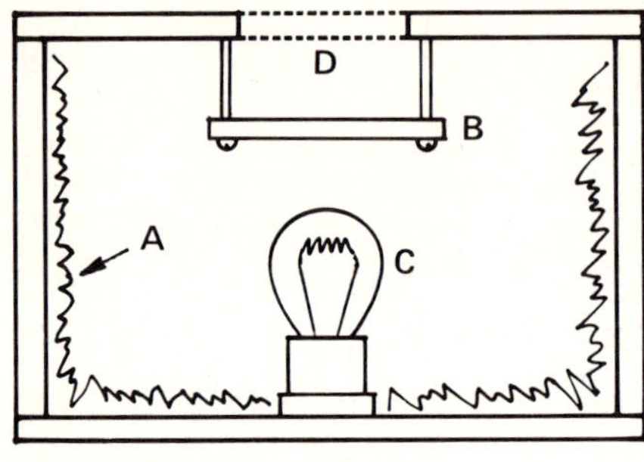

Fig. 52. Cross-section of home-made
viewing box for geldiffusion and immunoelectrophoresis. *A* Wrinkled alu-
minium foil lining; *B* Flat black metal
plate; *C* Light bulb; *D* Opening, where
the gel plate will be placed

it there in about 200-300 ml physiologic saline for approximately
three days, changing the saline twice daily, at room temperature.
For staining, the same Coomassie blue dye, or any other protein
stain, can be used. Staining for carbohydrate-containing components with periodic acid-Schiff reagent can also be done. The
procedure described in Exercise No. 45 can be used.

Quantitative estimation of the components in the analyzed
mixture is possible if the area under the peaks is measured by
planimetry. Quite dependable information can be obtained about
the relative amounts of the components present in the mixture. To
collect similar information about the absolute amount of a certain
component in various body fluids, isolated and purified components
must be available. Various amounts of these must be taken and
subjected to crossed immunoelectrophoresis in a fashion completely identical to the procedure described above. Planimetric
data of the areas shown by the pure compounds can be plotted
against the amounts of these compounds applied. A graph obtained
by this procedure can be used for estimation of the amount of the
components in question in the analyzed body fluid.

Use and Limitations

Serum, serum fractions (as obtained in Exercise No. 1, No. 2,
No. 5 or No. 7), ascites fluid, or any other body fluids can be
subjected to the above analysis, provided that a good and specific
antiserum is available. Here, as in any other immunodiffusion or
immunoelectrophoretic method, only those components will be detected to which specific antibodies in sufficient quantities can
be obtained. Nonimmunogenic components and contaminants in the
analyzed samples will remain undetected. Excellent resolution of
erythrocyte membrane components was achieved by Bjerrum and
Lundhal (1974), using crossed immunoelectrophoresis.

The limitation of crossed immunoelectrophoresis is in its quantitation. The height of the peaks shown by individual components is dependent not only upon the quantity present, but also on the electrophoretic mobility of the components studies. For example, IgG, which barely moves at this pH, will give a smaller area under the peaks, as measured by planimetry, than a much smaller amount of prealbumin, which has a high mobility.

References

Bjerrum, O.J., Lundhal, P.: Biochem. Biophys. Acta. *342*, 69 (1974)
Laurell, C.-B.: Anal. Biochem *10*, 358 (1965) and ibid.*15*, 45 (1966)
Ressler, N.: Clin. Chem. Acta *5*, 795 (1960)

Exercise No. 75

Counter Immunoelectrophoresis

The principle of this method is that at pH 8.4 the immunoglobulins will migrate to the negative end of a capillary system if an electric field has been applied to the capillary system. They will do so in spite of the fact that their isoelectric point is below pH 8.4. The migration is due to electro endosmosis which makes the fluid within the capillary system stream towards the negative pole. In the same capillary system, many antigens will migrate towards the positive end at that pH despite electro-endosmosis. The result is that they move towards each other in an electric field and soon will form a visible precipitate.

This phenomenon was used by Lang and Haan for the detection of antibodies in 1957. The same principle was applied for rapid and very sensitive detection of hepatitis antigen by Bedarida et al. (1966). The procedure described here uses commercially available positive human antiserum to the so-called Australia or serum hepatitis antigen. Other samples to be tested, hepatitis positive as well as normal human sera, will be collected from hospitals.

Materials and Equipment

    Normal and hepatitis serum samples from hospitals
    Positive anti-Australia antigen serum (available from IBL Labs.,
      Rockville, MD 20850)
    Agarose (Pharmacia Fine Chemicals, Piscataway, NJ 08854)
    Sodium barbital buffer, pH 8.4. Dissolve 8.2 g sodium barbital
      in 90 ml 0.1 *N* HCl. Add 3.92 g Na acetate (anhydrous) and
      dilute the solution to 1000 ml
    Microscope slides
    Electrophoresis chamber, same as in Exercise No. 72
    Plate stand, which can be leveled by adjusting screws, equipped
      with a level indicator

Procedure

1. Dissolve 0.5 g Agarose in 40 ml Na barbital buffer in a
boiling water bath. Before it solidifies add merthiolate to a
0.01% final concentration. This preservative will prevent the
growth of microorganisms. Its addition is not essential unless
the agar plates are to be stored in a moist chamber for several
months.
2. Take clean microscope slides, place them on the leveled
stand. Carefully pipette 2 ml melted Agarose on each slide. Do
not let the agar run off the slides. Wait until it solidifies.
3. Punch 3 ml diam holes into the agar according to the
pattern shown in Figure 53. Fill the wells indicated by AS in
Figure 53 with antiserum. Fill the wells indicated by numbers
with the sample to be tested.

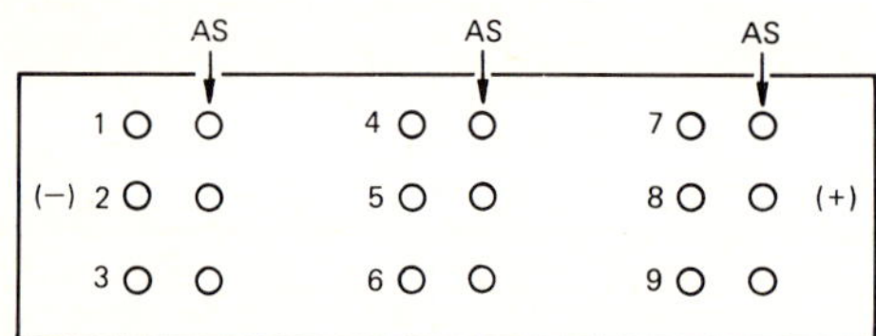

<u>Fig. 53.</u> Pattern for counter-immuno-
electrophoresis on microscope slides

4. The electrophoresis must be carried out at cold room temperature. Place the slides into the electrophoresis equipment and
connect the positive end, as indicated in Figure 53, with the
positive electrode chamber. Use wet filter paper strips to make
the connection. Do the same with the negative end. Cover the instrument, make sure that no excessive evaporation can occur.
Apply 200 Vdc from a power supply and let the electrophoresis
proceed for 60 min. Measure and record the millamperes going
through the system.
5. Disconnect the electrophoresis chamber from the power suppl
Put the slides into a moist chamber and keep them refrigerated.

Evaluation

Observing the area between the sample and antibody wells, precipitation line (or lines) will be visible if the sample contains
Australia antigen. This band is even more visible if the slides
are placed on a dark-field viewing box as shown in Figure 52.
Sometimes faint bands will be more visible if the plates are
immersed for 20 min in a 15% trichloracetic acid solution. Before
doing this, the nonprecipitated proteins must be eluted completel
from the gel by soaking the plates for 48 h in phosphate-buffered
saline, changing the fluid twice daily. The use of a magnifying
glass or a low resolution microscope is recommended for the
evaluation.

Use and Limitations

To get the best precipitation line, the proper concentration of
both antiserum and sample must be established. In antibody excess
or in antigen excess false negative results may be obtained. To
establish the optimal antiserum and sample concentration ratios,
a few two-fold dilutions of the antiserum should be prepared and
used in counter immunoelectrophoresis against the undiluted
samples. If the antibody titer in the antiserum is too high, a
precipitate will be formed in the sample well. Old sera, with
precipitate or with visible turbidity, may form a halo around
their wells. Such sera should not be used without removing the
turbidity, either by sharp centrifugation at 5000 to 8000 g for
30 min or by filtration through a 0.45 µ filter.
   The same system can be used for the detection of all kind
of soluble antigens or antibodies which will form precipitation
bands in gels.

References

Bedarida, G., Trinchieri, G., Carbonara, A.: Haematologica *54,* 591 (1969)
Lang, N., Haan, J.: Int. Arch. Allergy 10, 305 (1957)

Exercise No. 76

Semiquantitative Microprecipitin Assays

Quite often only very limited amounts of reactants are available
for immunochemical studies of certain antigens or antibodies. In
such cases the investigator has to be satisfied with less accurate methods if they can be carried out with small amounts.
The method described here in *Part A* allows the comparison of different antigen- or antiserum-containing preparations on a microliter scale, using precipitation in gel layers (Nowotny et al.,
1963). Another procedure described in *Part B* of this Exercise
allows the determination of the optimal antibody/antigen zone
using an approach similar to the one used by Everhart and Shefner
(1966).

Part A. Microtitration in Gel

Materials and Equipment

    1 mg/ml antigen solution in pH 7.4 saline
    Undiluted antiserum
    1% ion agar, pH adjusted to 7.4 with NaOH
    Physiologic saline, pH 7.4
    Takatsy microtitrator, commercially available (Fig. 40). If
      the laboratory is not equipped with this microtitrator,
      20 micro test tubes (3 mm i.d. and 40 mm length), and a few
      20 µl capacity micropipettes will also be useful to carry
      out this Exercise.

Disposable capillaries pulled out in a flame as shown in Fig. 5
Pipettes
Moist chamber for agar plates
Gel cutter tool (same as in immunoelectrophoresis or Fig. 55
Microscope slides

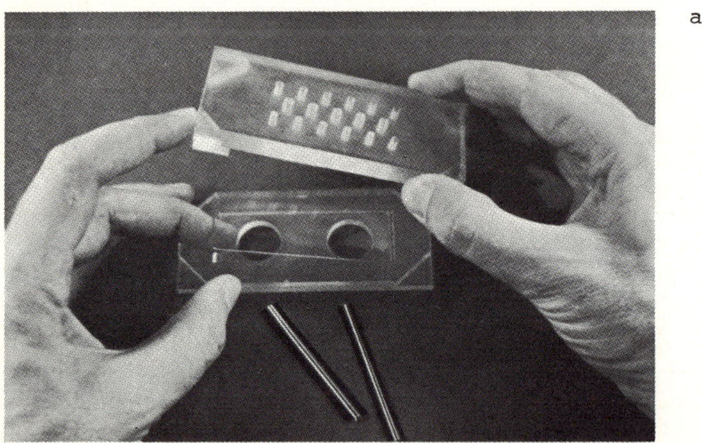

Fig. 54. Pulled-out disposable capillary

Procedure

1. Pipette 2 ml 1% ion agar, pH 7.4, on clean microscope slide:
and wait until it solidifies. Take the tool shown in Figure 55,
place a microslide in it. Put on the upper part of the tool and
by using the cutter tubes, punch holes in the agar layer. Remove
the slide and lift the small pieces of agar punched out of the

a

b

Fig. 55.a and b. Gel
cutter for microscopic
slide gel diffusion,
laboratory made

layer with the aid of a sharpened toothpick. The pattern seen in
Fig. 55 will be obtained. (It is also convenient to use the procedure applied to prepare microscope slides for immunoelectrophoresis, but for the purposes described here, six wells must be
cut on each side of the central trough equidistant from it and
from each other, as shown in Figure 56.) Place the agar plates
in a moist chamber until further use.

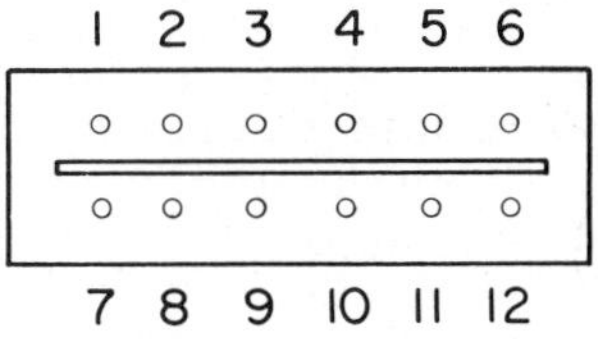

Fig. 56. Gel pattern for semi-quantitative
antigen-antibody reaction studies

   2. If the antigen contents of different preparations are to
be compared, serial double dilutions have to be prepared from
them. If the antibody titer of two or more sera are to be analyzed, the serum samples are to be diluted as in Exercise No. 69.
   The dilution of the solutions may be prepared very conveniently
and rapidly by the Takatsy microtitrator, which is described in
detail in Exercise No. 68. Take a Takatsy plate with cups 5 mm
i.d. and, with the calibrated dropper, place one drop of saline
in each cup. With the same size dropper, add to the first cup
of the first row one drop of the antigen or antibody solution,
whichever is to be diluted. Take a calibrated loop with 0.025 ml
capacity and mix the contents of the first cup by rotating the
loop rapidly in it for 10 s. Then lift the loop and transfer its
content to the next cup. Proceed to make the dilutions as before.
   3. When all dilutions are made, remove the agar plate from the
moist chamber. Take a fine glass capillary and remove a drop of
diluted sample from the 12th cup of the first row and fill the
12th well on the agar plate with it. Blow out the leftover liquid
from the capillary into a tissue paper, rinse the capillary by
drawing up into it a physiologic saline solution, blow out the
capillary again. Pick up a small sample from the 11th cup with the
same capillary, and fill the 11th well on the agar plate with it.
Proceed in this way from the highest toward the lowest dilution,
filling all 12 outer wells with diluted samples. If antigen was
placed in the outer wells, fill the central wells with the homologous antiserum. Return the plates to the well-sealed moist chamber
and incubate them at 37°C overnight.

Evaluation

The next morning, read the results and record the highest dilution which still gives visible precipitin band formation in the
agar.
   Use and Limitations see *Part B*.

Part B. Determination of Optimal Antigen/Antibody Ratio by Gel Diffusion

Materials and Equipment (Same as in *Part A*).

Procedure

1. For the determination of optimal antibody/antigen ratio by gel diffusion, make a serial double dilution of the antiserum in a row of seven cups on the Takatsy plate as described above with the calibrated dropper and loop. Place one drop of a 1 mg/ml antigen-containing solution in each cup of the same row. Take a fine glass rod and mix the contents of each well, rinsing and wiping the rod between each cup. Cover the plate with another one and incubate the plates in a moist chamber at 37°C for 1 h.

2. In the meantime, cut a microslide agar plate in the pattern shown in Fig. 57. Use the tools described for the immunoelectrophoresis Exercise (No. 72). Fill the upper channel or trough with antigen and the lower trough with antiserum solution.

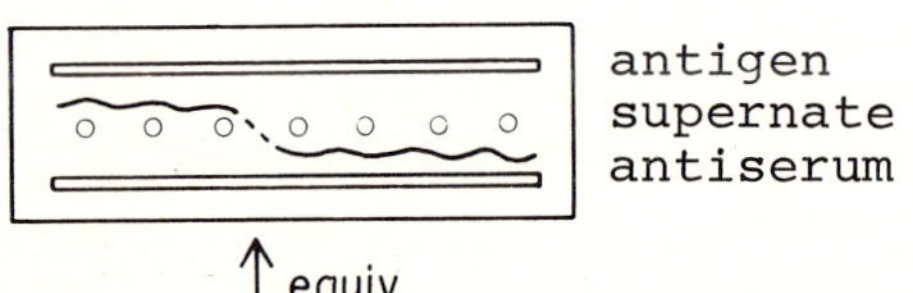

Fig. 57. Gel pattern to determine equivalence point

3. After the 60 min incubation of the antibody/antigen mixture, take the plate and by using a glass capillary similar to that used before, remove a sample from the seventh cup and fill the seventh well on the microslide. It does not matter if precipitate is also transferred into the well, because the precipitate will not diffuse into the agar. Rinse the capillary with saline, blow out its contents into a tissue paper, and transfer sample from the sixth cup into the sixth well on the agar plate. Proceed as above. Place the plates in a sealed moist chamber and incubate them at 37°C overnight.

Evaluation

In those wells where antigen excess is present, a precipitin line will be formed between the well and the antibody-containing trough. This will be the case at the end of the dilution series. In the first few wells, antibody excess will probably be visible, which means that precipitin lines will be on the other side of the central wells. The optimal antibody/antigen ratio will not give precipitate with either side.

Observe the precipitin lines and make sketches in your laboratory records.

If you find that all precipitin lines are on the side where the antigen filled channel is, your antiserum was not diluted out far enough. If the precipitation lines were formed between the wells and the antiserum filled channel either the titer of your

antiserum is not high enough or your antigen solution is very
reactive. Isolate and concentrate the immunoglobulins from your
serum, as described in Exercise No. 1 or 8. You can also repeat
the experiment with a 10- or 100-fold diluted antigen solution.

Use and Limitations

These two methods are convenient, they do not require more than
one drop of the solutions, and at the same time they give useful
information about the approximate activity of the preparations.
They may serve as useful guides for quantitative precipitin assays,
but they do not replace them.

References

Everhart, D.L., Shefner, A.M.: J. Immunol. *97,* 231 (1966)
Nowotny, A.M., Thomas, S., Duron, O.S., Nowotny, A.: J. Bacteriol. *85,* 418
   (1963)

Exercise No. 77

Quantitative Precipitation

This procedure will determine the amount of specific antibodies
in the rabbit serum immunized with human serum according to Exercise No. 65. The procedure consists of three parts. The first is
a preliminary test to find optimal antigen-antibody ratios for
maximal precipitation. Next is the determination of the curve of
precipitin reaction with the biuret test. The last part is the
measurement of the amount of antibody nitrogen in the maximal
precipitate. The method given here is essentially identical with
the procedure developed by Heidelberger and Kendall (1935).

Materials and Equipment

> Human serum albumin
> Rabbit antihuman serum albumin
> Physiologic saline
> Biuret reagent
> Laboratory centrifuge
> Conical centrifuge tubes, 10 ml capacity
> Small test tubes, 5 × 80 mm
> Calibration curve (obtained from Exercise No. 50)
> Spectrophotometer

Procedure

   1. A preliminary test must be performed to find the optimal
condition of maximal precipitation. Pipette 0.25 ml of undiluted
antiserum into each of five small tubes. Add 0.1, 0.25, 0.5, 1.0,
and 2.0 ml of antigen solution containing 10 mg/ml of antigen.

Make up all tubes to 2.5 ml with saline. Mix, let stand 15 min,
then centrifuge. Do not remove the sediment. Add two drops of
antiserum to each tube, gently overlaying them upon the supernatant. Observe this carefully, without shaking the tubes, for
the appearance of a ring of precipitate indicating the area of
excess antigen. If a precipitate forms in all five tubes, lower
amounts of antigen should be used and the test repeated. If no
precipitate forms in any tube, antigen excess has not been reached, and higher aliquots of antigen should be used (2.0, 2.5,
3.0, etc.). The optimal zone can also be found by gel diffusion
as described in Exercise No. 76.
   2. The remainder of this Experiment will be carried out in
duplicate in conical centrifuge tubes. The quantity of antigen
needed for maximum precipitation was determined in step 1. Add
twice this amount of antigen to the first tube. Add the amount
needed for maximum precipitation to the second. Add one-half
this amount to the third, and one-quarter to the fourth. Add
0.5 ml undiluted antiserum to each tube and fill all tubes to
5 ml with saline. Include a serum control consisting of 0.5 ml
of serum and saline. Mix all tubes thoroughly and incubate at
37°C for 1 h, then place in the refrigerator for at least 48 h.
Thoroughly mix the contents of the tubes twice daily. This facilitates complete reaction between antigen and antibody.
   3. Centrifuge in the cold at 3500 g for 30 min and carefully
decant the supernatant from the packed precipitate, being careful
not to lose any particles or sedimented precipitate. Save the
supernates. Wash and centrifuge the precipitates thoroughly twice
in cold saline, perform all operations in the cold. The precipitates are now ready for analysis.
   4. At this point the supernates from the first centrifugation
must be checked for the presence fo a slight antigen excess. The
supernates must be divided into two aliquots. Use the small test
tubes again and to 0.25 ml aliquot add 0.25 ml of the 10 mg/ml
antigen solution and to another 0.25 ml add 0.25 ml of antiserum.
Incubate all tubes for 30 min at 37°C and find the tube containing
a slight excess of antigen. If the assay was properly set up, the
first tubes have to give precipitate with added antigen while the
added antiserum will give precipitate in the last tubes.
   5. The quantitative measurement of the specific precipitate
in the conical centrifuge tubes is made with the biuret test.
1.5 ml of biuret reagent must be added to the specific precipitate
and the tubes shaken carefully until the precipitate dissolves.
Add water to make up the volume to 2.5 ml. Incubate at 37°C for
30 min and read at 550 nm in the spectrophotometer.

Evaluation

Plot the amount of antigen against the optical density read in
the spectrophotometer. Determine the maximum of the precipitation
curve. Read the optical density at the maximum of the curve. From
the calibration curve obtained in Exercise No. 50 calculate the
amount of protein nitrogen present in the maximum precipitate.
Any nitrogen found in the serum control should be subtracted from
each reading. Calculate the amount of antibody nitrogen in the
serum.

In the biuret test, the protein content of the specific precipitate was measured. The specific precipitate consists of antibody and antigen. The amount of antigen nitrogen has to be deducted from the nitrogen content of the maximum precipitate, in
order to find the amount of precipitable antibody nitrogen in
1 ml of serum. This correction is of course, unnecessary if the
antigen is nitrogen-free. It has been shown in experiments carried out with dye or isotope labeled antigens that at the equivalence zone the total amount of added antigen is precipitated.

Use and Limitations

The biuret reaction cannot be used for all antigen-antibody
determination. For example the assay will give very high antibody-antigen ratios if the optical transmittance of the reaction
mixture is reduced due to the low solubility of the antigen. In
such case the quantitative nitrogen determination method as
described in Exercise No. 48 are recommended.
It is very important to be sure that the antigen used in this
test is completely soluble and does not contain any particles
which can be sedimented by the centrifugation involved in this
determination. Such insoluble antigen preparations will settle
together with the specific precipitate and may falsify the results of the biuret determination unless necessary corrections
are made.
If the system contains several antigenic components, the optimal zone may be wide, and the results will be unreliable.
Several other factors which may influence this measurement are
thoroughly discussed by Kabat and Mayer (1961).

References

Heidelberger, M., Kendall, F.E.: J. Exp. Med. *61,* 563; *62,* 467 and 697 (1935)
Kabat, E.A., Mayer, M.M.: In: Experimental immunochemistry. 2nd edition.
    Springfield, Ill.: C.L. Thomas, Publ. 1961, p. 22

Exercise No. 78

Quantitative Radial Immunodiffusion

Mancini et al. (1965) developed a very simple method to quantitate
antigens using gel diffusion. The principle is this: The antigen
is placed in a well punched into an agar layer. The agar was
mixed while melted at 45°-50°C with the corresponding antibody.
The antigen will diffuse into the agar and will form a radial
precipitate around the well with the antibody already in the agar.
The more antigen we put into the well, the further it will diffuse
into the agar and the greater will be the diameter of the circular
precipitin zone. Keeping the antibody concentration in the gel and
the volume of antigen solution pipetted into the well constant,
the amount of antigen in the solution can be determined by measuring the diamter of the precipitation ring. This Exercise describes
the procedure for quantitative determination of complement C3

components in sera based on the publications of Fahey and McKelvey
(1965) as well as Shanbrom et al. (1967).

Materials and Equipment

  Antihuman C3 antibody produced in goats (available from Cordis
    Laboratories, Miami, FL 33137, or from Hyland, Costa Mesa,
    CA 92626)
  Human C3 (available from Cordis Laboratories or from Hyland)
  0.1 M Na barbital buffer. Dissolve 9 g Na barbiturate, 65 ml
    0.1 N HCl and 0.5 g Na azide in a total of 1000 ml distilled
    water
  Noble agar (Difco, Detroit, MI 48232)
  Glass plates, 11 × 9 cm
  Plastic or brass U-shaped mold as shown in Figure 48. Outer
    dimensions 105 × 90 mm, width of the branches 15 mm, thickness
    2 mm
  Clothes pins or clamps
  Silicon
  Gel puncher, 3 mm o.d.
  Semilog graph paper with mm divisions
  10 µl capacity microsyringe (Hamilton Company, Reno, NV 89510)
  Moist chamber. Any container which can be closed airtight can
    be used. Desiccators with water at the bottom will serve well
  Water bath at approximately 50°C
  Air incubator, 37°C

Procedure

  1. Take a nonsiliconized glass plate and place the U mold on
it. Take a siliconized glass plate, same size, and place it on
top. Put on six clothes pins or clamps to hold the plates together
  2. Dissolve 15 g Noble agar in 100 ml barbital buffer. Distribute the agar into 10 ml aliquots and keep it at cold room temperature.
  3. The undiluted anti-C3 serum must have a minimum of 0.1 mg/ml
anti-C3 immunoglobulin content. Prepare a ten-fold dilution of
serum using the barbital buffer. Mix 2 ml of this serum heated
at 50°C with 2 ml 1.5 Noble agar melted and cooled to 50°C. Take
a pipette, warmed in an air incubator (37°), and quickly fill
the gap between the two glass plates with the agar and serum
mixture, without introducing bubbles. It is recommended to warm
the glass plates in the air incubator also before filling. Now
let it cool to room temperature, then carefully slide the siliconized glass plate off, and remove the U-shaped mold, leaving
the solid agar on the glass.
  4. Place the gel with the glass plate on a sheet of millimeter
graph paper and, with the gel puncher, cut holes into the agar
15 mm apart. One can make 16 holes in one plate as given here if
the centers of the holes are 15 mm away from the edges and from
each other. Remove the cut-out agar plugs with a pulled-out glass
tube (approximately 2 mm diam) and gentle suction.
  5. Make six dilutions of the C3 preparation with known concentration. These should be 1, 0.6, 0.3, 0.1, 0.05, and 0.03 mg/ml

6. Using the microsyringe, put exactly 2 μl of the various antigen dilutions into the wells numbered 1 to 6. Start with the highest dilution. Rinse the syringe very thoroughly with Na barbital buffer after each dilution. Wells 1 to 6 will serve as calibration standards.

7. Take the fluid to be analyzed for C3 content and make the following dilutions: 1:1, 1:2, 1:4, and 1:8, again using the Na barbital buffer. Take the most dilute sample first, and pipette 2 μl into well No. 11. Well No. 10 will receive 2 μl of the 1:4 dilution. Continue with the samples and add 2 μl undiluted fluid to well No. 7

8. Incubate the plate in a moist chamber at room temperature (8-16 h). Make sure that the moist chamber is well closed and has enough water in it to prevent drying out.

Evaluation

Determine the extension of the precipitation ring. Measure the distance in 0.1 millimeters between the edge of the well and the outer edge of the circular precipitation zone. A very useful instrument for this purpose is the Precision Viewer, made by Hyland. Plot the reading obtained from wells No. 1-6 on the arithmetic scale of a semilog paper versus the amount of antigen C3 standards used. These latter values should be plotted on the log scale. Draw a line of best fit. The preparation of the above standard curve has to be repeated with every measurement. A rather linear correlation should be the result. Now measure the diameter of the various dilutions of the unknown and read their C3 content from the calibration curve. Multiply the values found in well No. 8 by 2, in well No. 9 by 4, in well No. 10 by 8, and in No. 11 by 16 (dilution factors) and compare these data with the values obtained from well No. 7.

Use and Limitations

The antibody concentration in the gel is critical. If it is too high, the diameters will be small, if it is too dilute, the precipitation ring will be diffuse or not visible. Mancini et al. used an antiserum which could precipitate all antigens from an equal volume of an 0.1% solution of human serum albumin. Good results were obtained if this serum was diluted 1:6. In order to save serum, the highest useful dilution should be mixed with the agar. This can be found empirically by making agar plates with 1:2, 1:4, 1:8, or 1:16 or even higher dilutions of the available sera. If only a few samples are to be analyzed, smaller agar plates will be sufficient. This can be obtained by changing the position of the U-shaped mold between the glass plates.

The antigen concentration is less important. According to Mancini et al. as little as 0.01 μg antigen can be measured quantitatively in the human serum albumin-antiserum system they described. This sensitivity could be further increased by diluting the anti-HSA serum 1:100.

If more precise determination is required, incubate the slide for 48 h in the moist chamber at 37°C. Under these conditions the reaction is virtually complete. One can plot the squared diameter of the precipitation on a linear graph paper against the

known amount of antigen standard in the wells using the evaluation method of Mancini. Determine the squared diameter of the
unknown sample wells, and read the amount of antigen in them from
this calibration curve. This curve should be linear.

The greates value of the system lies in its simplicity and
versatility. If monospecific antisera to an antigen are available,
any kind of body fluids or solutions can be diffused into the plat
and their antigen content can be determined (within certain limits
of molecular weight and antigenicity) with considerable precision
and with high specificity.

If the circles are faint, they can be made better visible using
15% trichloroacetic acid, by the procedure described in Exercise
No. 73. Of course, the excess antibody must first be removed from
the agar layer by thorough soaking in PBS for 48 h, changing the
PBS twice daily.

The volume of antigen placed in the wells should be kept constant (2 or 3 µl). Pipetting the same amount of antigen, but this
time dissolved in larger volume of solvent, will give a slightly
higher ring diameter than a more concentrated solution of the
same amount of antigen.

This method has been successfully used for the quantitative
determination of IgG, IgM, IgA, IgD, transferin, haptoglobin,
ceruloplasmin, macroglobulin, and numerous other antigens. Commercially prepared agar plates containing specific antibody are
now available for the above determinations from several sources.
One of these is Hyland, Costa Mesa, CA 92626.

References

Fahey, J.L. McKelvey, E.M., J. Immun. *94*, 84 1965
Mancini, G. Carbonara, A.O., Heremans, J.F., Immunochem. *2*, 235 (1965)
Shanbrom, E., Khoo, M., Luo, K., Clin. Res. *15*, 114 (1967)

Exercise No. 79

Radioimmunoassays for Cyclic AMP and Cyclic GMP Determinations

Radioimmunoassays (RIA) utilize antigen-antibody reactions. A
standard amount of antibody is mixed with a standard amount of
radio-labeled antigen, and the amount of antigen-antibody complex
formed will be measured quantitatively. If to the above two components of the reaction a sample is added which contains unlabeled
antigen, it will compete for the reactive site of the antibody wit
the labeled antigen. As a result of this competition between labeled and unlabeled antigens, the amount of the labeled antigen-
antibody complex will be reduced. There is a direct relationship
between the amount of the unlabeled antigen in the unknown sample
and the reduction in the amount of labeled antigen-antibody complexes. This is the basic principle which has been used by Yalow
and Berson to determine plasma insulin concentrations in patients
(1960)

In this Exercise the quantitative determination of cyclic
AMP (cAMP) and cyclic GMP (cGMP) has been selected to demonstrate

the use of the highly sensitive radioimmunoassays. cAMP plays a
key regulatory role in the function of mammalian cells. It medi-
ates the action of a variety of hormones and acts as an intracellular "second messenger". It has been discovered by Sutherland
and co-workers in 1957. cGMP is similarly widely distributed in
nature. It occurs in all animal tissues as well as in prokaryotes.
It has been discovered by Ashman and associates (1963). The biologic role of cGMP is not firmly established. It could be shown
that cAMP and cGMP function together via dual opposing actions
to control cellular functions (Goldberg 1975).

   RIA kits are available from a number of manufacturers. In this
Exercise we use the products obtained from New England Nuclear
(Boston, MA 02118).
        Part A of this Exercise describes the cAMP determination
based on the method of Steiner et al. (1972) as adapted by New
England Nuclear (NEN) for the use of their RIA kit. In *Part B*
the cGMP procedure is described based on the same reference.

Part A. Cyclic AMP

Material and Equipment

   Labeled antigen: succinyl tyrosine- [125 I] -methyl ester deriv-
      ative of cAMP. This preparation will be designed throughout
      this exercise as "hot cAMP". The NEN kit contains a vial of
      lyophilized hot cAMP. Add to this 5 ml distilled water. The
      kit also contains vials of lyophilized normal rabbit serum.
      Take one and redissolve its content in 5 ml distilled water.
      Add the 5 ml normal rabbit serum to the hot cAMP vial. Mix
      well. This vial will contain the hot cAMP in 1% normal rabbit
      serum and 0.05 M Na acetate buffer pH 6.2. This solution is
      stable at refrigerator temperature for two weeks, or for
      several months when frozen
   0.05 M pH 6.2 phosphate buffer. Mix 81.5 ml 0.05 M KH_2PO_4
      (6.80 g KH_2PO_4 in 1000 ml water) and 18.5 ml 0.05 M
      $Na_2HPO_4 \cdot 2H_2O$ (8.90 g $Na_2HPO_4 \cdot 2H_2O$ in 1000 ml water)
   Test tubes, approximately 12 × 120 mm
   Centrifuge tubes, conical or round bottom, approximately 10 ml
      capacity
   cAMP standard containing 5000 pmol/ml. Since this is not la-
      beled the term "cold cAMP" is used in this Exercise for the
      preparation. The cAMP standards are lyophilized and have to be
      redissolved in 2 ml distilled water
   Antiserum to cAMP. Prepared in rabbits by immunizing them with
      succinylated cAMP complexed with serum albumin. This is the
      "primary" antibody. A "secondary" antibody is used to precipi-
      tate the "primary" antibody plus cAMP complex. This "secondary"
      antibody is an antirabbit gamma-globulin prepartion prepared
      in sheep. The kit from NEN has these two antibodies already
      complexed and lyophilized. The vial has to be reconstituted
      with 21 ml distilled water. Do not shake it, just invert it
      gently
   Normal rabbit serum
   Acetic anhydride
   Triethylamine

Refrigerated centrifuge with swinging buckets for 10 ml centrifuge tubes
Gamma isotope counter

<u>Procedure</u>

1. In this Exercise the cAMP content of spleen cells will be
determined. Take 100 mg spleen (or any amount around 100 mg, but
determine the weight of the sample accurately). Grind it up in a
tissue homogenizer for 10 min in the presence of 0.9 ml 6%
trichloracetic acid. Centrifuge at 3000 g for 15 min at 2°-4^oC.
Lift 0.25 ml of the supernate, transfer it into a 10 ml test
tube provided with a glass stopper. Add to the sample 2 ml diethyl
ether (saturated with water), close the stopper very tightly and
shake it. Release the pressure by lifting the stopper. Wait for
5 min, then discard the ether phase. Use a Pasteur pipette and a
rubber bulb for this purpose. Repeat the extraction four more
times. Dry the extracted water phase by blowing air or nitrogen
on the surface and by warming the tube from the outside with a
hairdryer. Dissolve the dry residue in 0.4 ml acetate buffer and
take 0.1 ml aliquots of this (representing 25% of the fresh spleen
sample taken) for cAMP determination. Add to it 1.9 ml buffer.
2. Preparation of the dilutions from the cold cAMP standard
is the next step. These standard dilutions are to be prepared
freshly every day. Take eight disposable glass test tubes with
approximately 10 ml capacity. Mark them with A, B, C, etc. Pipette
into the first tube the reagents as indicated in Table 7 line A.
Mix it thoroughly by Vortexing. Transfer 1.0 ml from this into
tube B and mix it with 1.5 ml buffer. Proceed with the preparations of the standard curve as described in Table 7. The same
Table gives the amount of cAMP in these standards as picomols
(pmol) per 0.1 ml aliquots.
3. Take now 24 centrifuge tubes, of approximately 10 ml capacity. Number them consecutively. These will be used for the
determination and the calibration curve preparation. Table 8
shows how to proceed. Tubes No. 1 and 2 will have only hot cAMP,
nothing else. Tubes No. 3 and 4 will have the same amount of hot
cAMP plus buffer added. Tubes No. 5 and 6 will have hot cAMP plus
buffer plus antiserum to cAMP. Tubes No. 7 to 22 will give the
calibration curve. Increasing amounts of cold cAMP (all in 0.1 ml)
will be pipetted into these from the tubes A-H from step No. 1
above. They will compete with the hot cAMP for the anti-cAMP immunoglobulins. Tubes No. 23 and 24 will have the unknown samples.
No cold cAMP is added to these, and the cAMP content of the samples will compete here with hot cAMP for the antibody.
4. After everything is pipetted into the tubes at room temperature, cover each tube with aluminum foil or with parafilm. Mix
them very thoroughly without spilling their contents. Transfer
the tubes to cold room for approximately 16-18 h.
5. Add to each tube (except No. 1 and 2) 2 ml chilled pH 6.2,
0.05 M phosphate buffer. Vortex the tubes at cold room temperature
and centrifuge them at 3000 g in a refrigerated centrifuge for
15 min. Take a Pasteur pipette attached to a 2000 ml filter flask
under vacuum. By carefully immersing the tip of the pipette into
the tubes, remove the supernatant. Be aware that this is radioactive. Do not remove any of the pelleted antigen-antibody pre-

Table 7. Preparation of dilutions from the cold cAMP standard

A) Pipette 0.1 ml cAMP std. (5000 pmol/ml) + 1.9 ml buffer into tube A (25.0 pmol/0.1 ml)

B) Pipette 1.0 ml cAMP std. from tube A   + 1.5 ml buffer into tube B (10.0 pmol/0.1 ml)

C) Pipette 1.0 ml cAMP std. from tube B   + 1.0 ml buffer into tube C (5.0 pmol/0.1 ml)

D) Pipette 1.0 ml cAMP std. from tube C   + 1.0 ml buffer into tube D (2.5 pmol/0.1 ml)

E) Pipette 1.0 ml cAMP std. from tube D   + 1.5 ml buffer into tube E (1.0 pmol/0.1 ml)

F) Pipette 1.0 ml cAMP std. from tube E   + 1.0 ml buffer into tube F (0.50 pmol/0.1 ml)

G) Pipette 1.0 ml cAMP std. from tube F   + 1.0 ml buffer into tube G (0.25 pmol/0.1 ml)

H) Pipette 1.0 ml cAMP std. from tube G   + 1.5 ml buffer into tube H (0.10 pmol/0.1 ml)

Table 8. Procedure for the determination of the cAMP content of spleen cells and preparation of the calibration curve

Tube No.	Buffer	cAMP Std.		Sample	ScAMP-TME-[^{125}I]	Antiserum	Purpose	
1,2	–	–		–	0.1	–	Total Counts	
3,4	0.2	–		–	0.1	–	Blank	
5,6	0.1	–	*	–	0.1	0.1	"O" Standard	
7,8	–	0.1	(A)	–	0.1	0.1	0.10 pmol std.	(A)
9,10	–	0.1	(B)	–	0.1	0.1	0.25 pmol std.	(B)
11,12	–	0.1	(C)	–	0.1	0.1	0.50 pmol std.	(C)
13,14	–	0.1	(D)	–	0.1	0.1	1.0 pmol std.	(D)
15,16	–	0.1	(E)	–	0.1	0.1	2.5 pmol std.	(E)
17,18	–	0.1	(F)	–	0.1	0.1	5.0 pmol std.	(F)
19,20	–	0.1	(G)	–	0.1	0.1	10.0 pmol std.	(G)
21,22	–	0.1	(H)	–	0.1	0.1	25.0 pmol std.	(H)
23,24	–	–		0.1	0.1	0.1	Samples	

cipitates. Tilt the tubes to remove the last drop of supernate, but do not disturb the sediment.

6. Count all tubes in a gamma counter, including No. 1 and 2.

Evaluation

If the counting time was everywhere 1 min, the total count will give the count per minute or cpm. Wherever duplicates (or triplicates) were made, calculate the average cpm.

Tubes No. 1 and 2 will have the total hot cAMP content. Tubes No. 3 and 4 received only buffer and hot cAMP. These will not form a sediment during steps 4 and 5, but they will have residual radioactivity which could not be removed by siphoning off the liquid from the tubes in step 5. The average of the cpm values from tubes No. 3 and 4 should be therefore deducted from the cpm values of all tubes from No. 5 to 24.

The cpm of tubes No. 5 and 6 will give you the total antibody bound hot cAMP (B_T). The cpm of tubes No. 7 to 22 will be less, since the standard cold cAMP competed with the hot standard. Indicate these cmp values as B_S. Calculate the percent bound hot cAMP as follows:

Percent bound = $B_S/B_T \times 100$.

Take a semilog graph paper and plot the percent bound value on the linear scale vs pmol cAMP plotted on the log scale. A slightly S-shaped curve will be obtained. Take now the cpm of the unknown sample (B_U) and calculate the percent bound value ($B_U/B_T \times 100$) and read the pmol cAMP content of the unknown sample. Calculate the cAMP content of 100 ml spleen.

Part B. Cyclic GMP

Materials and Methods

Cold cGMP standard, 2000 pmol per ml
Hot cGMP derivative (succinyl tyrosine-[125 I]-methyl ester of cGMP)
Anti-cGMP serum
Besides these above reagents the same materials and equipment is used for the cyclic GMP determination as were listed in *Part A* of this Exercise.

Procedure

1. The extraction of cGMP is done by the same procedure as described in *Part A*, step 1 for cAMP.

2. Preparation of cGMP standard curve is very similar to the one used for cAMP standards. Pipette the amounts shown in Table 9. The only difference is in the amount of cyclic nucleotide content in the tubes. Please note that the cold cGMP standard has a two and a halftimes lower concentration than the cAMP had, and so will all the dilutions made from it.

Table 9. Preparation of cGMP standard curve

A) Pipette 0.1 ml cGMP std. (2000 pmol/ml) + 1.9 ml buffer into tube A (10.0 pmol/0.1 ml)

B) Pipette 1.0 ml cGMP std. from tube A   + 1.0 ml buffer into tube B (5.0 pmol/0.1 ml)

C) Pipette 1.0 ml cGMP std. from tube B   + 1.0 ml buffer into tube C (2.5 pmol/0.1 ml)

D) Pipette 1.0 ml cGMP std. from tube C   + 1.5 ml buffer into tube D (1.0 pmol/0.1 ml)

E) Pipette 1.0 ml cGMP std. from tube D   + 1.0 ml buffer into tube E (0.50 pmol/0.1 ml)

F) Pipette 1.0 ml cGMP std. from tube E   + 1.0 ml buffer into tube F (0.25 pmol/0.1 ml)

G) Pipette 1.0 ml cGMP std. from tube F   + 1.5 ml buffer into tube G (0.10 pmol/0.1 ml)

H) Pipette 1.0 ml cGMP std. from tube G   + 1.0 ml buffer into tube H (0.05 pmol/0.1 ml)

3. From this point on, all steps, including evaluation, are the same as given in *Part A* for cAMP determination. Use Table 8 as a guide.

Use and Limitations

Most of the reagents can be prepared in the laboratory but the preparation of some of them requires considerably greater experience than it is expected from students. The major difficulty lies in the proper standardization of the preparations, so that results obtained with different batches of reagents will be comparable. The great value of commercially available kits is that they are standardized, ready for use by anybody properly trained in basic laboratory techniques.

It is recommendable to have a control on the efficiency of the extraction procedures, as described in steps 1 in *Part A* and *B*. This can be achieved by adding known amounts of ^{3}H-labeled cyclic nucleotide to a sample aliquot. One has to carry out the extraction process as described above and determine the amount of the extracted ^{3}H cyclic nucleotide. The recovery of the added amount must be over 90%. ^{3}H standards can be obtained from New England Nuclear and they are parts of the kits for the cAMP and cGMP determinations.

A normal rabbit serum control should be included in the determination, replacing rabbit antiserum to the cyclic nucleotide. This control will determine the amount of cyclic nucleotides nonspecifically absorbed by serum proteins.

The sensitivity of the determination of both nucleotides can be greatly enhanced by using acetylated samples. Steiner et al. (1972) showed that nucleotides substituted in 2'O position are more efficiently competing with hot nucleotides for the recognition sites on antibody molecule. Cailla et al. (1976) used acetylation for the substitution at 2'O position. An approximately 20-fold increase in the sensitivity of both determinations can be achieved by acetylating the cyclic nucleotides in the unknown samples. The acetylation of the cold cyclic nucleotide standards as well as of the extracted unknown samples can be carried out as follows:

Mix freshly distilled acetic anhydride with 2 vol of also freshly distilled triethylamine. This acetylating reagent must be mixed just before use. Pipette 5 µl of this into the standard sample tubes, which contain known amount of cyclic nucleotide. Since the sensitivity of the determination is greatly enhanced use only 1/20 of the amounts you used for nonacetylated samples (see Tables 7 and 8). This 5 µl acetylating reagent should be added directly to the solution and mixed immediately by Vortexing for a few seconds. Pipette the 5 µl acetylating reagent into the unknown samples also. From here proceed as with nonacetylated samples, as described in step 2 in *Part A*.

References

Ashman, D.F., Lipton, R., Melicow, M.M., Prise, T.D.: Biochem. Biophys. Res. Comm. *11,* 330 (1963)
Cailla, H.L., Vannier, C.J., Delaage, M.A.: Anal. Biochem. *70,* 195 (1976)

Goldberg, N.D.: Advances in cyclic nucleotide research, Vol. II, p. 307
   Drummond, Greenbard, Robinson (eds.), New York: Raven Press 1975
Steiner, A.L., Parker, C.W., Kipnis, D.M.: J. Biol. Chem. *247*, 1106 (1972)
Sutherland, E.W., Rall, T.W.: J. Am. Chem. Soc. *79*, 3608 (1957)
Yalow, R.S., Berson, S.A.: Diabetes *9*, 254 (1960)

Exercise No. 80

Solid-Phase Radioimmunoassay

This method has been developed by Catt and Tregear (1967) who
coated polymers with antibodies and observed the reactivity of
this coat with radio-labeled antigens. This reaction was partially
inhibited if free, nonlabeled antigen-containing preparations or
body fluids were incubated first with the antibody-coated polymers.
Measurement of radioactivity on the polymers gave quantitative
information about the amount of nonlabeled antigen in the samples.
   A modification of this procedure has been applied to quantitation of Australia antigen (hepatitis antigen), as developed by
Purcell and associates (1973). This Exercise follows their description.

Materials and Equipment

   Hyperimmune guinea pig serum, immunized with hepatitis B
     antigen according to the method of Purcell et al. (1970).
     If hyperimmune human serum is available, this is preferred
     to guinea pig serum
   Normal human serum
   Normal guinea pig serum
   1% bovine serum albumin (BSA) in phosphate-buffered saline
   V-bottom polyvinyl microtiter plate, with lid available from
     several companies distributing plastic laboratory ware
   Eppendorf micropipettes
   Phosphate-buffered saline (PBS), see Exercise No. 1

Procedure

   1. Prepare IgG from an aliquot of the hyperimmune guinea pig
serum by ammonium sulfate precipitation (Exercise No. 1) followed
by chromatography on a DEAE cellulose column (Exercise No. 3).
   2. Label the isolated IgG with ^{125}I as described in Exercise
No. 13.
   3. Pipette 75 µl of the nonfractionated, 1:1000 diluted guinea
pig anti serum into each well of a polyvinyl V-bottom microtiter
plate provided with a lid. Leave the plates in a refrigerator or
cold room at 4°C for 4 h. The serum can be collected from the wells
by pipetting. Save it, since it can be used to coat several other
microtiter plates. Wash the plates by immersing them in PBS.
Discard the PBS wash, and repeat washing. Shake the plates to
remove as much PBS as possible and pipette into the wells 200 µl
1% BSA. Incubate the plates again at 4°C overnight and then wash
twice with PBS as above.

4. Next add the sample which will be checked for hepatitis B
antigen content. Pipette 25 µl aliquots into three or four wells
coated with guinea pig antiserum. Incubate the plates for 24 h
at 4°C. Remove the sample from the wells by aspiration and rinse
the wells five times with 200 µl PBS.

As negative control, pipette 25 µl normal human serum, free
of hepatitis B antigen, into another three or four wells. Treat
these in the same way as the positive control samples. It is recommended to run a few twofold dilutions of a hepatitis B positive
serum on the same plate. This can serve as your positive control.

5. Add 50 µl of ^{125}I-labeled IgG. Cover the plate with the
lid and incubate it in a humidified box at 37°C for 4 h.

6. Remove the lids and aspirate off the radioactive serum.
Follow the instructions of the radiation safety regulations for
the disposal of the serum as well as of the washing fluids which
follow. Wash the plates five times with 200 µl PBS as in step 4
above.

7. Cut the wells apart, one by one, with scissors or a hot
spatula. Drop them individually into gamma counting tubes and
measure the radioactivity.

Evaluation

Determine the mean of the residual counts on the unknown sample
wells (U) and on the negative control wells (N) also. Divide the
mean of the unknown sample wells by the mean of the negative
wells. This will give the U/N ratio. If the unknown sample did
not contain hepatitis B antigen, no ^{125}I-labeled IgG will adhere
to the coated plate, U/N will be near 1. If the sera are positive,
the U/N values will be greater than 1.

As additional control, one may add nonlabeled anti hepatitis
B IgG to the wells before adding labeled antihepatitis B IgG.
Purcell et al. (1973) considered a test positive if the addition
of nonlabeled hepatitis B antibody before labeled hepatitis B
IgG reduced the U/N value by 50% or more.

Use and Limitation

Nonspecific binding of ^{125}I IgG to the polymer is one of the
major sources of error in this assay. Incubation with 1% BSA
can reduce it to an acceptable minimum.

The great advantage of this assay is that we do not need
purified antibodies for coating the polymer surface. High antibody titer-containing convalescent sera can be obtained from
hospitals and used for this purpose. The use of human serum as
opposed to guinea pig serum has the advantage that the solid-phase
assay set up with human serum and with isolated human IgG is more
specific for the detection of hepatitis B antigen in human serum
samples than the assay set up by using guinea pig antibodies to
the same antigen. However, the specificity of the test with hyperimmune guinea pig serum can be improved by diluting the ^{125}I-
labeled guinea pig IgG reagent with a mixture of normal human and
normal guinea pig sera. Dilution with human serum will block
guinea pig antihuman antibodies which are present in the guinea
pig IgG. The addition of normal guinea pig serum will neutralize
anti-guinea pig antibodies present in the human serum samples

which will be analyzed for hepatitis B antigen. (Personal communication from R.H. Purcell.)

The newest findings indicate that there are several hepatitis antigens. The latest review on this topic has been published by Purcell (1978).

Finally, we should emphasize the wide applicability of the solid-phase radioimmunoassay. The procedure can also be used for the detection of hepatitis B antibodies according to the method of Ling and Overby (1972). In this case the microtiter wells are coated with hepatitis B antigen. ^{125}I-labeled hepatitis B antigen must also be prepared. The steps for the procedure are very much the same as those described above. Rosenthal et al. (1973) used such a procedure for the detection of various viral antibodies and antigens. Purcell and coworkers elaborated a procedure for the detection of hepatitis B core antigen in serum and hepatitis A antigen in feces. Similar assays can be developed for a great number of other systems, provided either that purified antigen which can be labeled with isotope is available, or that high titer antiserum to the antigen can be obtained. This antiserum must be fractionated and similarly labeled with iodine as described in this Exercise.

The most modern version of solid-phase immunoassays uses beads coated with antibodies. Such preparations are commercially available from Bio-Rad Laboratories, Richmond, CA 94804.

Very careful handling of the human serum samples is imperative to avoid hazards of infection with hepatitis!

References

Catt, K., Tregear, G.W.: Science *158*, 1570 (1967)
Ling, C.M., Overby, L.R.: J. Immunol. *109*, 834 (1972)
Purcell, R.H.: Hospital Practice, July, 1978
Purcell, R.H., Gerin, J.L., Holland, P.V., Cline, W.L., Chanock, R.M.: J. Inf. Dis. *121*, 222 (1970)
Purcell, R.H., Wong, D.C., Alter, H.J., Holland, P.V.: Appl. Microb. *26*, 478 (1973)
Rosenthal, J., Hayaski, D.K., Notkins, A.L.: Appl. Microbiol. *25*, 567 (1973)

3.2.3   Other Reactions of Antibodies and Immunocytes

Exercise No. 81

Complement Fixation

The term "complement" is used for those heat-labile protein materials which are present in the serum of different animal species and according to our present knowledge they are not products of immunization. Complement consists of several components which are different in their chemical nature, heat sensitivity, and biologic function. Early investigators observed that if the antigen participating in an antigen-antibody reaction is an erythrocyte, after it has been combined with corresponding antibodies, lysis of the erythrocytes will result if complement is

present in the mixture. These two facts, (a) the "fixation" of
complement by antigen-antibody complexes, and (b) the lysis of
cells combined or "sensitized" with their corresponding antibodies
in the presence of complement, were used by Bordet and Gengou
(1901) to design the complement fixation assay for quantitative
determination of antibodies in the serum of immunized animals.
    The complement fixation reaction consists of two parts. In
the first part, the antigen and antibody complex is formed, and
this is incubated with the complement. Depending upon the amount
of antigen and antibody present in this mixture, different
amounts of the complement will be fixed. This reaction is not
visible, therefore in the second part of the reaction the amount
of unused complement is measured. For this purpose an indicator
system is used which consists of sheep erythrocytes sensitized
with the corresponding antibodies. Sheep erythrocytes and their
antibodies will combine with the unused complement and will undergo
lysis. The amount of hemoglobin liberated from the lysed cells is
measured photometrically and the values obtained are in direct
relationship with the amount of unused complement.
    Complement fixation can be used for the determination of anti-
gens or antibodies in different biologic preparations. If the
antibody content of a serum is to be determined, obviously the
proper antigen preparation must be available in the laboratory.
If a known antiserum is available, the complement fixation test
using this antiserum may be applied to detect antigens in different materials.
    For a quantitative complement fixation reaction, the proper
amount of every reagent used in the assay must be determined.
It is important to know the optimal amount of hemolysin, the ap-
propriate dilution of complement, and the proper amount of antigen. It is also important to know whether the components participating in the reaction have a nonspecific anticomplementary
effect. Certain materials are able to fix or inactivate complement in the apparent absence of the proper antigen or antibody.
A series of preliminary experiments must be set up before the
quantitative complement fixation test can be carried out, which
will be described in this Exercise.

Materials and Equipment

    Washed sheep erythrocyte suspension
    Hemolysin (commercially available)
    Complement (guinea pig serum, commercially available)
    Glycerin
    Phenol (crystalline)
    Sodium diethyl barbiturate
    Sodium chloride
    Veronal-buffered saline solution (see Procedure)
    Magnesium chloride · 6 H_2O
    Calcium chloride · 2 H_2O
    Centrifuge
    Spectrophotometer
    Electrometric pH meter
    Pipettes, capacity 1 ml and 5 ml
    12 × 100 mm test tubes

Procedure

Part A. Preparation of Reagents

Veronal-buffered saline is prepared as follows: Dissolve 83 g NaCl
and 10.19 g sodium diethyl barbiturate (Veronal) in 1500 ml
distilled water. Add to this solution 34.6 ml 1 *N* HCl. Also
prepare a stock solution containing magnesium and calcium by
dissolving 20.33 g of $MgCl_2 \cdot 6\ H_2O$ and 4.4 g of $CaCl_2 \cdot 2\ H_2O$ in
100 ml distilled water. Pipette 5 ml of this stock solution into
buffered saline and make up the volume to 2000 ml. The Veronal
buffered saline is five times more concentrated than will be
required for the Exercise. Keep this standard solution in the
refrigerator and, unless otherwise directed, dilute 1 vol with
4 vol of distilled water each time before using. The pH of the
solution must be between 7.3 and 7.4.

3% Sheep red blood cell (SRBC) suspension is prepared by centrifuging
10 ml of sheep blood and discarding the supernate. Wash the cells
at least three times with buffered saline in order to remove the
plasma proteins. Take 3.0 ml of this packed and washed sediment
and pipette it into 97 ml buffered saline. Rinse the pipette by
drawing up the red cell suspension into it a few times. It is
important to transfer all the cells from the pipette into the
suspension.

Hemoglobin color standard is necessary for the calibration curve
in the quantitative evaluation of the degree of hemolysis. First
prepare a hemoglobin solution by mixing 2 ml of the above SRBC
suspension with 14 ml distilled water which will lyse them completely. Add 4 ml of the *undiluted* buffered saline stock solution
to restore the isotonic salt concentration of this hemolysate.
Mix well. Next, prepare a 0.3% sheep erythrocyte suspension by
mixing 2 ml of the 3.0% suspension thoroughly with 18 ml buffered
saline. After these have been prepared, place 11 small test tubes
in a rack. Do not put hemoglobin solution in the first tube, but
pipette 0.3 ml in the second, 0.6 ml in the third, 0.9, 1.2, etc.
into the remaining tubes. The 11th tube will contain 3.0 ml hemoglobin solution. Now take the 0.3% SRBC suspension and fill each
tube up to 3 ml, which means that the first tube will receive
3 ml of the 0.3% SRBC suspension, the others will have decreas-
ing amounts, and none will be added to the 11th tube. Mix the
contents of the tubes well, centrifuge them at 1000 g for 10 min,
and store them in the refrigerator. These hemoglobin color standards may be used for the establishment of a calibration curve.
Read the optical density of the supernates at 541 nm in a spectrophotometer. For less accurate analysis, these color standards
can be used merely for visual comparison of the hemolysis obtained in the mixture containing unknown samples.

Complement is prepared from a commercially available lyophilized
product. The lyophilized complement is dissolved in the proper
diluent, usually supplied or indicated by the manufacturer.
Pipette 0.5 ml samples of the undiluted complement solution into
test tubes and keep them in the freezer. Complement, which is
very sensitive to heat and will be inactivated rather rapidly at
room temperature, will be relatively stable for several weeks in
the frozen state, although a slight decrease in activity may be
observed during prolonged storage even in the freezer. For the

complement fixation assays, 0.2 ml of accurately pipetted complement must be added to 7.8 ml of cold buffered saline. This
1:40 diluted complement must be kept in the refrigerator or in
an ice bucket during the experiment and must be freshly prepared
on the day of each assay.

   Hemolysin (also called amboceptor or antisheep erythrocyte serum)
solution is prepared from glycerinized hemolysin by diluting it
with buffered saline. Glycerinized hemolysin is obtained by mixing equal volumes of hemolysin and neutral glycerin, which sta-
bilizes the hemolysin. Mix well 94 ml of the buffered saline
with 4 ml saline containing phenol (5% crystalline phenol dissolved in 100 ml buffered saline), then add 2 ml glycerinized
hemolysin. This is the standard hemolysin stock solution, diluted
1:100. It keeps well in the refrigerator.

Part B. Hemolysin Titration

Prepare a 1:1000 dilution of the standard hemolysin by diluting
1 ml stock solution with 9 ml buffered saline. Pipette 1 ml of
this 1:1000 dilution into each of 8 tubes and prepare 1:2000,
1:3000, 1:4000, etc. dilutions of this hemolysin up to 1:8000
final dilution. Use buffered saline to make these dilutions. Take
another eight tubes and transfer 0.3 ml of each dilution from
1:1000 up to 1:8000 into the new set of 12 × 100 mm test tubes.
Pipette 0.3 ml of the 3.0% SRBC suspension into each tube and mix
properly. Let the tubes stand at room temperature for 20 min. To
these sensitized cells, first add 1.2 ml buffered saline and then
1.2 ml of a freshly prepared 1:40 dilution of guinea pig complement. Mix well and incubate the tubes at 37°C for one h.

   Centrifuge the tubes at 1000 g for 10 min and measure the
optical density of the supernate at 541 nm in a photometer. Use
the calibration curve prepared previously from the hemoglobin
color standards and determine the percent of hemolysis obtained.
It can be seen from the data that a certain hemolysin dilution
will result in maximum hemolysis, and a further increase in the
hemolysin concentration will not appreciably enhance the amount
of hemoglobin liberated. The purpose of the hemolysin titration
is to find the proper amount of hemolysin which should be used
in the complement fixation assay. Therefore, select the highest
dilution (lowest hemolysin concentration) which resulted in maximal hemolysis. This is one minimal hemolytic unit (MHU) in 0.3 ml
hemolysin. To prepare a 3% suspension of sensitized SRBC, take
1.5 ml of a washed erythrocyte sediment, suspend it in 18.5 ml
buffered saline and add 30 ml hemolysin solution which contains
2 MHU/0.3 ml.

Part C. Titration of the Complement

Since complement is very unstable, it is important to determine
the titer of the complement in each experiment. Wadsworth, Maltaner
and Maltaner (1931) were the first to elaborate standard condition
for quantitative measurements, by which one can determine the amou
of complement which will result in 50% hemolysis (one $C'H_{50}$ unit)
of the standardized SRBC in the presence of the previously determined optimal amount of hemolysin. This can be achieved by graphic

interpolation. First prepare dilutions of complement as shown in
Table 10 using 10 ml centrifuge tubes.

Table 10. Preparation of complement dilutions

Tube No.	ml 1:100 dil. complement	ml buffered saline	Dilution obtained	log dilution
1	2.70	0	1:100	2.0000
2	1.35	1.35	1:200	2.3010
3	0.90	1.80	1:300	2.4771
4	0.67	2.03	1:400	2.6020
5	0.54	2.16	1:500	2.6990
6	0.45	2.25	1:600	2.7782
7	None	2.70	–	
8	None	+2.70 ml distilled water		

Add 0.3 ml sensitized SRBC to each tube, mix well, and incubate at 37°C for 60 minutes. Centrifuge the mixtures at 500 g
for 10 minutes to obtain a clear supernate and read the optical
density of the hemolysate at 541 nm. Tube 7 is the cell control
(no hemolysis); tube 8 is the 100% hemolysis control.
W.H. Taliaferro and L.G. Taliaferro (1950) developed details
of the 50% titration method. In this procedure the degree of
hemolysis is determined photometrically and the amount of complement which would give 50% lysis of the given amount of erythrocytes will be determined graphically. The titer of the dilution
is identified by the logs of the reciprocals. For example, the
log of the reciprocal of a 1:2000 dilution is 3.30, and a 1:200
dilution is 2.30. Therefore, it is convenient to use dilutions
which are spaced at relatively close log intervals. The degree
of hemolysis is calculated by the term log Y (1-Y) where Y is
the degree of hemolysis, determined photometrically, when 100%
hemolysis is 1.0. By plotting the degree of hemolysis as log Y
(1-Y) on the abscissa and the corresponding logs of guinea pig
serum dilution on the ordinate, a straight line relationship
will be found. The intercept of this line which connects the
readings with the ordinate is the dilution of the serum necessary
for 50% hemolysis. A typical graphic determination of complement
titer is shown in Fig. 58. This is one complement ($C'H_{50}$) unit.

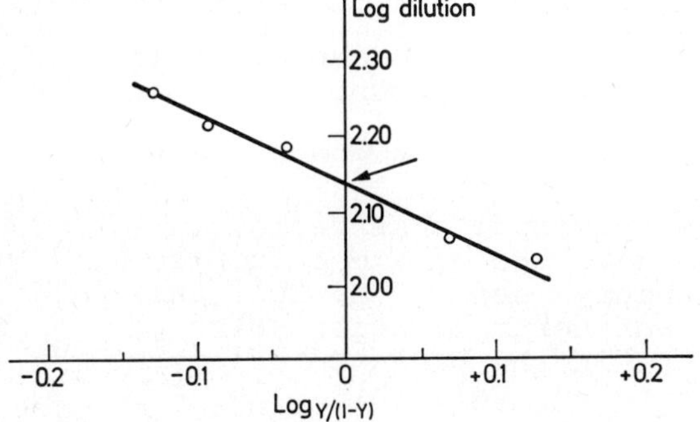

Fig. 58. Determination of one
complement ($C'H_{50}$) unit

If 50% hemolysis is not obtained, repeat the experiment with
higher (or lower) serum dilutions.

For the quantitative complement fixation assay, 5 C'H$_{50}$ units
in 0.5 ml vol will be added. Calculate the dilution which will
have 10 C'H$_{50}$ units per milliliter.

Part D. Qualitative Complement Fixation (Determination of the Proper Antigen Dose for Quantitative Complement Fixation Test)

It is essential to use the proper antigen dose if the antibody
content of an unknown serum is to be determined. Several antigens,
such as viral antigens, are strongly anticomplementary, a fact
which must also be determined in control experiments. The most
adequate antigen dose must fulfill several criteria: It must be
in a sensitive dose range which can detect minute amounts of
antibodies; it should not be hemolytic; it should be able to
bind all complement given to the system (the optimal complement
dilution has already been titrated) if enough homologous antibody
is present.

For this purpose, the so-called checkerboard or block titration is used in which a known high titer homologous antiserum
will be used in dilution series. Several rows of dilutions will
be prepared, and to each row a different amount of antigen will
be added. After the addition of complement and sensitized SRBC,
the degree of hemolysis will be observed merely by visual comparison with the hemoglobin color standard. It will be seen that
the different antigen doses will show different antibody titers
of the antiserum. The aim of this preliminary assay is to find
the highest amount of antigen (lowest dilution) which is not
anticomplementary and measures the highest titer of the antiserum.

Take 64 tubes, make eight rows. Pipette 0.3 ml of the 1:10
diluted antiserum into the first tube of each row. Make serial
twofold dilutions in all eight rows using buffered saline.

Prepare 5 ml of each of the following twofold dilutions of
the antigen solution which contains 0.1 mg/ml dry weight substance dissolved in buffered saline: 1:1, 1:2, etc. up to 1:128.
This will give you eight different antigen dilutions. Add 0.3 ml
of the first antigen dilution to each tube of the first row of
antiserum dilutions. Add the second antigen dilution to the second row of antiserum dilutions and proceed with the further dilutions as above. After all 64 tubes have received 0.3 ml of the
different antigen and 0.3 ml of the different antiserum dilutions,
incubate them at room temperature for 15 min. The protocol which
students are to follow is shown in Table 11.

Pipette 0.3 ml of 1:40 diluted guinea pig complement into each
tube. Mix the contents of the tubes, then incubate them at cold
room temperature overnight. Add to each tube 1.8 ml buffered
saline and 0.3 ml 3% suspension of sensitized SRBC. Incubate at
37°C for 30 min.

Read the degree of hemolysis by visual observation only. Find
the antigen concentration which gave the greatest complement fixation without being anticomplementary (see *Part E*). Call this proper
antigen dose 1 Ag. The highest antiserum dilution which gave very
definite adsorption of complement in the presence of optimal
amount of antigen has also been found. Call this 1 Ab. The

Table 11. Protocol for "checkerboard" titration

0.3 ml dilution of a 0.1 mg/ml antigen sol'n	Dilution of antiserum (0.3 ml)							
	1:10	1:20	1:40	1:80	1:160	1:320	1:640	1:1280
1:1								
1:2								
1:4								
1:8								
1:16								
1:32								
1:64								
1:128								

reciprocal of this dilution, therefore, is the titer of the antiserum used, determined with the complement fixation assay.

Part E. Anticomplementary Controls

It is recommended to determine the degree of anticomplementary
effects of the antigen and antibody dilutions in a separate assay.
Using the same antigen and serum dilutions used in the preceding
section, prepare the following sets of tubes according to the
protocol in Table 12. Incubate all tubes at cold room temperature
overnight. The next day, add 0.3 ml sensitized SRBC to each tube
and incubate at 37° for one-half hour. Read the degree of hemolysis.
Those antigen or antiserum dilutions which "fixed" more than 1/5
of the added complement have high anti-complementary effect,
therefore they cannot be used in the quantitative complement fixation test.

Part F. Quantitative Complement Fixation

From the previous sections not only the optimal amount of antigen
but also the optimal amount of serum dilution can be determined.
Select the highest dilution of serum which gave the least hemolysis with the optimal amount of antigen. If the presence of antigens is to be determined by complement fixation, use the next
lower dilution of this serum in the quantitative assay.
   The optimal amount of antigen was determined in one of the
previous sections, and if the complement fixation assay is used
to detect antibodies present in different sera, use twofold
amounts of this antigen dissolved in the same volume for quantitative purposes.
   After mixing a standard amount of serum to different dilutions
of antigens (or standard amount of antigen to different dilutions
of unknown serum), make up the volume to 2.2 ml with buffered
saline and add to them 5 $C'H_{50}$ complement units in 0.5 ml. It is
absolutely essential to include the necessary controls in the
same assay. The procedure for antigen titration is according to
the protocol shown in Table 13.

Table 12. Preparation of tubes for determining degree of anticomplementary effects of antigen and antibody dilutions

		1	2	3	4	5	6	7	8	9	10	11	12	13	14	15	16	17	18
Antigen dilution	1:2	0.3	0.3	0.3	.....	.....	.....	.....	.....	.....	.....	.....	.....	.....	.....	.....	.....	.....	.....
	1:4	.....	.....	.....	0.3	0.3	0.3	.....	.....	.....	.....	.....	.....	.....	.....	.....	.....	.....	.....
	1:8	.....	.....	.....	.....	.....	.....	0.3	0.3	0.3	.....	.....	.....	.....	.....	.....	.....	.....	.....
Antiserum dilution	1:2										0.3	0.3	0.3	.....	.....	.....	.....	.....	.....
	1:4										.....	.....	.....	0.3	0.3	0.3	.....	.....	.....
	1:8										.....	.....	.....	.....	.....	.....	0.3	0.3	0.3
Buffered saline		1.9	2.2	2.3	1.9	2.2	2.3	1.9	2.2	2.3	1.9	2.2	2.3	1.9	2.2	2.3	1.9	2.2	2.3
5 C'H$_{50}$ complement		0.5	0.2	0.1	0.5	0.2	0.1	0.5	0.2	0.1	0.5	0.2	0.1	0.5	0.2	0.1	0.5	0.2	0.1

Table 13. Protocol for quantitative complement fixation

Tube No.	Antiser. ml (2 Ab/0.3 ml)	Antigen ml (1 mg/ml, diluted)	Buffered Saline ml		Complement ml (10 C'H$_{50}$/ml)	
1	0.3 ml	1.0 (1:1)	0.9		0.5	
2	0.3	1.0 (1:10)	0.9		0.5	
3	0.3	1.0 (1:100)	0.9		0.5	
4	0.3	1.0 (1:1000)	0.9	Incubate at R.T. 15 min	0.5	Incubate overnight at +5°C
5	0.3	1.0 (1:10000)	0.9		0.5	
6	0.3	–	1.9		0.5	
7	–	1.0 (1:1)	1.2		0.5	
8	–	1.0 (1:1)	1.4		0.3	
9	–	1.0 (1:1)	1.6		0.1	
10	–	–	2.2		0.5	
11	–	–	2.7		–	

Following overnight incubation, add 0.3 ml 3% sensitized SRBC
to each tube. Incubate for 30 min at 37°C, centrifuge at 500 g
for 15 min, and read the optical density of the supernate at
541 nm.

The determination takes place in the first five tubes. Tube
No. 6 is the antiserum control, tube Nos. 7, 8, and 9 the antigen
controls. Tube No. 10 is the 100% hemolysis control and No. 11
should not show any hemolysis. If tube Nos. 7-9 show anticomplementary effect, lower concentrations of antigen should be
tried.

If the antibodies are to be determined in an unknown serum,
pipette 2 Ag in 0.3 ml dilution (next lower than optimal dilution
found in the "checkerboard" titration) into all tubes, and add
different serum concentrations. It is usually satisfactory to
start with a 1:10 dilution, and make double dilutions. The anticomplementary effect of the lowest serum dilution must be measured
in the same way as was done with the antigen in Table 12.

Evaluation

If the anticomplementary effect of the antigen or antibody resulted in less than 20% inactivation of the added 5 $C'H_{50}$, the
results may be evaluated. Often the assay which measures the
titer of the antiserum is satisfactory. If more accurate figures
are needed, the amount of antibody or antigen which caused 50%
hemolysis (see *Part C*) has to be determined.

Use and Limitations

The complement fixation test is useful where no antigen-antibody
precipitate is formed which can be measured by the quantitative
precipitin assay or by other simpler procedures. The best known
application of the complement fixation test is the Wasserman-
Kolmer test to detect syphilitic antibodies in patient serum. A
very wide use for the complement fixation assay can be found in
the literature of virology. (See *Viral and Rickettsial Infections in
Man* edited by Horsfall and Tamm, 4th edition, Lippincott.) Complement fixation detects incomplete antibodies. It is used in
forensic medicine to identify species.

The sensitivity of the complement fixation reaction is several
orders of magnitude lower than passive hemagglutination in the
detection of antibodies, but better than the sensitivity of the
conventional precipitin reaction. It measures 0.05 µg antibody
nitrogen with an accuracy of 5%. The accuracy of the procedure
is remarkable because lysis of red blood cells may be measured
easily. However, it has to be reemphasized that the quantitative
complement fixation reaction requires thoroughly controlled and
well standardized experimental conditions. The procedure is described here so that all tubes in all determinations will have a
final volume of 3.0 ml, which makes regular spectrophotometric
cuvettes suitable to make the measurement. Naturally, if smaller
photometric cuvettes are available, such as 0.5 ml capacity,
everything may be scaled down accordingly. If test tubes of identical size are used throughout the experiments, very often satisfactory estimation of the degree of hemolysis can be achieved

merely by visual comparison of the obtained tubes with the
hemoglobin color standards which may be kept in the refrigerator
for a relatively long time.

It must also be mentioned that not all antibodies fix complement, therefore neither the measurement nor the detection of
antigens by using noncomplementfixing antibodies can be carried
out by these assays. Horse antisera do not always fix complement;
pneumococcal antisera from man, dog, and mouse do not fix complement either. Nonspecific fixation of complement is usually due
to contaminants in the antigen preparations. Reliable complement
fixation assay should require purified antigens, which are not
easily available in most investigated immunologic systems.

An outstanding chapter covering complement and complement
fixation has been written by M.M. Mayer (1961). A critical review
of the complement fixation assay with a brief summary of the
achievements in the kinetics of immune hemolysis has been published by H.J. Rapp (1964).

References

Bordet, J., Gengou, O.: Ann. Inst. Pasteur *15,* 289 (1901)
Mayer, M.M.: In: Experimental immunochemistry. Kabat and Mayer, p. 13
    Springfield, IL.: C.L. Thomas 1961
Rapp, H.J.: In: Immunological methods, p. 1. Ackroyd, J.F. (ed.), Philadelphia
    Davis & Co. 1964
Taliaferro, W.H., Taliaferro, L.G.: J. Infect. Dis. *87,* 37 (1950)
Wadsworth, A.: Standard methods of the division of laboratories and research
    of the New York State Dept. of Health. 2nd ed. Baltimore: Williams and
    Wilkins 1939
Wadsworth, A., Maltaner, E., Maltaner, F.: J. Immunol. *21,* 313 (1931)

Exercise No. 82

Total and Viable Cell Counts

It is essential for almost all experiments involving cells to
count the exact cell number per milliliter suspension. In addition, it is frequently necessary to determine the number of viabl
cells in a cell suspension. Various procedures are available for
this purpose. One of them, described here, requires a hemocytometer and a simple microscope.

Materials and Equipment

Spleen cell suspension (Exercise No. 66)
Hemocytometer (Neubauer type available from A.H. Thomas,
  Philadelphia, PA. 19105)
Microscope with 2oo- to 300-fold magnification
Turk's blood diluting fluid (available from A.H. Thomas,
  Philadelphia, PA. 19105) or dissolve 0.01% gentian violet
  in 3% acetic acid
Trypan blue, 0.5% in distilled water

        Disposable Pasteur pipettes
        8 × 100 mm test tubes
        4.25% NaCl solution

<u>Procedure</u>

    1. Determine first the total leukocyte number. Take 0.1 ml
cell suspension and mix it with 0.9 ml Turk's solution. Let is
stand a few minutes at room temperature. The Turk's solution will
lyse the erythrocytes, leaving the leukocytes intact.
    2. Take a Pasteur pipette and fill the hemocytometer with one
drop cell suspension. Place it under a regular microscope and
count the cells. The Neubauer type hemocytometer has nine large
squares delineated by double lines. One such square corresponds
to 0.1 mm^3 suspension. These large squares (1 mm^2) are shown in
Figure 59 (A, B, C and D). Count the cell number in four large
squares. Use the equation in "Evaluation" to determine the total
cell number.

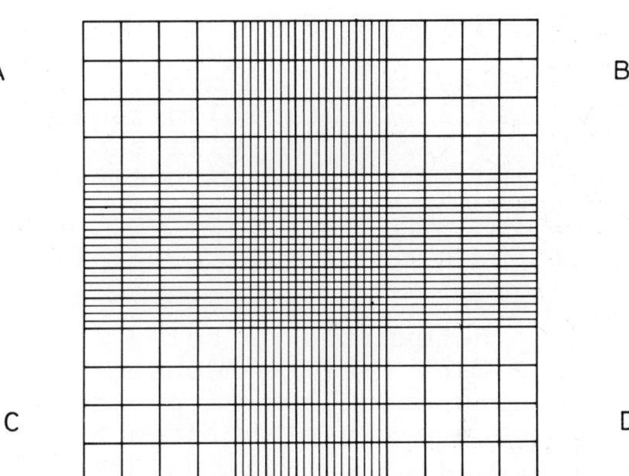

<u>Fig. 59.</u> Subdivisions etched into
glass in the hemocytometer

    3. To determine the viable cell number take 2 ml 0.5% Trypan
blue solution and add, just before use, 0.5 ml 4.25% NaCl solution. Mix well. Pipette 0.1 ml cell suspension into a test tube
and add 0.9 ml of the diluted Trypan blue. Mix well and wait for
exactly 5 min.
    4. Take a Pasteur pipette and place one drop of stained sus-
pension in the hemocytometer. Dead cells will pick up Trypan blue
and will be distinctly stained. Count the number of blue cells
in the four large squares as above.

<u>Evaluation</u>

Use the following equation to determine N, which is the total
number of cells per milliliter.

$$N = \left[\frac{n \text{ (dilution factor)}}{4} \right] 10^4$$

Where n = number of cells counted in four large squares. The dilution factor in the above example was 10. If the cell suspension is too concentrated, dilute it again ten-fold by adding 0.1 ml to 0.9 ml phosphate buffered saline. In this case the dilution factor will be 100.

To find the percent viable cells in the total suspension, use this equation:

$$\text{\% viable cells} = \frac{\text{total dead cells in four large squares} \times 100}{n \text{ (as determined above)}}$$

Use and Limitation

Trypan blue is cytotoxic, therefore do not wait longer than 5 min after mixing the cells with the dye. Start counting, and the whole process should be finished in another 5 min. The reason why we do not dissolve Trypan blue in saline is that the dye slowly precipitates in the presence of salts. The mixing of four parts of dye with one part of 5 X saline (4.25%) should be carried out daily, before the experiment.

Some cells may be dead without showing intensive blue color. Such cells are morphologically different from the others, they are slightly larger and will show ruffled contours. These can be recognized after some experience in this procedure.

The Trypan blue exclusion test is an insensitive measurement of cell viability. A better procedure has been described by Bodmer et al. (1967) who used fluorescein diacetate which will be converted to fluorescein by viable cells only. Their number can be determined by using a fluorescence miscroscope equipped with a dark field condenser.

Fluorescein-labeled proteins are also used to detect dead cells. A procedure was elaborated by Riggs et al. (1958) by which tissue culture cells as well as damaged tissue cells can be stained in vivo. The above described test can be used with some modifications for cytotoxicity assays. It has been applied to antibody mediated cytotoxic reactions by Harris et al. (1954).

References

Bodmer, W., Tripp, M. Bodmer, J.: In: Histocompatibity Testing, p. 341 (1967)
    See also Transplantion *10* 135 (1970)
Harris, S., Harris, T.N., Farber, M.: J. Immunol. *72*, 148 (1954)
Riggs, J.L., Sewiwald, R.J., Burckhalter, J., Downs, C.M., Metcalf, T.J.: Am. J.
    Pathol. *34*, 1081 (1958)

Exercise No. 83

Direct and Indirect Staining of Bacteria With Fluorescein-Labeled Antibodies

The two most frequently used fluorescent procedures for the localization of antigens in different preparations are the direct and indirect methods. In direct staining, the antibody which has been produced in experimental animals or isolated from human serum is directly labeled with fluorescein. In the indirect method, the antibody which will react with the antigen present in the preparation is not labeled with any tracers. To make the antigen-antibody complex visible, antibodies have been produced in another animal species against the antibodies participating in the antigen-antibody reaction. These anti-immunoglobulin antibodies are labeled with fluorescein dyes. The latter procedure is not only convenient but also it gives a more intensive fluorescent staining. The use of fluorescein-labeled antibodies has recently been made extremely easy, since several fluorescein-labeled antirabbit or antihuman immunoglobulins are commercially available.

Both of the above procedures are described in this Exercise. For the direct staining of *E. coli* bacteria, anti-*E. coli* rabbit sera which are labeled with fluorescein are used. For the indirect procedure, the *E. coli* bacteria are reacted with nonlabeled rabbit antiserum and this complex made visible under fluorescent microscope by reacting it with antirabbit goat serum, labeled with fluorescein tracer (see Exercise No. 12).

Materials and Equipment

      24-h old broth culture of *E. coli* (10 ml)
      E.coli antibody produced in rabbits
      E.coli antibody produced in rabbits, conjugated with FITC
      Antirabbit antiserum produced in goats, labeled with fluoro-
        chrome (commercially available preparation)
      Anti-*Pseudomonas aeruginosa* rabbit serum
      Rhodamine B isothiocyanate
      Soil sample
      Trypticase soy broth
      Methanol
      Phosphate buffered saline, pH 7.4
      Chromic acid washed microscope slides
      Coplin jars
      Moist chamber (Petri dish with wet paper lining)
      Fluorescent microscope

Procedure

   1. Bring approximately 1 g of fresh soil into the laboratory, mix it with 50 ml of broth and filter it through two layers of paper. Incubate it overnight at 37°C. Take a tube of 24-h old broth culture of the *E. coli* strain which has been used as the antigen to produce antiserum in rabbits. In a third test tube mix 5 ml of soil extract with 1 ml of *E. coli* broth. Take 5 ml of all preparations and wash the cells three times with saline.

Suspend the sedimented cells in 1 ml saline and place one drop
on a microscope slide. Spread this suspension by making a smear.
Prepare three slides from each and fix the preparations by slip-
ping them into a Coplin jar containing methanol. Keep the slides
in the jar for 10 min, then dry them at room temperature. It must
be mentioned that other fixatives have been found to be more
useful for the fixation of different smears or tissue sections,
but methanol is one of the most commonly used.

Take one slide from each preparation and place one drop of
fluorochrome-conjugated anti-*E. coli* rabbit serum preparation on
each. Place the slides in the moist chamber and incubate them
at 37°C for 60 min. Rinse the slides with phosphate-buffered
saline in a Coplin jar for 10 min. Replace the saline with fresh
solution, repeat the washing by carefully shaking the jar; three
or four washings usually remove the nonbound antibody conjugates
from the preparation. Examine the slides under the fluorescent
microscope.

2. The indirect or "sandwich" technique is perhaps the most
popular among the fluorochrome tracer methods. Take three other
smears prepared from the different microorganism suspensions
and place on each of them a drop of anti-*E. coli* rabbit serum
which has not been labeled with fluorescein dye. Incubate the
slides with the rabbit serum in the moist chamber at 37°C for
1 h. Using the same washing procedure as before, rinse the slides
in phosphate-buffered saline at least three times for complete
removal of the nonreacted antibodies from the preparation. Remove
the slides from the Coplin jar and blow the surfaces with a fan
until they dry. Avoid excess heat or prolonged drying of the
slides. Some moisture may remain on the surface of the smear.
Place one drop of antirabbit goat serum, labeled with FITC, on
the same place where the nonlabeled anti-*E. coli* rabbit serum
was placed previously. Put one drop of the same goat serum on
another part of the smear and incubate the slides in the moist
chamber for 1 h as before. After the incubation wash off the
excess antiserum from the slide with several rinsings in the
phosphate-buffered saline, and examine the slides under the
fluorescent microscope. The spot where the anti-*E. coli* rabbit
serum and the antirabbit goat serum were placed will show fluorescence of the *E. coli* cells. The control drop, where only
antirabbit goat serum was incubated with the smear, should not
show fluorescence.

3. For control purposes, take an unused smear and stain it
with Ziehl's carbolfuchsin procedure, in order to demonstrate
the presence of a large number of different microorganisms. If
anti-*Pseudomonas aeruginosa* rabbit serum is available in the laboratory, it is worth while to combine it with Rhodamine B isothiocyanate. Using this serum on the slide which was developed
by the direct procedure using FITC-conjugated rabbit antiserum,
a double staining may be achieved. To apply a second staining to
the mixture using the Rhodamine conjugate, the procedure is exactly the same as described in the direct procedure. In this
double staining procedure, it is irrelevant in which sequence
the two different conjugates were applied. If both *E. coli* and
P. aeruginosa cells were present in the smear, the *E. coli* will show
a bright green fluorescence while the *Pseudomonas* cells will appear
as a less intense orange fluorescence.

Evaluation

Observe the preparation under the fluorescent microscope. For
the proper adjustment and use of such microscopes, the instructions must be carefully studied. Students should not use the instrument without supervision.

Use and Limitations

Identical with those discussed regarding the labeling techniques.

Exercise No. 84

Unlabeled Antibody-Enzyme Method to Localize Antigens

A method has been developed to detect antigens or antigenic receptors on cell surfaces with great sensitivity, using a peroxidase-antiperoxidase (PAP) soluble antigen-antibody complex.
The procedure requires specific antisera but no covalent coupling between the reagents. They react with each other as antigens and antibodies and the whole complex formation can be detected with ease and great sensitivity because the catalytic
site of the peroxidase enzyme is not inhibited by antiperoxidase
antibodies. This is one of the few enzymes in which antibodies
to the enzyme do not interfere with the enzymatic activity.
   The following reaction steps will lead to the detection of
antigen, to which rabbit antiserum can be produced. First: The
rabbit antiserum will react with its antigen, for example with
the H2 histocompatibility antigen on spleen cells of strain A/J
mice. Second: the Fc portion of the rabbit immunoglobulin will
react with an excess amount of antirabbit immunoglobulin produced in sheep or other animal. Third: Peroxidase-rabbit antiperoxidase soluble antigen-antibody complex will also react
with the sheep antirabbit immunoglobulin, since the latter has
been applied in excess and the complex formed in the second
step will have free receptors for further rabbit Fc portions.
The peroxidase-rabbit antiperoxidase will now be attached to
the complex through the Fc portion of the rabbit immunoglobulin.
Fourth: The entire complex is detected by an enzymatic peroxidase
reaction using diamino benzidine and H_2O_2, which will give a
colored, insoluble oxidation product, suitable for light microscopy (Graham and Karnowsky, 1966). Exposure of the preparation
to OsO_4 vapors or solutions will form a heavy, electron-dense
deposit on the complex, thus rendering it highly visible in electron microscopy.
   The unlabeled antibody-enzyme method was developed independently by Mason et al. (1969) and by Sternberger and co-workers
(1969). The Exercise described here is based on the procedure of
Sternberger and associates (1970).

Materials and Equipment

Rabbit PAP complex. Commercially available from several sources
for example, Bionetics Laboratory Products, Litton Industries,
Kensington, MD 20795
Rabbit antiserum to Treponema pallidum. Commercially available
from Cappel Laboratories, Downington, PA 19335. The serum can
also be prepared by injecting anesthetized rabbits with approximately 10^5 live microorganism intratesticularly. Serum
can be harvested six to eight weeks after the injection
Sheep antirabbit immunoglobulin. Available commercially from
several companies. It can be prepared in the laboratory by
injecting the animals with rabbit gamma globulin IM mixed with
complete Freund's adjuvant
Tris buffer stock solution. Dissolve 60.57 g Tris (hydroxymethyl) aminomethane in 500 ml water. Add 1 N HCl until the
pH reaches 7.4. Make up the solution to 1000 ml with water
Tris buffer will be made from the above stock solution by
diluting it 1:10 with water. Tris-buffered saline; dilute
the stock solution tenfold with 0.9% saline
3,3'-diaminobenzidine tetrahydrochloride (DBT)
2% solution of OsO_4 in phosphate-buffered saline

Procedure

1. Spread one drop bacterial suspension obtained from cultures
or from clinical specimens on microscope plates with the aid of
a loop. To facilitate the spreading, add one drop of Tris-buffered
saline to the suspension. Spread a drop of *T. pallidum* (American
Type Culture Collection 27087) suspension on a second microscope
slide and subject it to the same treatments as described below.
This will serve as a positive control. Prepare four microscope
slides from the clinical specimen as well as from the positive
controls.
2. Prepare 1:4, 1:16, 1:64, 1:256 and 1:1024 dilutions of the
anti-*T. pallidum* rabbit serum and add 15 drops of the antiserum
to the specimens spread on the micro slides. Let them stand at
room temperature in a covered Petri dish for half an hour. Rinse
the microslides with a few milliliters of Tris-buffered saline
and three times with Tris buffer.
3. Cover the slides with approximately five drops of sheep
antirabbit immunoglobulin. Incubate them for 30 min at room
temperature, and rinse them with Tris-buffered saline.
4. Take the PAP reagent and add approximately five drops to
each micro-slide. Incubate again for 30 min at room temperature.
Rinse them off twice with Tris-buffered saline and submerge the
slides into Petri dishes containing Tris buffer.
5. After about 10 min, take the slide out of the Petri dishes
and cover them with a layer (approximately ten drops) of freshly
prepared solution containing 0.05% DBT and 0.001% H_2O_2 in water.
Wait 5 min and rinse the slides with distilled water at least
three times.
6. Let the slides dry. They are now ready for light microscopic
observation.
7. For electron microscopic studies, expose the slide to OsO_4
vapors. Take a screw cap jar which has a Teflon line in cap. The

jar should be at least 60 × 120 mm in size. Place a few crystals
of OsO$_4$ in the bottom of the jar and put the microslide into it.
Close the jar and keep it at room temperature for 6 h or at 50°C
for 30 min. Osmium labeling of the bacteria can be done also by
placing a few drops of a 2% OsO$_4$ solution on the microslide.

Evaluation

The preparation can be observed either by light or electron microscopy. Approximately 500 X magnification will clearly show the
T. pallidum morphology. The deposition of OsO$_4$ on the PAP-reacted
and DBT-developed microorganisms will render the spirochetes
highly visible at this magnification. Control slides where the
anti-*T. pallidum* rabbit serum was replaced with normal rabbit serum,
but otherwise treated entirely as above should reveal no microorganisms.
   The PAP method can also be used in electron microscopy as
mentioned above. The preparations will be reacted as described
above but in centrifuge tubes instead of on microscope slides.
The OsO$_4$-stained and embedded preparations can be sectioned for
electron microscopic purposes. The resolution reported in the
literature is excellent.

Use and Limitations

The same procedure can be used for detection of antibodies, toxins
or other immunogenic substances. It has been used not only for
bacteria and viruses but also for any other cell types or histological tissue sections of plant or animal origin, provided a
specific rabbit antiserum can be produced to an antigenic component present in them. Obviously, the better purified the rabbit
antiserum, the more specific the staining. Highly specific antiserum can be obtained if highly purified immunogen is available
and is used to produce antisera. Having a rabbit PAP reagent and
a good antirabbit sheep (or goat) serum, anything can be detected
to which specific rabbit antisera can be obtained. If the antigen
is not very immunogenic in rabbits but is immunogenic in guinea
pigs and such guinea pig antiserum has been obtained, a guinea
pig PAP is needed and anti-guinea pig serum produced in another
species. In addition to the great sensitivity, the flexibility
of the PAP procedure makes it a very valuable methodology.
   Soluble PAP preparations are available commercially, but one
can also prepare them in the laboratory by the procedure described
in the publication of Sternberger et al. (1970). The same paper
describes the proper analytic procedures which can be used to
evaluate the quality of such preparations.
   Exercise No. 8 in this Manual describes the isolation of horse
radish peroxidase antibodies using affinity chromatography. The
immunization schedule given in the present Exercise for the production of antirabbit gamma globulin serum in sheep can also be
used to produce antisera to enzymes.

References

Graham, R.C., Karnowsky, M.J.: J. Histochem. Cytochem *14*, 291 (1966)
Mason, T.E., Phifer, R.F., Spicer, F.F., Swallow, R.A., Dreskin, R.B.:
    J. Histochem. Cytochem. *17*, 190 (1969)
Sternberger, L.A., Cuculis, J.J.: J. Histochem. Cytochem. *17*, 190 (1969)
Sternberger, L.A., Hardy, P.H., Jr., Cuculis, J.J., Meyer, H.G.: J. Histochem.
    Cytochem. *18*, 315 (1970)

Exercise No. 85

Isolation of Human Lymphocytes by Sedimentation

Böyum separated white blood cells from erythrocytes by aggregating red blood cells and sedimenting them (1964). The same procedure was further elaborated by Böyum and applied for bone marrow
cell separations (1968). Several modifications of this procedure
have been published. The principle is aggregation of erythrocytes
by chemicals which do not affect white blood cells. Sedimentation
of aggregated erythrocytes can be achieved without centrifugation
at room temperature, and the cells remaining in the supernatant
can be sedimented by low speed centrifugation. In the procedure
described here, the aggregating agent is Ficoll, a polysaccharide
preparation. Sodium metrizoate is added to Ficoll to increase
its density, thus preventing sedimentation of nonaggregating
white blood cells while permitting the sedimentation of aggregated
erythrocytes. The blood sample is layered in a tube on the mixture
of Ficoll and sodium metrizoate. The aggregated cells will rapidly
settle to the bottom of the container. In less than an hour at
room temperature, without centrifugation, the supernate will contain only white blood cells, accumulated at the interface. A modification of this procedure is described here, using low speed centrifugation to facilitate the sedimentation of aggregated erythrocytes as proposed by Ting and Morris (1971), and as applied for
lymphocyte separation.

Materials and Equipment

Fresh blood, mixed with ACD solution at the time it is taken.
 A proper ratio is 8.5 ml blood to 1.5 ml ACD
ACD solution: 2.2 g sodium citrate $\cdot$ 2H$_2$O + 2.5 g glucose + 0.8 g
 citric acid dissolved in 100 ml glass distilled water
Tris (hydroxymethyl) aminomethane
Balanced salt solution (BSS). Prepare the following two stock
 solutions: A. 1 g D-glucose + 7.5 mg CaCl$_2 \cdot$ 2H$_2$O + 199.2 mg
 MgCl$_2 \cdot$ 6H$_2$O + 403 mg KCl and 17.56 g Tris dissolved in approxi-
 mately 900 ml distilled water. Adjust the pH at this point to
 7.4, using a pH meter, and make up the solution to 1000 ml.
 B. Dissolve 8.2 g NaCl in 1000 ml water. For the balanced
 salt solution, mix 1 part of A with 9 parts of B
1% silicon (Dow Corning Z4141)
Ficoll-Paque solution (available from Pharmacia Fine Chemicals
 Piscataway, NJ 08854) or Lymphoprep solution (available from
 Nyegaard and Co. AS, Oslo, Norway)

        10 ml centrifuge tubes
        Pasteur pipettes
        5 ml sterile syringe with needle
        Table top centrifuge, adjustable to 100 and 400 g

Procedure

1. All glassware in this experiment should be siliconized. Immerse every glass item into 1% silicon solution for a minute. Remove the glassware and let them drip upside down for 30 min. Dry the pieces in an oven at around 100°-110°C. Siliconization is not needed if plastic labware is used.

2. Take blood by collecting 8.5 ml into 1.5 ml ACD solution. Mix thoroughly. Dilute this blood 1:1 with BSS. Mix again.

3. Do not open the rubber septum of the Ficoll-Paque flask, but take a 5 ml syringe and withdraw 3 ml solution from the bottle under sterile conditions. Put it into a 10 ml siliconized centrifuge tube.

4. Pipette carefully 4 ml diluted blood on the top of the Ficoll-Paque solution. Do not mix. Centrifuge the tube at 400 g for 30 min at room temperature.

5. The supernate contains plasma. Remove as much supernatant as you can, using a Pasteur pipette, without disturbing the milky middle layer, which contains the lymphocytes and platelets.

6. Take a clean Pasteur pipette and lift the middle layer. Transfer it to a 10 ml centrifuge tube. Add 8-9 ml BSS, mix gently and centrifuge at 100 g for 10 min at room temperature. Discard the supernatant which contains platelets. Resuspend the sedimented cells in 2.5 ml BSS. The cells are ready to be counted (Exercise No. 82) and used either in the lymphoblast assay (Exercise No. 89) or in the cytotoxicity test (Exercise No.90), or in any other suitable system, using the proper dilution.

Evaluation

The purity of the preparation should be checked by making a smear of the suspension on a microslide and staining it with Wright stain or a similar method using the routine hematologic procedures for white blood cells. The percent of lymphocytes in the total cell population should be around 90% or higher. The purity of the lymphocyte preparation depends upon the care of separation of the lymphocyte layer from the others. If the lymphocyte layer was not completely washed, thrombocytes will be present in the lymphocyte preparation. If the lower layer has also been admixed to the milky lymphocyte-containing layer, granulocytes will be found in a higher percentage.

The viability of the cells can be tested by using the Trypan blue method, as described in Exercise No. 82. The viability of the lymphocytes, if freshly drawn blood cells were separated should be close to 100%.

Use and Limitations

Cadaver blood or blood conserves (with ACD solution) can also be fractionated. The viability of the cells as well as the recovery of lymphocytes in these cases will be significantly lower. Heparinized or defribrinated blood can be used instead of ACD-mixed blood samples. The procedure for the separation of lymphocytes from such blood samples is the same as described above. Spleen or bone marrow cell suspensions can also be subjected to the same procedure to obtain a lymphocyte-enriched cell suspension.

Both smaller or larger blood samples can be used. A microprocedure has been elaborated by Fotino et al. (1971). If larger blood samples are to be separated, use centrifuge tubes with greater diameter, so that the height of the blood as well as of the Ficoll-Paque layers will be very similar to those described above. If this cannot be done, the proper centrifugation time must be established for the new containers.

Sodium metrizoate is a high density compound with high iodine content. This is the main component of several commercial preparations sold under trade names such as Isopaque, Triosil, and others. It is available from Nyegaard & Co., AS, Oslo, Norway or from their distributors (in the U.S.: Accurate Chemical and Scientific Corp., Hicksville, NY 11801). Mixing this preparation in various ratios with Ficoll can give solutions with the desired density. Such solutions may be needed for the isolation of different white blood cell populations. Both sodium metrizoate and the Ficoll-Paque or the Lymphoprep preparations should be stored in the dark at +5°C.

References

Böyum, A., Nature London *204*, 793 (1964)
Böyum, A., Scand. J. Clin. Lab. Invest. *21*, Suppl. 97. p. 9 (1968)
Fotino, M., Merson, E.J., Allen, F.H.: Ann. Clin. Lab. Sci. *1*, 131 (1971)
Ting, A., Morris, P.J.: Vox Sang. *20*, 561 (1971)

Exercise No. 86

Isolation of Adhering Leukocytes

Macrophages will adhere to glass or to some synthetic polymer surfaces. The adhesion is even increased if the macrophages are activated. The simplest method to remove adhering cells from leukocytes or from spleen cell suspension is to incubate a thinly spread cell suspension in a tissue culture bottle or Petri dish and rinse the nonadhering cells off gently. *Part B* of this Exercise describes such a procedure. The recovery of adhering cells requires some careful treatment which will dislodge them from the surface without affecting their viability. A more precise and elaborate method for the isolation of adhering cells as well as 99% pure lymphocytes uses column chromatography, where glass beads are used to remove adhering cells. *Part A* of this Exercise describes such a procedure, based on the method of Garvin (1961) as modified by Rabinowitz (1964).

<u>Part A</u>

<u>Materials and Equipment</u>

   50 ml heparinized human blood or 25 ml mouse spleen cell
    suspension adjusted to approximately 10^7 cells/ml
  Medium 199 for tissue culture. (Available from Microbiological
    Associates, Bethesda, MD 20016 or from several other suppliers)
  1% silicon in water (use Dow Corning Silicon Z 4141)
  EDTA buffer with fetal calf serum. Dissolve 0.2 g disodium
    ethylenediamine tetraacetic acid (EDTA) + 8.0 g NaCl + 0.2 g
    KCl + 1.15 g Na_2HPO_4. $2H_2O$ + 0.2 g KH_2PO_4 and 0.2 g glucose
    in 800 ml glass distilled water. Use only reagent grade
    components. Adjust the pH, if necessary, to 7.4. Add 20 ml
    fetal calf serum and make up the buffer to 1000 ml with
    distilled water. Sterilize by filtration
  1 N HCl
  50 g KOH dissolved in 500 ml methanol (KOH-methanol). Caution:
    This solution is very caustic. Wear protective goggles while
    handling the solution. Wash your skin with 10% acetic acid
    in water if the solution is spilled on it
  Glass beads. Use the product of Minnesota Mining & Manufactur-
    ing Co. (3M Center, St. Paul, MN 55101) catalog No. 100-15
    "Superbright" beads
  Chromatographic column, 20 × 300 mm, with side arm (Fig. 2)

<u>Procedure</u>

   1. All glassware should be siliconized. Simply rinse these
pieces with 1% silicon, let them drip for 15-20 min, and dry them
in an oven at 110°C.
   2. Take 50 ml heparinized blood, let the red blood cells
sediment for 60 min at room temperature. The supernatant will
contain the white blood cells.
   If adhering cells are to be isolated from spleens, spleen
cell suspension should be made as described in Exercise No. 66.
Disperse the cells in Medium 199 and adjust the cell concentration to approximately 10^7/ml.
   3. Wash the glass beads first by pouring the beads under constant stirring with a blunt glass rod into 1 liter 1 N HCl. Let
them stand at room temperature for about an hour. Stir occasionally
with the glass rod. Discard the supernatant. Fill the container
with approximately 1000 ml distilled water, stir again, and discard
the supernatant in a few minutes. Repeat the washing three more
times. Add 500 ml KOH-methanol to the beads, let it stand overnight.
Decant the supernatant on the next day and wash the glass beads
at least five times with glass distilled water until the pH of
the water is close to neutral. The last wash should be in 1%
silicon. Filter the beads on double layer filter paper in a
Büchner funnel, and dry them in an oven at 110°C.
   4. Take the chromatographic column, and clean it with chrome
sulfuric acid. Wash it very thoroughly with glass distilled water
and siliconize it. Fill it up half way with water and close the
effluent stopcock. Build up approximately 1 cm high hard-packed
siliconized glass wool plug in the bottom of the chromatographic
column. Use a long, heavy glass rod to tamp down the plug. Take

a spoonful of the siliconized glass beads and, with the aid of
a funnel, transfer them to the column. They will quickly settle
to the bottom. Repeat the process until the column is filled
with glass beads up to approximately 75%-85% of its total capacity. Wash the column with approximately 200 ml glass distilled
water. Autoclave it in upright position, wrapped in aluminum
foil.

5. Transfer the column into a 37°C incubator room. If such
room is not available, use a jacketed column and circulate 37°C
water in the jacket. Adjust the flow rate of the column to approximately 1-2 ml/min. Replace the water by adding 100 ml sterile
Medium 199 to the column.

6. Pour the blood plasma (diluted 1:1 Medium 199) or the
spleen cell suspension into a siliconized, sterile 200 ml separatory funnel. This should be positioned above the chromatographic
column so that its contents will drip directly onto the surface
of the glass beads in the column. Adjust the flow rate of this
separatory funnel to approximately 30-60 drops per minute. Collect the cells leaving the column in a siliconized glass container
use a 500 ml sterile and siliconized Erlenmeyer flask. After all
plasma or cell suspension has been applied to the column, pour
100 ml Medium 199 into the separatory funnel and let it go through
the column. Pool all the effluents.

7. To elute the adhering cells from the glass beads, use the
EDTA buffer. Replace the 500 ml Erlenmeyer flask in which the
nonadhering cells and the washing fluids were collected with
another sterile and siliconized 500 ml Erlenmeyer flask. Now
pour 200 ml sterile filtered EDTA buffer into the separatory
funnel and let it drip through the column at the same flow rate
as used above. This eluate will contain the adhering cells.

Part B

Materials and Equipment

    Petri dishes, glass, 10 cm diam
    Medium 199 for tissue culture
    0.25% trypsin in phosphate-buffered saline (see Exercise No. 1)
    5% CO_2 incubator

Procedure

1. Suspend the mixture of lymphocytes and monocytes in Medium
199 and adjust the concentration to 10^6 to 10^7 ml. Pipette 5 ml
of this cell suspension into a Petri dish and transfer to a CO_2
incubator at 37°C. Leave the Petri dishes there overnight to allow
the monocytes to adhere to the glass surfaces.

2. Using a Pasteur pipette, remove the supernatant from the
Petri dishes. Replace it with 5 ml Medium 199, tilt the plate
gently several tymes to loosen the nonadhering cells. Take the
Pasteur pipette and remove the washing fluid with the cells.
Repeat the rinsing twice more, using the same procedure. The
washing of the adhering cell layer should also be carried out
at 37°C. The monocyte purity in these layers varies between 70
and 95%. If higher purity is needed, the cells should be culti-

vated for 24 or 48 h and the tubes washed twice more with Medium
199. The purity of the adhering cell layer will be high, consisting of 95%-98% monocytes.
   3. The adherent cell layer can be removed from the Petri
dishes by incubating them with 0.25% trypsin for 10-15 min at
37°C. Complete removal of the cells can be facilitated using a
rubber policeman.

Use and Limitations

Shortman and associates (1971) further elaborated the procedure
of Rabinowitz. As mentioned in the introduction, synthetic fibers
can also be used to retain adhering cells. Nylon wool was used by
Greenwalt et al. (1962).
   For the elution of adhering cells several procedures are used
in addition to EDTA. 0.25% Trypsin dissolved in saline may be
added to the properly washed adhering cells. A 15 min exposure
to diluted trypsin at 37°C will release the cells rapidly.
   Regarding the purity of the cells obtained by this fractionation procedure, Alter and Bach (1970) reported that passage of
peripheral blood leukocytes on a Nylon wool column will give a
nonadhering cell preparation which consists of 70%-90% lymphocytes.
If this cell filtrate is now passed through a glass bead column
as described above, a large percentage of the cells will adhere
to the glass beads and the first 30% which can be collected in the
column effluent will be uniformly small cells which may be up
to 99.9% pure small lymphocytes. The adhering cells may contain
up to 90% blood monocytes, as determined by morphological criteria
of Wright's stained smears.

References

Alter, B.J., Bach, F.H.: Cell Immunol. *1*, 207 (1970)
Garvin, J.E.: J. Exp. Med. *114*, 51 (1961)
Greenwalt, T.J., Gajewski, M., McKenna, J.L.: Transfusion *2*, 221 (1962)
Rabinowitz, Y.: Blood *23*, 811 (1964)
Shortman, K., Williams, N., Jackson, H., Russel, P., Byrth, P., Diener, E.:
   J. Cell Biol. *48*, 566 (1971)

Exercise No. 87

Rosette-Forming Cell Assay

Cells actively synthesizing and releasing antibodies can be
detected by the plaque assay, as described in Exercise No. 66.
Other lymphocytic cells will have antigen-recognizing immunoglobulin groups on the surface, held by their Fc portion in the
membrane. Such cells will react with specific antigens which they
can recognize. If this antigen is a surface immunogen of the sheep
red blood cell (SRBC), the recognition-capable immune lymphocyte
may hold several erthrocytes attached, thus forming a rosette.
If mice are immunized with SRBC, in ten days their spleens will
contain rosette-forming cells (RFC). Immune lymphocytes can also

form such rosettes if the SRBC has been reacted first with anti-SRBC antibodies. In that case, the lymphocyte will receive and hold the FC portion of the already reacted anti-SRBC immunoglobulins, and form a rosette. According to the overwhelming majority of experimental data, the cell which is capable of forming such rosette is an immune B lymphocyte (Greaves and Raff, 1971).

The enumeration of RFC is considered to be a useful parameter to estimate cell-mediated immune response levels, although the exact immunologic role of cells capable of rosette formation is not known. The procedure described here is based on the method published by Zaalberg (1964) as modified by Behling et al. (1976).

Materials and Equipment

Mice (inbred or outbred mice can be used from healthy and in-
fectious disease-free colonies)
Sheep blood in Alsever's solution. To prepare Alsever's solution dissolve 20.5 g dextrose, 8.0 g Na citrate· $2H_2O$, 55 g
citric acid · H_2O and 4.2 g NaCl in 1000 ml glass distilled
water. Mix two parts of blood with one part of Alsever's
solution
Phosphate-buffered saline (PBS) (see Exercise No. 1)
Hank's balanced salt solution (HBSS), commercially available
from Gibco, Grand Island, NY 14072
Glass tissue homogenizer
Gauze, 10 × 10 cm squares
1 M $NaHCO_3$. Dissolve 8.4 g in 100 ml distilled water
Hemocytometer (Neubauer type, available from Fisher Scientific
Co., King of Prussia, PA 19406)
Vortex

Procedure

1. Take 5 ml sheep blood, centrifuge it, discard the plasma and wash the red blood cell sediment three times with PBS. Resuspend the cells in 10 ml PBS and Vortex briefly. From this primary SRBC suspension, take a 1 ml aliquot and dilute 1:500 using PBS. By means of a hemocytometer determine the cell concentration of the diluted SRBC suspension. Now calculate the cell concentration of your original SRBC suspension and adjust to 2.5×10^9 cells/ml.

2. Inject six mice IP with 5×10^8 SRBC suspended in 0.2 ml saline.

3. Ten days later, kill the mice by cervical dislocation and remove their spleens. Homogenize the spleens one by one in 3 ml HBSS using a glass tissue grinder and applying only five to six strokes. Pool all the cell suspensions and filter them through two layers of gauze. Adjust the pH of the suspension to 7.2 using 1 M $NaHCO_3$.

Determine the spleen cell number by the procedure given in Exercise No. 82. Adjust the white blood cell count to 6×10^7 per ml.

4. Take triplicate samples of 100 µl spleen cell suspension aliquots and mix them with 100 µl SRBC suspension. This latter

should contain approximately 3×10^8 cells per ml. Add O.8 ml
HBSS to the cells and Vortex for a full minute before incubation
at 4°C for 3 h.

Evaluation

To determine the number of RFC in the above incubation mixture,
a hemocytometer is used. First rotate the tube containing 1 ml
cell mixture gently to lossen up the sedimented cells. Very
gently swirl the tube by hand to insure a homogeneous cell suspension with minimum stress before analysis in a hemocytometer. Use
a 430-fold magnification and count the lymphocytes which have
more than 5 SRBC's attached. Lymphocytes with less than 5 SRBC's
are not considered to be rosettes. For statistical accuracy count
four large squares of the hemocytometer three separate times.
Each large square corresponds to a total volume of O.OOO1 ml.
From the average number of rosettes per large square, the spleen
cell concentration, and the total volume of the spleen cell
suspension, the number of rosettes can be quantitatively expressed as (1) rosette forming cells (RFC) per 10^6 spleen cells and/or
(2) RFC's/spleen.

Use and Limitations

The existence of T-cell rosettes is still an open question, while
investigators agree that mostly B cells are involved in rosette
formation. Electronmicroscopic pictures show that the rosette-
forming plasma cells have extensive endoplasmic reticulum, indicating that they are actively secreting antibodies. The majority
of the cells in the center of a rosette, however, have only very
little of such organels, indicating lack of antibody secretion,
and these cells also manifest a different ultrastructure morphology. According to McConnel et al. (1969) a few of the early
plaque-forming cells are also capable of making rosettes.
   As has been mentioned in the Introduction, sensitized erythrocytes will form rosettes with either B cells or macrophages. It
is a well-established fact that B cell and macrophage membranes
can accept Fc portions of immunoglobulins or immunoglobulin an-
tigen-complement complexes. These rosettes can be separated by
low speed centrification and the supernatant will be enriched in
T cells.
   One may use antigen coated erythrocytes in the RFC assay. $CrCl_3$
has been used to couple antigens to SRBC by Ling et al. (1977).
Carbodiimide coupling, as described in Exercise No. 69 can be
also used for such purposes.
   The above described reaction should not be confused with
spontaneous rosette formation between some human peripheral
lymphocytes and SRBC which was described by Bach et al. (1969).
According to Wybran et al. (1972) primary T cells participate
in these so-called E rosettes.

References

Bach, J.F., Dormont, J., Dardenne, M., Balner, H.: Transplantation *8,* 265
   (1969)

Behling, U.H., Campbell, B., Chang, C., Rampf, C., Nowotny, A.: J. Immunol.
   117, 847 (1976)
Greaves, M.F., Raff, M.C.: Nature (New Biol.) *233*, 239 (1971)
Ling, N.R., Bishop, S., Jefferis, R.: J. Immunol. Meth. *15*, 279, 1977
McConnell, I., Munro, A., Gurner, B.W., Coombs, R.R.: Int. Arch. Allergy
   35, 209 (1969)
Wybran, J., Carr, M.C., Fudenberg, H.H.: J. Clin. Inv. *51*, 2537 (1972)
Zaalberg, I.B.: Nature, London *202*, 1213 (1964)

Exercise No. 88

Passive Cutaneous Anaphylaxis (PCA)

It has been shown that during antigen-antibody interactions,
vasoactive substances are liberated, the most important among
them being histamine. Histamine increases capillary permeability.
This increase is a transient reaction of relatively short duration, therefore its observation is difficult, and even semi-quantitative evaluation is impossible. A much more permanent result of
the permeability increase can be achieved if large molecular
weight dyes are injected into the circulation together with the
antigen. Due to the increased permeability of the capillary walls
these dyes may enter the tissues and remain for a relatively long
time in those areas where histamine has been liberated through
the antigen-antibody interaction.

The test is very sensitive, being able to detect less than
0.01 µg antibody nitrogen per injected sample in the proper
animal. It has been shown that not all types of immunoglobulins
are able to elicit the PCA reaction. The method described here
follows the procedure developed and used by Ovary (1952, 1964).

Materials and Equipment

    Guinea pigs weighing approximately 250 grams
    Rabbit antiserum, obtained from Exercise No. 65, or immuno-
     globulin fractions obtained from Exercise No. 2 or No. 3
    0.2 mg/ml homologous antigen solution in saline. (The antigen
     which was applied for the production of rabbit antibodies is
     to be used in this assay)
    2% Evans blue dye in saline
    Diethyl ether (or sodium pentothal)
    Physiologic saline
    0.5 ml syringe
    26 gauge needles, short bevel, for intradermal injection
    Electric hair clipper (Oster 40)

Procedure

1. Prepare solutions of antiserum and isolated immunoglobulin
preparations in saline. The nonfractionated antiserum must be
diluted 1:100, giving an antibody nitrogen concentration between
1 and 10 µg/ml. The nitrogen content of the isolated immunoglobul
preparations or their breakdown products have to be measured ac-

cording to Exercise No. 48. Adjust the protein nitrogen content
of these samples to 1 µg/ml in saline.

2. Take three guinea pigs. It is not necessary to anesthetize
the animals. According to Ovary, deep anesthesia may result in
negative reactions. One student should hold the animal while
another student carries out the experiment. Clip the fur from the
animal's backs with an electric clipper as close as possible
without exposing the animals to strenuous long handling. Four
sites for injection may be selected on the back of each animal.
These must be approximately 1 ½ cm from the middle line and approximately 3 cm from each other. Skin areas above the shoulders
or below the sacrum should not be used.

3. Inject intradermally 0.1 ml of the antibody-containing
solutions diluted as described above. This volume will contain
0.1 µg nitrogen from the isolated antibody fractions and approximately 0.1-1.0 µg antibody nitrogen from the whole immune serum.
Be sure that the injection is given intradermally. Proper application of 0.1 ml volume will result in a clearly visible bleb
approximately 6-8 mm in diameter.

4. Wait 4 h before the injection of the antigen. In the meantime, prepare the antigen solution, which should contain 0.2 mg/ml
dissolved or suspended in saline. Mix with an equal volume of 2%
Evans blue dye and inject 1.0 ml of this preparation intravenously
in one of the hind foot veins.

Intradermal, intravenous, or intracardial injections require
practice. Inexperienced students will fail to do these injections
properly, which will result in a negative reaction. It is necessary to practice the injections on animals which will not be
used later for bioassays. Several handbooks describe most pre-
cisely the necessary equipment and also the handling for such
exercises.

5. Observe the skin of the animals after 30 min. At the site
of the injection of antibodies a blue coloration will be seen
which reaches its maximum intensity generally in a few minutes.
Weak reactions can be seen better if the animals are killed,
skinned, and the subcutaneous surface of the skin investigated.

Evaluation

Semi-quantitative evaluation of this reaction is possible only
if the proper dose ranges for both antibody and antigen have
been established in preliminary experiments. The range in which
relatively linear increase in the intensity of the skin reaction
can be observed with increasing antibody doses injected must be
found. Identical doses of different preparations within this
linear response range may be compared. For orienting numerical
values, some authors use the size of the colored skin area expressed in square millimeters, as described in Exercise No. 91.

Use and Limitations

The application of the PCA test to mice is described by Ovary
(1958). Detailed methodological instructions as well as application possibilities and limitations are reviewed by Ovary (1964).

References

Ovary, Z.: Int. Arch. Allergy *3*, 164 (1952)
Ovary, Z.: Immunological methods. Ackroyd (ed.) p. 259 Philadelphia: F.A.
   Davis Co. 1964
Ovary, Z.: Progr. Allergy *5*, 460 (1958)
Ovary, Z.: J. Immunol. *81*, 355 (1958)

Exercise No. 89

Lymphoblast Assay

The existence of blast cells among circulating lymphocytes was
first described in 1902 by Maximow, but thorough studies of
lymphoblast formation did not start until Nowell's publication
in 1960, reporting the initiation of mitosis of human lymphocytes
in vitro by phytohemagglutinin (PHA). Soon thereafter Pearmain
et al. (1963) reported that lymphocytes from tuberculin-sensitive
individuals showed blast formation if they were exposed in vitro
to tuberculin preparations. Similar reports using other agents
were published by Hirschhorn et al. (1963) and by Schrek (1963).
   Dutton and Bulman showed first in 1964 that in vitro exposure
of sensitized lymphocytes to their in vivo sensitizing agents
enhances DNA synthesis in the cells. This was elaborated for a
quantitive determination of lymphoblast formation by Adler and
associates (1970). Exposure of viable human leukocytes to killed
histo-incompatible leukocytes will also enhance the rate of DNA
synthesis in the viable cells, as measured by the incorporation
rate of tritiated thymidine. This is also called mixed lymphocyte
culture or MLC assay and it is used extensively in transplantation
immunology. Lymphoblast assay, mitogenic response assay, or
lymphocyte transformation assay are synonyms describing the same
procedure.
   This Exercise uses bacterial lipopolysaccharide and phytohemagglutinin to demonstrate lymphoblast formation of mouse spleen
cells. The procedure as described here is a modification of the
method of Adler et al. (1970).

Materials and Equipment

   Swiss albino or ICR mice
   Heat-inactivated fetal calf serum (Flow Labs., Rockville,
    MD 20852)
   Medium RPMI-1640. Dissolve 10.43 g powdered Medium RPMI-1640
    and 2 g $NaHCO_3$ in 900 ml distilled water. Add 10 ml penicillin
    and streptomycin solution containing 10,000 U penicillin and
    10,000 µg streptomycin (commercially available from Difco
    Laboratories, Detroit, MI 48232. Adjust the pH to 7.0
    and bring the volume to 1000 ml with glass distilled water.
    Sterilize the Medium through Millipore filtration. The final
    pH should be between 7.2 and 7.3
   Phytohemagglutinin solution containing 100 µg/ml in pyrogen-free
    saline. It is highly recommended to prepare an approximately

500 ml solution of 100 µg/ml phytohemagglutinin. This should
  be subdivided into 1 ml aliquots in small tubes and kept
  frozen. The activity of such standard solution will not
  change unless it is thawed and frozen repeatedly. Take one
  tube for every experiment and include it as a standard positive control
Lipopolysaccharide (as obtained in Exercise No. 21) dissolved
  in saline at a concentration of 100 µg/ml
10% trichloracetic acid in water
Methanol
Protosol (available from New England Nuclear, Boston, MA 02118)
Liquid scintillation fluid. Dissolve 2.5 g PPO (2,5-diphenyl-
  oxitol) and 50 mg POPOP (2,2'-paraphenylene-bis-5-phenyloxitol)
  in 500 ml toluene. Keep the solution in the cold room
^{3}H Thymidine (New England Nuclear, Boston, MA 08118)
Trypan blue, commercially available, or dissolve 0.5% in
  distilled water
Turk's stain, commercially available, or dissolve 0.01% gentian
  violet in 3% acetic acid
Pyrogen-free physiologic saline, commercially available
Tissue culture tubes with caps, disposable, size 12 × 75 mm
Dissecting tools
CO_2 incubator (Hotpack International, Philadelphia, PA 19135)
5 cm diam funnel with two layers of gauze
Glass fiber scintillation pads (A.H. Thomas, Philadelphia,
  PA 19105)
Stainless steel filter unit (Hoefer Sci. Instr., San Francisco,
  CA 94107) model FH 224
Vortex mixer
Note: All glass and metal tools and containers must be rendered
  pyrogen-free by heating them to 180°C for 3 h. If plastic
  is used, purchase pyrogen free products

Procedure

1. Prepare a spleen cell suspension as described in Exercise
No. 66 from four mouse spleens. Filter and centrifuge the cells,
discard the supernate. Resuspend the cell sediment in 10 vol of
RPMI containing 10% fetal calf serum (mix nine parts of Medium
RPMI-1640 with one part of heat-inactivated fetal calf serum).
Count the number of viable cells as follows: First take 0.1 ml
cell suspension and add 0.9 ml Turk's stain. Wait 10 min at room
temperature, shake the suspension a few times. Determine the
total cell number. Take another 0.1 ml cell suspension and determine the viable cell count using trypan blue, as described in
Exercise No. 82. Adjust the cell concentration to 2×10^6 viable
cells/ml.
2. Prepare triplicate tissue culture tubes for each sample
containing 0.1 ml sample (LPS or saline or PHA) and 1 ml spleen
cell suspension. Incubate at 37°C for 68 h in a 5% CO_2 incubator.
3. Each tube should now receive 1 µCi/100 µl ^{3}H-thymidine.
Incubate the cell suspension for 4 h.
4. Take the tubes out of the incubator and place them in a
melting-ice bath. First add 4 ml cold 0.15 M NaCl to each tube.
Centrifuge the tubes in the cold for 10-12 min, at 1000 g speed.
Carefully remove the tubes and aspirate off the supernatant. Do
not disturb the pellet. Keep the tubes in a melting-ice bath.

Add 1 ml 10% TCA to each tube and Vortex carefully for approximately 15 s. Add 4 ml more of 10% TCA. Again Vortex carefully
without spilling radioactive fluids. Keep the tubes in ice for
1 h.

5. Take the tubes one by one and filter them through glass
fiber paper (called "scintillation pads"). Use the Hoefer filter
unti model FH 224. Place the glass fiber pad into the filter
with its rough side up, apply the vacuum, and filter the tube
contents through. Wash the tube with 5 ml TCA and filter again.
Wash the glass fiber with 5 ml cold methanol, again use vacuum.
Remove the pad with tweezers and place it in a scintillation vial
Add 1 ml 100% Protosol to the vial and make sure that it covers
the filter pad.

Clean the filter unit with Kimwipe and dispose of the paper
in a radioactive waste container. Take the second tube and filter
this as described above. Proceed with all the tubes, one by one.
Store the vials at 56°C overnight.

6. Transfer the vials to the cold room for approximately 30 min
Now add 5 ml scintillation fluid to each vial. Tightly close the
screw caps and Vortex the scintillation vials vigorously. Keep
the vials in the dark in the scintillation counter for an additional hour before counting.

Evaluation

Determine the counts/minute (cpm) values of the samples in a
liquid scintillation counter. Calculate the stimulation index
for each sample. The stimulation index equals cpm in sample/Cpm
in saline control.

Use and Limitations

While products of tubercle bacteria are capable of inducing
sensitization of lymphocytes in man as well as in several experimental animals, other bacteria and their products do this
to a lesser extent and to a greatly varying degree. Furthermore,
the degree of immune response not always parallels the degree
of lymphoblast formation. On the other hand, there are natural
products as well as synthetic compounds which can induce high-
level lymphoblast formation with no or barely detectable specific
immune response to them. Examples of these are Concanavalin A,
phytohemagglutinin, or synthetic glycolipids (Behling et al.,
1976). The correlation between mitogenic and adjuvant effect is
similarly absent using some preparations. Phytohemagglutinin,
a potent mitogen, given before or together with immunogen, is
immunosuppressive or is without effect. Other compounds are adjuvants without inducing lymphoblast formation (Behling and Nowotny
1977, Frank et al., 1977). In spite of these, the lymphoblast
assay is widely used and is considered by many investigators as
a useful parameter to measure the responsiveness of man and animals to some immunogens.

The intracellular events in stimulated lymphocytes are sub-
jects of intensive studies in several laboratories. A review on
this topic has been published recently by Cogoli and Pearson
(1976).

References

Adler, W.H., Takiguchi, T., Marsh, B., Smith, R.T.: J. Exp. Med. *131,* 1049
  (1970)
Behling, U.H., Campbell, B., Chang, C.M., Rumpf, C., Nowotny, A.: J. Immunol.
  117, 847 (1976)
Behling, U.H., Nowotny, A.: J. Immunol. *118,* 1905 (1977)
Cogoli, A., Pearson, T.W.: Exp. Cell. Biol. *44,* 235 (1976)
Dutton, R.W., Bulman, H.N.: Immunology *7,* 54 (1964)
Frank, S., Specter, S., Nowotny, A., Friedman, H.: J. Immunol. *119,* 855
  (1977)
Hirschhorn, K., Bach, F., Kolodny, R.L., Firschein, I.L., Hashem, N.: Science
  142, 1185 (1963)
Maximow, A.A.: Beitr. Pathol. Anat. Allg. Pathol. *32,* Suppl. 5. (1902)
Nowell, P.C.: Cancer Res. *20,* 462 (1960)
Pearmain, G., Lycette, R.R., Fitzgerald, P.H.: Lancet *i,* 637 (1963)
Schrek, R.: Am. Rev. Respir. Dis. *87,* 734 (1963)

Exercise No. 90

T Cell-Mediated Cytotoxicity of Influenza Virus-Infected Cells

As discussed by Cerottini and Brunner (1974), and Doherty et al.
(1976), cytotoxic T lymphocytes (CTL) constitute part of the
immunologic response to tissue grafts, tumor cells, and virus
infections. It is thought that CTL interact directly with the
surface of the foreign or virus-modified cells in a process
which results in lysis of the target cells. This phenomenon can
be demonstrated in vitro using a modification of the ^{51}Cr release
assay described by Lightbody (1971 and 1976). In this Exercise
cultured fibroblast cells will be infected with influenza virus,
and exposed to spleen cells of mice immunized to influenza virus.
The modification of the cytotoxicity assay as described here has
been elaborated and made available for this Manual by Ann Marschak.

Materials and Equipment

Tissue culture of human fibroblast cells has to be provided
  by the instructors
Na$_2$^{51}CrO$_4$ (5 mCi/ml) available from New England Nuclear,
  Boston, MA 02118, and from several other vendors of radio-
  active chemicals
Hank's balanced salt solution (HBSS). Available from Gibco,
  Grand Island, NY 14072
0.25% trypsin solution
RPMI-1640-culture medium (Available from Gibco)
Influenza virus (Hong-Kong virus, WHO Bulletin Vol. 41, 643)
Mice previously immunized with influenza virus (Inject IP 1:10
  HBSS diluted chick embryo allantoic fluid, which has been
  infected with the virus)
Dissecting instruments
Glass Tissue Homogenizer
Cetrimide detergent, 1% in water
"Lysing solution", 1% NH$_4$Cl in water

Flat-bottom microtiter plates (Falcon product, available from
  Becton, Dickinson and Co., Cockeysville, MD 21030)
Plastic gloves, disposable
Hemocytometer
Trypan blue, stock solution, 1% in water
Microscope
Centrifuge

Procedure

1. The ^{51}Cr labeling of the cultured fibroblast is carried
out by the following simple procedure: $Na_2{}^{51}CrO_4$ (100 µCi) dissolved in 1 ml RPMI-1640 is added to each milliliter of tissue
culture. Incubation continues overnight.

2. Fibroblast target cells must be trypsinized to obtain a
single cell suspension. Aspirate the supernatant from the monolayers into a flask marked radioactive. Rinse the monolayers
with HBSS by adding 5 ml to the flask, tilting the flask and
removing the supernate. Then rinse with 5 ml of trypsin, remove
4 ml and let the flask sit at room temperature for several minutes
until the cells fall off the plastic. Add 10 ml RPMI to the
flasks, gently pipette the cells to break up clumps and remove
to a 50 ml test tube. Spin down the cells at 500 g for 10 min,
discard the supernatant and resuspend the sediment in 5 ml of
RPMI.

3. Take 2.5 ml of the target cell suspension and add 0.25 ml
of virus solution. Add 0.25 ml medium to the other half of the
target cell suspension. Incubate both for 1 h shaking the tubes
every 15 min. Centrifuge again as above, and resuspend in 10 ml
RPMI.

4. Count the number of viable cells using trypan blue and a
hemocytometer as described in Exercise No. 82. Adjust to a final
concentration of 1.5×10^5 cells/ml.

5. Preparation of effector cells is the next step. Remove the
spleen from immune mice, drop it into tissue homogenizers containing 5 ml of HBSS. Make a single cell suspension by about
four to six strokes, filter the homogenate on two layers of gauze.
Spin down the cells in 50 ml centrifuge tubes and resuspend in
HBSS. Spin them down again and this time resuspend them in 5 ml
of cold "lysing solution" to remove red cells. Leave test tubes
on ice for 10 min and then quickly fill tubes with HBSS and centrifuge. Wash the cells one more time, resuspend them in 5 ml of
RPMI and count. Adjust to 1.5×10^7 cells/ml and make three twofold dilutions.

Do the same with the spleens of nonimmunized mice.

6. The microtiter plate should be filled as follows: The first
wells in eight rows will receive 100 µl cetrimide detergent.
(This will lyse the cells completely and give values for maximum
levels of ^{51}Cr release.) The second wells in all eight rows will
receive 100 µl medium RPMI 1640. (Readings from these wells will
give you the level of spontaneous ^{51}Cr release.) The third, fourth
fifth, and sixth wells will receive 100 µl of the four dilutions
of the normal spleen cell suspension. Start with the highest
dilution (1:4) in the third wells of the eight rows. Undiluted
spleen suspension (1.5×10^7 cells/ml) is added to the sixth wells.
Seventh, eighth, ninth, and tenth wells of all eight rows will

receive the four dilutions of the immune spleen cell suspension.
Start again with the most diluted suspension in the seventh
wells.

Add 100 µl of ^{51}Cr labeled but influenza virus-uninfected
target cells to the first four rows (4 × 10 wells). Add 10 µl of
^{51}Cr labeled and virus-infected target cells to the second four
rows of the plate. The target to effector cell ratio is 1 to
100 in this set up.

For easier understanding of the plating procedure observe the
pattern in Table 14.

Table 14. Plating pattern for the cytotoxicity assay

	Detergent	Medium	1:4 dil.	1:2 dil.	1:1 dil.	undiluted	1:4 dil.	1:2 dil.	1:1 dil.	undiluted	
			normal spleen cell dilutions				immune spleen cell dilutions				
	1	2	3	4	5	6	7	8	9	10	
A	o	o	o	o	o	o	o	o	o	o	
B	o	o	o	o	o	o	o	o	o	o	^{51}Cr labeled normal target cells
C	o	o	o	o	o	o	o	o	o	o	
D	o	o	o	o	o	o	o	o	o	o	
E	o	o	o	o	o	o	o	o	o	o	
F	o	o	o	o	o	o	o	o	o	o	^{51}CR labeled virus infected target cells
G	o	o	o	o	o	o	o	o	o	o	
H	o	o	o	o	o	o	o	o	o	o	

7. Cover the plates, and incubate the effector and target
cell mixtures (100 to 1 ratio) in a moist 5% CO_2 incubator overnight.

8. Take the plates out of the incubator without shaking them.
Take 80 tubes, for isotope counting, number them consecutively.
Very carefully, without touching the sedimented cells, pipette
100 µl from each well into a tube. Follow the same sequence as
in Step 6: First take the content of the first wells in the eight
rows.

9. Determine the cpm in each sample of supernatants.

Evaluation

Calculate the percent specific lysis for each dilution according
to the formula:

$$\frac{x - s}{m - s} \times 100$$

Where x is the cpm of the experimental ^{51}Cr release, m is the
cpm of the maximal release (due to cetrimide lysis) and s is the
spontaneous release.

Use and Limitations

Cerottini and Brunner (1974) discussed the advantages and limitations of the ^{51}Cr release assay as compared to other cytotoxicity
measurement procedures. Unquestionably, till today this is the
simplest, the most reliable and the most quantitative of all
methods. It is applicable to studies of cell-mediated immune
response against allogeneic normal cells as well as against
tumors.
   One should also point out that difficulties are often encoun-
tered. One of these is the sometimes too high spontaneous release
of the ^{51}Cr label, particularly with freshly explanted tumor
cells. Another difficulty frequently found is the inefficiency
of the effector cells at the here used 100 to 1 target cell ratio
to release significant amount of ^{51}Cr. In such cases 200:1 or
300:1 ratios may be needed. The occurence of these two difficultie
depend almost always upon the nature of the target cells used,
provided that the above described experimental conditions were
properly maintained.
   It may be desirable to increase or decrease the ^{51}Cr label
concentration on the target cell surface. This can be done by
changing the mCi^{51}Cr content of the sodium chromate added to the
culture.

References

Cerottini, J.C., Brunner, K.T.: Adv. Immunol. *18*, 67 (1974)
Doherty, P.C., Blanden, R.V., Zinkernagel, R.M.: Transpl. Rev. *29*, 89 (1976)
Lightbody, J., Bernoco, D., Miggiano, V.C., Ceppellini, R.: J. Bact. Virol.
    Immunol. *64*, 243 (1971)
Lightbody, J.: In: Manual of clinical immunology. Rose, Friedman (eds.)
    p. 851, American Society for Microbiology Publication 1976

3.2.4   Miscellaneous Biologic Reactions

Exercise No. 91

The Local Shwartzman Phenomenon

Sanarelly (1924) observed that two intravenous injections of
bacterial culture, 24 h apart, will result in extensive hemorrhage
in different organs of the animal. Shwartzman (1928) reported that
if an intracutaneous injection of bacterial lipopolysaccharide
into rabbits is followed in 8 to 24 h by an intravenous injection
of the same material or similar bacterial products, the initial
skin site becomes inflamed in 2 to 4 h, and subsequently undergoes
hemorrhagic necrosis. In this Exercise different doses of bacterial lipopolysaccharides will be injected intracutaneously,
and the size of the hemorrhage elicited by an intravenous injection of the same endotoxin given 24 h later will be compared in
a semiquantitative manner.

Materials and Equipment

    White New Zealand rabbits, weighing between 2 and 3 kg each
    0.2 mg/ml endotoxic lipopolysaccharide solution in saline ad-
      justed to pH 7.2
    1 ml syringe with 26 gauge needle, short bevel
    Hair clipper
    Rabbit cages
    Marker, felt pen

Procedure

    1. Shave the abdomen of the animal with the electric hair
clipper and mark it into six sectors. Number the sectors with
the felt pen.
    2. Make five twofold dilutions of the endotoxic lipopolysac-
charide in saline. Undiluted solution will contain 40 µg in 0.2 ml,
the last dilution will contain 2.5 µg in the same volume.
    3. Using a 1 ml syringe and 26 gauge needle, inject intra-
cutaneously 0.2 ml of each dilution of the bacterial endotoxin
in the center of five of the six sectors. Into the sixth sector
inject sterile saline as a control. These are the preparatory
doses. Observe the injection site for evidence of reaction.
Besides slight erythema, there should be none.
    4. After 24 h inject intravenously in an ear vein 20 µg of endo-
toxin dissolved in 0.2 ml saline. This is the provocative dose.

Evaluation

Observe skin test sites a few hours after the provocative injection for evidence of inflammation. Also observe them on the following day, at which time hemorrhagic necrosis may develop.
    If an unknown preparation is tested for Shwartzman reactivity,
it is essential to include at least three different doses from

a standard endotoxin in the same rabbit. It is usually observed
that one out of three rabbits does not react in the same degree
to the control endotoxin standard as do the orther rabbits. Some
rabbits may show an increased sensitivity against the endotoxin.
These cases are easily detectable by comparing the size of the
skin lesions elicited by the control standard endotoxin injections
The results observed in these abnormally reacting rabbits must
be excluded from the evaluation.

For numerical evaluation of the hemorrhagic lesions, measure
the diameter of the lesions in millimeters at various angles to
each other. Calculate the average radii and the areas of the
lesions, express the latter in square centimeters. These give
indicative figures about the relative potency of the preparations.
See Fig. 60.

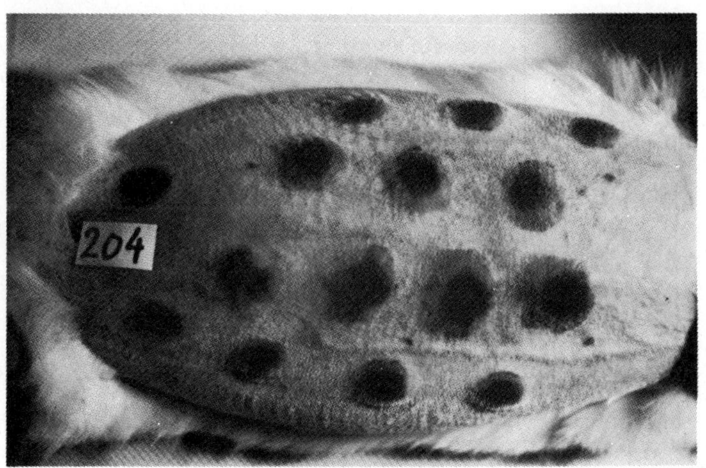

Fig. 60. A typical
Shwartzman skin
test

Use and Limitations

The Shwartzman reaction resembles hypersensitivity reactions
(Stetson, 1951 and 1964), although serologically unrelated substances may also be used in the provocative injection. For more
information on this problem, students should study publications
by Thomas and co-workers (1957 and 1959). While many events involved in the elicitation of hemorrhagic necrosis of the skin have
been observed, the local Shwartzman assay is not fully understood.
Nevertheless, the assay is one of the parameters frequently used
in comparing the effects of different bacterial products. With
proper distribution of the injection sites on the abdomen of a
rabbit, 16 injections may be given (Radvany et al., 1966). The
total amount which is given in 16 different intradermal injections
of one rabbit is several times higher than the lethal dose for
the same rabbit if given intravenously.

References

Radvany, R., Neale, N.L., Nowotny, A.: Ann. N.Y. Acad. Sci. *133,* 763 (1966)
Sanarelli, G.: Ann. Inst. Pasteur *38,* 11 (1924)
Shwartzman, G.: J. Exp. Med. *48,* 247 (1928)
Stetson, C.A., Jr.: J. Exp. Med. *94,* 347 (1951)
Stetson, C.A., Jr.: In: Bacterial endotoxins. Landy and Braun (eds.), p. 658.
   New Brunswick, N.J.: Rutgers University Press 1964
Thomas, L.: In: Cellular and humoral aspects of the hypersentitive states.
   Lawrence, H.S. (ed.), p. 451. New York, N.Y.: Hoeber-Harper 1959
Thomas, L., Zweifach, B.W., Benacerraf, B.: Trans. Ass. Am. Physicians *70,*
   54 (1957)

Exercise No. 92

Measurement of the Activity of the Reticuloendothelial System

Innate immunity is largely influenced by cellular factors in the
animal body. The most important of these is phagocytosis by the
reticuloendothelial system (RES). The cells and tissues of the
RES constitute a vast and widely disseminated apparatus which
effectively filters foreign materials from the circulating blood
as it passes through organs containing RE tissue (Aschoff, 1924).
In this experiment, living bacteria will be used as an indicator
to determine where in the body they are removed from the circulation of the mouse (Bull, 1915) (Kerby et al., 1950).

Certain natural and synthetic substances have the capacity
to enhance or block the RES activity. Bacterial endotoxins stimulate the removal of bacteria from the circulation, and colloidal
carbon (India ink) or ThO_2 (Thorotrast), if given in large doses,
may completely block the RES.

In this Exercise a simple method is given which will demonstrate the functioning of the RES and give a semiquantitative
procedure to measure its activity. One group of mice will receive
10 µg endotoxin from *E. coli* intraperitoneally, followed 24 h later
by an injection of living *Serratia marcescens* cells. The other group
will receive saline before the intravenous injection of viable
bacteria. Blood will be taken at various time intervals and the
number of bacteria will be determined by colony count on agar
culture. For purposes of demonstration, a few organs will be
homogenized and the number of viable cells in the organs will be
compared.

Materials and Equipment

   10 Swiss albino mice, 18-20 g each
   Viable *Serratia marcescens* cells, 1×10^6/ml in saline
   Endotoxic lipopolysaccharide from *E. coli* 0111 (obtained in
     Exercise No. 21). Dissolve 50 µg/ml in saline
   Broth culture medium
   Trypticase soy agar pour plates
   Ether
   Physiologic saline

    Sterile Petri dishes
    Sterile dissecting tools
    Sterilized glass tissue homogenizers
    0.1 ml pipettes, rinsed with heparinized saline and dried

Procedure

1. Inject five mice intraperitoneally with 0.2 ml endotoxin.
Inject another five mice with nonpyrogenic saline.

2. After 24 h inject about 1×10^6 *Serratia marcescens* cells into
each of five mice intravenously.

3. Immediately after injection, cut off approximately 1 cm
from the end of the mouse tails, and pick up 0.01 ml blood with
the heparinized micropipette. Rinse this blood into 10 ml saline.
This is a 10^{-3} dilution. Prepare 1 ml 10^{-4} and 1 ml 10^{-5} dilutions
also.

4. Add 1 ml of each dilution to trypticase soy agar pour plates

5. Repeat 30 min after injection steps 3 and 4, using 10^{-3} and
10^{-4} dilutions.

6. Take another 0.01 ml blood, 60 min after injection, suspend
it in 1.0 ml saline, and prepare a single pour plate with the
resulting 1.0 ml vol (10^{-2} dilution).

Immediately after the blood for step 6 has been withdrawn,
kill your animal with ether anesthesia.

7. Remove the liver and spleen to separate Petri dishes.
Prepare tissue homogenates in 5 ml sterile saline using separate
sterile homogenizers. Prepare 10^{-2}, 10^{-3}, and 10^{-4} further dilutions from the homogenates. Prepare pour plates from the dilutions
Incubate all plates at 37°C for two days.

Evaluation

Some dilutions taken right after the injection of bacteria will
have too many colonies. Count the colonies on those plates where
it is possible. Record the count of the proper dilutions of blood
and compare these numbers with those obtained after different
time intervals. You will observe a relatively fast clearance of
cells from the circulation. The mice which were pretreated with
endotoxin will show a faster clearance.

Use and Limitations

For the measurement of blood clearance, isotope-labeled carbon
particles are better to use (Halpern et al., 1951). In this
Exercise only those cells which are viable and able to multiply
will be detected on the agar plates. Cells which were lysed or
killed by serum factors will not produce colonies, although they
may be still in the circulation. With isotope-labeled particles,
this problem is eliminated (Benacerraf and Miescher, 1960).

The procedure given here provides a simple assay for routine
analyses. In testing the stimulating effect of different natural
products on the RES, one has to bear in mind that the substance
may be contaminated with bacterial products, even after it has bee
autoclaved, which is known to leave the endotoxins still active.
Traces of this substance may be present in India ink or in Thoro-

trast also. Treatment of endotoxins with 0.1 N NaOH at room temperature for 24 h detoxifies them, but still does not destroy
their capacity to stimulate the RES. To eliminate the presence
of endotoxin from natural products is extremely difficult. The
best one can do is to work with a sterile starting material and
carry out all operations under aseptic and pyrogen-free conditions.
Glassware and metal equipment can be rendered pyrogen-free in a
hot sterilizer at 180°C for 2 h; some inorganic materials also
survive this treatment.

References

Aschoff, L.: Ergebn. inn. Med. Kinderheilk. *26*, 1 (1924)
Benacerraf, B., Miescher, P.: Ann. N.Y. Acad. Sci. *88*, 184 (1960)
Bull, C.G.: J. Exp. Med. *22*, 475 (1915)
Halpern, B.-N., Biozzi, G., Mene, G., Benacerraf, B.: Ann. Inst. Pasteur *80*,
    582 (1951)
Kerby, G.P., Holland, B.C., Martin, S.P.: J. Immunol. *64*, 123 (1950)

Exercise No. 93

Enhancement of Nonspecific Resistance to Infection

Bacterial lipopolysaccharides are able to stimulate the defense
mechanism of the host against infection (Rowley, 1956). Several
other natural products given in properly dispersed form and in
the proper doses may also show similar effects. This protection
is nonspecific. For example, if animals receive injections of a
lipopolysaccharide extracted from one species of bacterium, their
resistance will be raised to subsequent infection not only with
the homologous organism, but also with immunologically unrelated
strains.
    In this experiment, lipopolysaccharide from a Gram-negative
bacillus, *Serratia marcescens,* will be injected into mice to stimulate their nonspecific resistance. The animals will be challenged
with virulent *Salmonella typhi* O901 cells 24 h later, and the number
of survivors determined.

Materials and Equipment

    20 White Swiss albino mice, 18-20 g each
    S. typhi O901 viable cell suspension containing 10^9 cells/ml
    Bacterial lipopolysaccharide (endotoxin) from *Serratia marcescens*
      (obtained from Exercise No. 21), concentration 1 µg/ml in
      saline
    Physiological saline
    1 ml syringe with 26 gauge needle
    Mouse cages

Procedure

1. Using a sterile 26 gauge needle, inject ten mice intraperitoneally with 0.5 ml of endotoxin preparation. Inject ten mice in the same manner with saline; these are the controls.
2. Challenge all mice by careful intraperitoneal injection of 1 ml of a suspension of *Salmonella typhi* O901 containing 10^9 cells/ml. Handle the bacterial suspension with proper care.

Evaluation

Observe the results of the challenge injection. Record the total number challenged and the number of survivors, both in the test group and in the controls.

If quantitative figures are needed, use further dilutions of the endotoxin tested for nonspecific resistance enhancing effect and determine the dose which will give 50% protection. Use graphic interpolation, similar to Exercise No. 81, where the ordinate shows the log of the amount of material tested in µg/injection and the abscissa shows the log of S/D (where S is the number of survivors and D is the number of deaths). Connecting the two closest readings to log $S/D = 0$ (50% protection), the intersection on the ordinate will give the dose which elicits 50% protection (see Fig. 61). Naturally, for such determination a much larger number of mice must be used in each group, and at least five different concentrations must be injected.

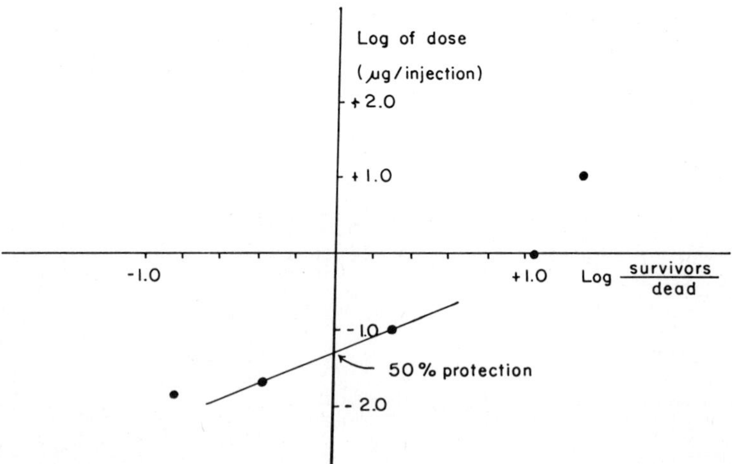

Fig. 61. Determination of 50% protection in the nonspecific resistance enhancement assay

Use and Limitations

This assay may be used to measure the effect of different natural
products on the resistance of mice against serologically unrelated
microorganisms. The choice of the microorganism used as challenge
and the route of application are up to the investigator. However,
the proper dose of the challenge has to be carefully standardized
and rechecked in each assay. The dose is the smallest number of
viable cells per injection which will result in close to 100%
mortality in the nonprotected animals. In this Exercise the 10^9
S. typhi 0901 viable cells in 1 ml saline, injected intraperitoneally,
give 80-90% mortality. If the challenge is too high, no or only
marginal protection will be seen.

Reference

Rowley, D.: Brit. J. exp. Path. *37*, 223 (1956)

Exercise No. 94

Determination of Toxicity

One of the most characteristic properties of bacterial O-antigen
preparations is their toxicity, which can be shown in humans as
well as in experimental animals. In the research of such products
as well as of antibiotics and several natural products, the measurement of their toxicity is often necessary.

In this Exercise the lethality of the endotoxic O-antigen
preparation will be measured. Calculation of the LD_{50} (50% of the
lethal dose) will also be made.

Materials and Equipment

    25 Swiss albino mice, male, weighing 18-20 g each
    Endotoxic O-antigen
    1 ml sterile syringes with hypodermic needles
    Physiologic saline

Procedure

1. Divide the 25 mice into five groups and mark the cages,
containing five mice each, with the name or number of the preparation, the doses injected, and the date and hour of injection.

2. Prepare 10 ml of a 4 mg/ml solution of the endotoxic O-
antigen in saline. Take five small test tubes and pipette 4 ml
endotoxin into the first one. Measure 2.0 ml into the second and
add 2.0 ml saline. Pipette 1.0 ml into the third tube and make
it up with 3.0 ml saline. To the fourth tube add 0.5 ml endotoxin
and 3.5 ml saline, and to the last tube add 0.25 ml endotoxin and
3.75 ml saline. Mix the contents of the tubes with a glass rod.

3. Inject 0.5 ml from the first tube into each of five mice
in the first cage. Give the injection intraperitoneally. This

dose is 2 mg/mouse. Inject 0.5 ml from the second tube into each
of the next five mice. This dose is 1 mg/mouse. The third dose
will be 0.5 mg/mouse, the fourth 0.25 mg/mouse, and the fifth
0.125 mg/mouse. After the injection of each concentration, the
syringe should be rinsed with approximately 0.5 ml of the next
concentration before proceeding.

Evaluation

Count the survivors for three days at about the same time of
day as the injections were given. Record the data.
   Use the following equation to calculate the LD_{50}:

$$\log LD_{50} = \log \text{ (highest dose tested)} + (\log D) \left[\frac{1}{2} - \frac{\Sigma R}{N} \right]$$

where ΣR = Total number of dead animals
      N = Number of animals per dose
      D = D-fold difference between successive doses.
   We have to have data from doses where all animals died from
the highest dose and none from the lowest dose. If this was not
achieved, because a few of the total number of animals in one
dose survived or a few died, we may assume the next highest (or
lowest) dose which we did not test would give 100% mortality (or
100% survival). This assumption may be used only where it is re-
asonable. This calculation is based on the Spearman-Karber method
and was applied for this assay by Itkin (1964).

Use and Limitations

The quantitative determination of a biologic activity is usually
unrealistic if it is not repeated in a relatively large number
of experiments. The determination of the lethal effect of endotoxic O-antigens in mice is a relatively simple experimental
procedure. The endotoxins are injected in different doses into
the peritoneal cavity of mice and the number of survivors recorde
after a few days. However, the following points must be remembere
(Berry, 1966):
   a) Different mouse strains react differently to the same preparation.
   b) The age of the mice has a very important influence on the
sensitivity to endotoxin. Young animals are more resistant than
old ones.
   c) Variation of the temperature in the animal quarters affects
the sensitivity.
   d) The animals received from commercial animal farms should be
kept for at least one week in the new environment before using
them for experimental purposes.
   e) Deaths of animals in this assay cannot be recorded for
periods longer than 72 h. Any deaths after that may not be due
to the injection of endotoxin alone.
   f) It should also be mentioned that the animal cages, stands,
and the entire quarters must be maintained as clean as possible.
   It can be seen, therefore, that an apparently simple biologic
test requires great care if the aim of the experiment is to obtai
some quantitative information.

References

Berry, L.J.: Fed. Proc. *25*, 1264 (1966)
Itkin, A.: Personal communication (1964)

Exercise No. 95

Measurement of Pyrogenicity

Probably the most characteristic effect of bacterial lipopolysaccharides is the elevation of body temperature. Less than 1 µg
given intravenously raises the temperature of rabbits by several
degrees centigrade. The mechanism of this phenomenon is not fully
understood, although some experimental data show involvement of
the central nervous system in the reaction. In this Exercise a
stimplified version of this rather demanding measurement is given.
Please follow the instructions carefully.

Materials and Equipment

    Six New Zealand white rabbits
    1 µg/ml endotoxin dissolved in pyrogen-free saline
    Pyrogen-free saline
    1 ml syringe with 26 gauge needle (these must be hot-air
      sterilized at 180°C for 2 h)
    Six-channel electronic thermometer with six flexible thermo-
      couples
    Restraining cages for pyrogenic measurements (see Fig. 62)
    Air-conditioned animal experiment room, constant temperature
      (±0.5°C), constant light conditions. Sound isolation is re-
      commended

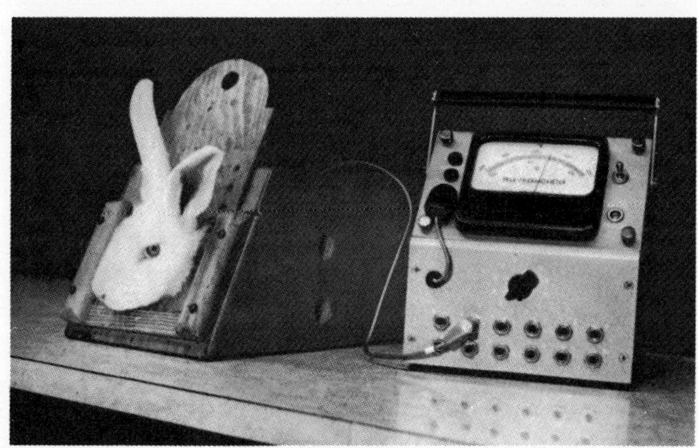

Fig. 62. Restraining
cage and thermometer
for pyrogenicity
measurement

Procedure

1. Rabbits must be acclimatized to the environment for a few days before the experiment starts. Only perfectly healthy rabbits may be used. Slight diarrhea may result in abnormal pyrogenic responses.

Each day, before the experiment, for at least three days, put the rabbits into their restraining cages, insert the flexible thermocouples approximately 3 in. into their rectum, and fix at this position with surgical tape. Read the temperature every hour. Only those rabbits which show a constant ($\pm 0.3^\circ$C) rectal temperature between 39° and 40°C are to be used. Rabbits which have an already elevated temperature ($+40^\circ$C or more) will not respond with pyrogenic response to endotoxin injection.

2. Carry out the actual measurement in the same room. Record the rectal temperatures at 15 min intervals 3 h before and 6 h after the test injection. Give the endotoxin injection into the ear vein of three rabbits. Inject 0.2 ml of the endotoxin solution, which amount will contain 0.2 µg dry material. As a control, inject three rabbits with commercial pyrogen-free saline.

Evaluation

Plot the temperature readings against time on a sheet of graph paper. Rabbits respond with a biphasic fever curve, the first peak being reached in one hour, the second in 3 h. Observe the height of the first and second peaks. Compare the fever responses after injection of endotoxin with the response elicited in the saline-injected controls.

Use and Limitations

There are several methods for quantitative evaluation of the pyrogenic effect. One is the determination of the minimum pyrogenic dose (MPD) which is the endotoxin dose that elicits a 0.2°C rise in body temperature of the rabbit, if given intravenously into the ear vein (Westphal et al., 1952). This can be obtained by measuring and extrapolating the effect of several different doses. Some authors require 0.4°C rise for MPD determination. Another method is based on planimetric measurement of the area under the entire fever curve (Bennett, 1948). Both methods have their shortcomings as well as advantages. For simple comparisons of two or more preparations, the measurement of the peak maxima is satisfactory, provided the comparison has been carried out in the proper dose range.

The above point deserves most careful consideration in the pyrogenic measurements. Because of the extreme sensitivity of the pyrogenic assay, only a very narrow dose range may be used for comparing the activity of different preparations. Outside of this range, the responses are not linear. Slightly active preparations, or materials which are merely contaminated with pyrogenic bacterial products, may show definite pyrogenicity if larger amounts are injected. Because of this fact, the measurements of the MPD seem to be the most reliable for quantitative studies, although this is the most laborious and time-consuming procedure.

For orientation about the proper dose range, it may be helpful
to know that the MPD of an average endotoxic lipopolysaccharide
lies between 0.01 and 0.001 µg/kg rabbit weight.

It is absolutely essential not to disturb the rabbits during
the determination. Harsh noises, sudden changes in temperature
or even in light, too frequent going in and out of the experiment
room may result in fluctuation in the temperature of the rabbits.

The pyrogen test, according to the U.S. Pharmacopeia, XVII.
edition, is designed for products that can be tolerated by normal
rabbits in a dose of 10 ml/kg. The injection must be given intravenously within a period of not more than 2 min. The material
must be tested in three rabbits weighing between 2 and 3 kg. If
no rabbit shows an individual rise in temperature of 0.6°C or
more above its respective control temperature, and if the sum of
the three temperature rises does not exceed 1.4°C, the material
under examination meets the requirements for the absence of
pyrogens.

References

Bennett, I.L., Jr.: J. Exp. Med. *88*, 267 (1948)
United States Pharmacopeia, XVII. ed. p. 863 1965
Westphal, O., Lüderitz, O., Eichenberger, E., Keiderling, W.: Z. Naturforsch.
    7b, 536 (1952)

Subject Index

ACD solution 280
Acetylation 203
Acidic compounds, detection on paper chromatograms 145
- polysaccharides, precipitation 77
Adhering leukocytes, isolation 282
Adjuvant effect 211
Affinity chromatography 19
Agar layer for gel diffusion 233, 246
Agglutination, bacterial 217
-, charcoal 230
-, hemagglutination 219
Allergens, isolation 90
Allergic reaction 92
Amboceptor 266
Amines, primary, detection 144
-, -, quantitative determination 169
Amino acids, C-terminal and N-terminal 147
-, -, detection 144
-, -, gas-liquid chromatography 124, 131
-, -, hydrolysis for determination 101
-, -, ion exchange column chromatography 120
-, -, paper chromatography 143, 146
-, -, quantitative determination 169
Amino sugars, detection 145
-, -, -, by preparative thin-layer chromatography 188
-, -, in phytohemaglutinin 101
-, -, quantitative determination with ninhydrin 169
-, -, -, -, with p-dimethylamino-benzaldehyde 178
Ammonium sulfate, precipitation with 1
Ampholyte 35
Anaphylaxis 288
Antibodies (see also Immunoglobulins)

Antibodies, bacterial 230
-, conjugation of 41
-, determination, see Antigen/Antibody reactions
-, labeled 41, 44, 48, 261, 275
-, to enzymes 20
-, viral 230
Antibody production, immunization and adjuvant effect 211
- -, in mouse ascites fluid 214
- -, at cellular level 215
-, titration 217, 224, 230, 245, 263
Antigen/Antibody reaction, bacterial agglutination 217
- -, charcoal agglutination 230
- -, hemagglutination and its inhibition 219
- -, passive hemagglutination and its inhibition 224
- -, complement fixation 263
- -, precipitation methods, double gel diffusion 232
- -, -, -, gel diffusion for optimal antigen/antibody ratio 248
- -, - - immunoelectrophoresis 235
- -, - - micro-precipitin assay 245
- -, - -, quantitative precipitation 249
Antigens, antibody-enzyme method 249
-, localization by unlabeled antibody 277
-, bacterial, H 66
-, -, K 76, 80
-, -, O (see also Lipopolysaccharides) 69
-, blood group, ABO 84
-, - -, activity 219
-, - -, differentiation 82
-, - -, extraction with detergent 85
-, - -, NN 86
-, - -, solubilization 84
-, flagellar 66
-, H-2 63
-, particulate 213, 217

Antigens, soluble 212
-, transplantation 63
Ascites tumors 57
Australia antigen 243, 261

Band centrifugation 94
Biuret method 168, 250
Blood groups 84
Bromphenol blue 145
Buffer, acetate 20
-, barbital (veronal) 24
-, borate 20
-, cacodylate 56
-, EDTA 14
-, phosphate 20
-, sodium acetate 170
-, sodium carbonate 149
Buffered saline 4

Capsular polysaccharides 76
Carbodiimide coupling 226
Carbohydrates, analysis by gas-
   liquid chromatography 127
-, determination by phenol-sulfuric
   acid 171
-, -, enzymatic 175
-, periodate oxidation 191
-, permethylation 156
-, reducing, detection 144, 173
-, -, determination 53
-, -, terminal 198
-, separation 127
-, staining in gels 153
-, sodium borohydride reduction 196
Carbon, radioactive 73
Carboxylic acids, analysis 188, 207
- -, methyl esters 125, 127, 209
Carboxymethyl cellulose 16
Cell counts, total 272
- -, viable 272
Cell membranes, isolation 57
Cell-wall polysaccharides (see
   Polysaccharides)
Cetylpyridinium chloride 77
Cetyltrimethylammonium bromide 84
Chaotropic agent 22
Charcoal agglutination 230
Chloroform-methanol extraction 72
Cholesterol liposome 63
Chromatography, affinity 19
-, column 120
-, -, adhering leukocytes 283
-, -, amino acids 120
-, -, bacterial O-antigens 70
-, -, conjugated proteins 42
-, -, equipment for 4
-, -, gas-liquid 124, 127, 131

-, -, IgG 8
-, -, separation of IgG fragments 16
-, -, serum proteins 8
-, paper, amino acids 151
-, -, analysis of an unknown 143
-, -, method 110
-, -, preparative 113
-, -, R_F values 146
-, -, solvents 110
-, -, staining reagents 144-145
-, thin-layer 135
-, -, preparative 185
Chromium, radioactive 294
Chromotropate reagent 194
Cold fingers 100
Complement fixation 263
Conjugation of antibodies with fluorescein and rhodamine 41
Coomassie blue 28
Crossed immunoelectrophoresis 240
Counter immunoelectrophoresis 243
Cyanogen bromide 20
Cyclic-AMP, determination by radio-
   immunoassay 254
Cyclic-GMP, determination by radio-
   immunoassay 254
Cytotoxicity, T-cell mediated 272,
   274, 293

DEAE cellulose, preparation 9
Degree of polymerization, determi-
   nation 196
Density-gradient centrifugation 94
Detergents 21, 31, 77, 84
Dinitrofluorobenzene 48
Dinitrophenyl (DNP) amino acids,
   preparation 148
- - -, paper chromatography 151
-, labeled erythrocytes 54
-, -, proteins, preparation of
   soluble 51
Disc electrophoresis 26
Dry weight determination 159
Dye exclusion test 272

Electrofocusing 35
Electron microscopy, immuno- 279
Electrophoresis, disc 26
-, free boundary 24
-, immuno 235
-, -, crossed 240
-, -, counter 243
-, -, quantitative 237
-, -, rocket 237
-, paper, advantaves of 26
-, -, amino acids 117

Electrophoresis, disc, for analysis of an unknown 143
-, -, carbohydrates 117
-, -, on cellulose acetate 23
-, -, equipment 25
-, -, high voltage 117
-, -, human serum 24
-, -, uses of 26
-, SDS gel 31
Elson-Morgan reagent 145
Endotoxins, adjuvant effect 211
-, ^{14}C-labeled 73
-, effect on non-specific resistance 301
-, extraction, chloroform-methanol
  72
-, -, phenol-water 71
- -, trichloroacetic acid 69
-, hydrolysis 102, 104
-, local Shwartzman assay 297
-, pyrogenicity assay 305
-, toxicity measurement 303
Enzymatic cleavage, IgG 14
- -, proteins or polysaccharides
  110
- determination, carbohydrates 175
- labeling 44, 48
Enzymes, isolation of 19
-, immobilized 22
-, unlabeled antibody method 277
Equilibrium centrifugation 94
Erythrocytes, human, ABO group
  antigens 84
-, -, SDS gel electrophoresis 31
-, -, agglutination by phytohemaglutinin 82
-, -, fixation of antigens to 224
-, -, NN group antigens 86
-, sheep, Forssman hapten 88
-, -, labeled 54
-, -, in rosette-forming
  assay 285
-, -, in complement fixation 265
Ester groups, determination by
  hydroxylamine 205
Esters, separation of fatty acids
  124
-, determination 209
Event marker 7

Fatty acids, see Carboxylic acids
Flagellin 66
Flazo orange 44
Fluorescein labeled antibodies 275
Fluorescent antibodies, preparation
  of 41

-, -, staining of E. coli bacteria
  275
Formaldehyde determination 194
Forssman hapten, isolation from
  erythrocytes 88
Fractionation of serum proteins 1
Fraction collector 5
Freeze drying 96
- -, equipment 89
Freund's adjuvant 211
Fuchsin, acid 25
-, -, for staining of serum pro-
  teins 24

Galactose, enzymatic determination
  177
Gamma golublin, isolation by ammonium sulfate 1
Gas-liquid chromatography, see
  Chromatography
Gel diffusion for determination of
  optimal antigen/antibody ratio
  248
-, -, double 232
- filtration, for isolation of
  immunoglobulins 3
Gels, staining for carbohydrates
  153
Glucose, enzymatic determination
  176
Glycolipid, endotoxic, analysis by
  preparative thin-layer chromatography 185
-, -, ^{14}C-labeled 73
-, -, extraction 72
-, synthesis 61
Glycosidic linkages 200
Gradient elution 16
- -, equipment 123

H-antigens, bacterial, isolation 66
H chain of IgG, isolation 17
Hapten, Forssman 88
Hemagglutination 219
-, degrees of 222
-, inhibition 219
-, passive 224
-, -, inhibition 224
-, by phytohemagglutinin 290
Hemocytometer 272
Hemoglobin color standard in complement fixation 265
-, separation 35
Hemolysin, titration 266
Hemolysis, inhibition by Forssman
  hapten 90

Hemolysis, passive 229
-, -, immunoplaque method 215
-, -, inhibition 229
Heptose, quantitive determination
    180, 188
Hexosamines, see Amino sugars
High voltage paper electrophresis
    (see Electrophoresis)
Horseradish peroxidase 20, 48
Human serum, double gel diffusion
    232
- -, immunoelectrophoresis 235
- -, paper electrophoresis 23
Hydrazinolysis 150
Hydrolysis, acid, O-antigenic
    lipopolysaccharides 102
-, -, phytohemagglutinin 99
-, alkaline 52
-, partial, endotoxic lipopolysac-
    charide 104
-, -, with ion exchanger 104
-, total, lipopolysaccharide 102
-, -, optimal conditions 99
Hydroxyl groups, quantitative de-
    termination 203
Hydroxylaminolysis, lipid deter-
    mination by 205

IgA 13
IgG, batch-type isolation using
    DEAE 11
-, chromatographic separation of
    fragments 16
-, enzymatic cleavage 14
-, H chain, isolation of 17
-, Isolation of 8
-, L chain, isolation 17
-, in immunoplaque assay 216
-, reductive cleavage 17
-, resistance against 2-mercapto-
    ethanol 14
-, -, against enzymatic cleavage
    14
IgM 3
- dissociation 12
-, separation on Sephadex 3
-, conjugation 41
-, fractions, enzymatic cleavage 14
-, precipitation with ammonium sul-
    fate 1
Immune precipitation 249
Immunization 211
Immunodiffusion, radial 251
Immuno electron microscopy 51, 279
Immunoelectrophoresis (see Electro-
    phoresis)

Immunogenicity 211
Immunoglobulins 3, 8, 12, 14, 17,
    19
-, chromatography on DEAE 8
Inorganic ash determination 159
Internal standards 126
Iodine, radioactive 44, 261
Ion exchangers 8, 16, 104, 107, 121
- -, column chromatography on 120
- -, hydrolysis with 104
Isoagglutinins, reaction with 84,
    219, 222
Isoelectric point 35
Isopycnic gradient centrifugation
    94
Isotopes, labeling with 73, 94,
    261, 290, 293

Jerne assay 215

K antigens, isolation 76
2-Keto-3-deoxysugars, detection 145
- -, determination 182
- - -, by preparative thin-layer
    chromatography 188
Kjeldahl method for nitrogen deter-
    mination 16

Lactoperoxidase 45
L chains of IgG, isolation 17
LD$_{50}$, determination 303
Lecithin liposome 63
Lectins (see Phytohemagglutinins)
Leukocytes, adhering, isolation 282
Limulus Lysate assay 190
Lipids, determination by hydroxyl-
    amine method 205
-, bacterial, extraction 136
-, -, separation by thin-layer
    chromatography 135
Lipopolysaccharides (see also en-
    dotoxins)
Lipopolysaccharides, isolation 69
-, partial hydrolysis 104
-, total hydrolysis 102
Liposomes 61
Lymophoblasts 290
Lymphocyte surface labeling 44, 49
Lymphocytes, blast formation 290
-, human, isolation 280
-, mitosis 290
-, sensitized 290
Lyophilization (see Freeze drying)

Macroglobulins, dissociation with
  2-mercaptoethanol 12
-, separation by gel filtration 3
Macrophages 228
Mancini technique 251
Medium RPMI-1640 290
Membrane, cell 27
-, tumor, isolation 57
2-Mercaptoethanol, dissociation
  with 12, 31
Methylated carbohydrates, chromato-
  graphy 158
- -, preparation 156
- -, spray reagent for 158
Micro-precipitin assay 245
Microtitration in gel 245
-, Takatsy method 220
Molar ratios 189
Molecular sieves 3, 8
Molecular weight determination 32
Motility agar 67
Moving zone centrifugation 94
Mucin, hog gastric 111

N-acetylneuraminic acid 182
N-acyl linkages 210
Nesslerization 165
Ninhydrin for determination of
  primary amino compounds 169
- for paper chromatographic
  analysis 144
Nitrogen decompression 59
-, microdetermination 162
NN blood group antigens 86
Non-specific resistance, enhance-
  ment of 301

O-acyl linkages 210
Oligosaccharides (see also carbohydrates)
-, high voltage paper electrophoretic separation 117
-, paper chromatographic separation
  113
-, preparation by partial hydrolysis 104
Ouchterlony method 232

PAP method 277
PAS reagent 154
Papain 14
Paper chromatography (see Chro-
  matography)
Passive cutaneous anaphylaxis 288

-, hemagglutination 224
- -, inhibition 227
Pentoses, quantitative determination
  181
Pepsin 220
Peptides, preparation by partial
  hydrolysis 104
-, separation 111
-, -, by paper chromatograhy 116
-, -, by high voltage paper electrophoresis 117
Periodate oxidation of carbohydrates 191
- -, measurement of products 194
Permethylation, carbohydrates 156
Peroxidase-antiperoxidase method
  277
Peroxidase labeling 48
Phenol-sulfuric acid, determination
  of carbohydrates 171
Phenol-water extraction procedure
  for bacterial O-antigens 171
Phosphate buffered soline (PBS) 1
Phosphorus, paper chromatographic
  identification 144
-, determination by thin-layer
  chromatography 188
-, micro determination 166
Phytohemagglutinins, amino acid
  content 101
-, hemagglutination 82
-, hydrolysis 99
-, isolation 82
-, mitosis, enhancement 292
-, reducing carbohydrate content
  101
Pneumococcus polysaccharide 80
Pollen, ragweed 90
Polyacrylamide gel 26
Polyglycerol phosphate 79
Polymerization, degree of, determination 197
Polyribitol teichoic acid (see
  teichoic acid)
Polysaccharides, Lipo- 69
-, -, extraction 69
-, -, hydrolysis 102, 104
-, capsular 76
Precipitation, quantitative 249
Precipitin reaction, micro 245
- -, quantitative 249
Proteins, enzymatic labeling 44
-, quantitative determination 168
-, structural study 147
-, serum 27
-, -, determination by Biuret method 168

Proteins, enzymatic, DNP-soluble 51
-, -, ethanol precipitation procedure 3
-, -, fractionation with ammonium
  sulfate 1
-, -, -, by gel filtration 3
-, separation by gel electrophoresis 27
-, staining with acid fuchsin 25
-, -, -, Coomassie Blue 30
Pyrogenicity, measurement 190, 305

Radial immunodiffusion 251
Radioimmunoassay (RIA) 254
-, solid phase 261
Ragweed pollen 90
- -, sensitivity 92
Red blood cells (see Erythrocytes)
Reducing terminal carbohydrates
  198
Reduction with NaBH$_4$ 196
Reductive cleavage 17
Reticuloendothelial system, mea-
  surement of activity 299
RFC assay 285
Rhodamine, conjugation of immune
  globulins with 41
Rocket immunoelectrophoresis 237
Rosette-forming cells 285

Saline, phosphate buffered 1
-, Veronal buffered 265
Schiff reagent 154
SDS gel electrophoresis 31
Sephadex, column chromatography 3,
  4
-, gel filtration 3, 4
Serum, human (see Human serum)
- proteins (see Proteins and Im-
  munoglobulins)
Schwartzman phenomenon, local 190,
  297
Sialic acid derivatives, detection
  145
- - -, determination 182
Silver nitrate, alkaline 144
Smith degradation 200
Sodium borohydride reduction 196

Solid-phase radioimmunoassay 261
Staining procedures, acid fuchsin
  25
- -, fluorescent antibodies 275
- -, proteins in gel 28
- -, for paper chromatography and
  electrophoresis 144-145, 153
Steam distillation 164
Stroma, red blood cell, extraction
  of antigens 84
Sucrose density gradient 58, 93
Sulfhydryl compounds 12

T-cell cytotoxicity 293
Takatsy microtitrator 220
Teichoic acid, extraction 78
Thin-layer chromatography 135
2-Thiobarbituric acid 183
Thymidine incorporation 290
Titer, antibody 218
TNP, labeling with 54
Tollens reagent 201
Total acetylation of carbohydrates
  203
- methylation of carbohydrates 156
Toxicity, determination of 303
Turk's solution 272
Tracers, fluorescent 41, 275
Transesterification 188, 208
Transplantation antigens 63
Trichloroacetic acid extraction of
  bacterial O-antigens 69
Trinitrophenol (TNP) labeled
  erythrocytes 54
Trypan blue 272
Tumor, membrane isolation 57

Ultrasonic disruption 93
- - of bacteria 93
- - of mamallian cells 95
Ultraviolet flow analyzer 7
- for viewing chromatographs 137

Vacuum distillation 96
Viable cell counts 272

Wasserman-Kolmer test 271

Antibiotics

Volume 5 (in 2 parts)

Antibiotics 5 is divided into two convenient-to-use volumes. Like its predecessors, **Antibiotics 1 and 3,** it is written by internationally recognized specialists who have made important contributions to research in their fields. **Antibiotics 5, Part 1** reviews the modes and mechanisms of action of antibacterial antibiotics and synthetic drugs with precision and scholarly thoroughness. They are either mentioned for the first time in this series, or have been included because more detailed information has now become available. **Part 2** looks at the actions of antieukaryotic and antiviral substances. For antieukaryotic substances of natural origin, the actions of certain bacterial toxins, as well as of toxins from higher plants are reviewed for the first time. This superlative reference will be welcomed by researchers, instructors, graduate students and scientifically interested physicians. It is an indispensable addition to the modern life sciences library.

Springer-Verlag
Berlin
Heidelberg
New York

Part 1

Mechanism of Action of Antibacterial Agents

Editor: F. E. Hahn

1979. 88 figures, 43 tables. Approx. 400 pages
ISBN 3-540-09342-7

Contents: Bacitracin. – Bicyclomycin. – Chloramphenicol. – Ethambutol. – Ionophore Antibiotics. – Isonicotinic Acid Hydrazide. – Kidamycin and Acetyl Kidamycin. – Lincomycin. – Moenomycin and Related Phosphorus-Containing Antibiotics. – Nalidixic Acid. – Nitrofurans. – Novobiocin and Coumermycin A_1. – Phenomycin and Enomycin. – Quinone Antibiotics. – Sparsomycin. – Streptomycin and Related Antibiotics. – Tiamulin und Pleuromutilin. – Tirandamycin. – Subject Index.

Part 2

Mechanism of Action of Antieukaryotic and Antiviral Compounds

Editor: F. E. Hahn

1979. 122 figures, 53 tables. Approx. 480 pages
ISBN 3-540-09396-6

Contents: Anisomycin and Related Antibiotics. – Antitumor Platinum Compounds. – 9-β-D-Arabinofuranosyladenine. – 8-Azaguanine. – Bleomycin. – Diphtheria Toxin and Exotoxin A from Pseudomonas aeruginosa. – Echinomycin, Triostin, and Related Antibiotics. – Ellipticine. – 2-Hydroxy-3-Alkyl-1,4-Naphthoquiones. – Hydroxystilbamidine-5-Iodo-2'-Deoxyridine. – Neocarzinostatin. – Nitracrine. – Phleomycin. – Polyene Antibiotics: Nystatin, Amphtericin B, and Filipin. – Protein and Glycoprotein Toxins that Inactivate the Eukaryotic Ribosome. – Quinine. – Showdomycin. – Streptonigrin. – Tilorone Hydrochloride. – The Vinca Alkaloids. – Virazole (Ribavirin). – Subject Index.